# Integrated Science

## The Energy Code

**Louis S. Campisi, Ph.D.**
**Warren Rosenberg, Ph.D.**
**Victor A. Stanionis, Ph.D.**

*Iona College*

KENDALL/HUNT PUBLISHING COMPANY
4050 Westmark Drive Dubuque, Iowa 52002

ISBN 0-7575-1300-X



Printed in the United States of America
10 9 8 7 6 5 4 3 2 1

# Brief Contents

Preface ix
Acknowledgments xi

Chapter 1 Introduction to Scientific and Technological Literacy 1
Chapter 2 Pattern Recognition 9
Chapter 3 Problem Solving: Blocks and Strategies 23
Chapter 4 Models and Matter 41
Chapter 5 Models and the Atom 53
Chapter 6 Systems 65
Chapter 7 Information and Coding 79
Chapter 8 Information and Coding in Living Systems 99
Chapter 9 Energy 109
Chapter 10 A Primer: Energy Sources, Their Nature and Use 123
Chapter 11 Energy Resources and Their Environmental Effects: The Carbon Cycle and More 139
Chapter 12 Metabolism and Nutrition 155
Chapter 13 Population and Growth 167
Chapter 14 Medical Screening Tests 181
Chapter 15 Technology and Risk 193

Appendix A Review of Basic Mathematics 205
Appendix B Scientific Notation and Significant Figures 213
Appendix C Functions 215

# Contents


PREFACE ix

ACKNOWLEDGMENTS xi

**CHAPTER 1:** INTRODUCTION TO SCIENTIFIC AND TECHNOLOGICAL LITERACY 1

- What Is Science Literacy? 1
- "Live Long and Prosper" 2
- Why Should We Study Science and Technology? 3
- Science as a Way of Knowing: Technology as an Act of Doing 3
  - *Science* 3
  - *Technology* 4
- What Is the Scientific Method? 4
- How to "Live Long and Prosper" 5

**CHAPTER 2:** PATTERN RECOGNITION 9

- Periodic Waveform Patterns 9
- Mathematics 10
  - *Fourier Analysis* 10
  - *The Spiral Spring* 11
  - *The Falling Body* 11
  - *The Fibonacci Sequence* 11
  - *The Fibonacci Constant* 12
  - *The Fibonacci Spiral* 12
- Music 13
- Crystallography 13
- DNA 14
  - *Pattern Recognition* 15
  - *The Unabomber* 16
  - *Periodic Table* 16
  - *Epidemiology* 18
  - *Evolution* 19
  - *Differential Medical Diagnoses* 19

**CHAPTER 3:** PROBLEM SOLVING: BLOCKS AND STRATEGIES 23

- Problem Solving 23
- Problem-Solving Methods 24
- Polya's Method 24
- Gap Model 24
- Scientific Method 26
  - *What Is the Scientific Method?* 26
- Hypothesis Testing 28
- Exponential Growth 29
- Blocks to Problem Solving 30
- Critical Thinking Skills 34

**CHAPTER 4:** MODELS AND MATTER 41

- Models 41
  - *Physical Models* 42
  - *Graphical Models* 43

*Mathematical Models* 43
*Analog Models* 43
*Verbal Models* 44
Building Models 44
*Mouse Model* 44
*Model of Matter* 45
Units 45
Metric System 45
Density 46
Matter 47
Chemical Structures 48
Compounds 48
*Solutions* 49
States of Matter 49
*Gases* 49
*Liquids* 49
*Solids* 50
*Changes in State* 50

**CHAPTER 5: MODELS AND THE ATOM** 53
Rutherford's Atomic Model 53
Photons 54
Light and Waves 54
Planck's Theory 55
Emission Spectra 56
Bohr's Atomic Model 57
The Atom 58
DNA 59
Mathematical Models 60

**CHAPTER 6: SYSTEMS** 65
Systems 65
Feedback 67
Systems Analysis 70
Equilibrium 70
*Hunting* 70
*Tacoma Narrows Bridge* 71
The Human Body 71
*The Pacemaker* 72
The Environment 74

**CHAPTER 7: INFORMATION AND CODING** 79
Codes 79
*Information* 80
*Zip Code* 80
*Pony Express* 81
*Morse Code* 81
*Typewriter* 81
Binary Code 82
*Binary Numbering Systems* 83
*Digital and Analog* 84
DNA 84
Musical Coding 85
Language Coding 85
*Russian* 86
*Navajo* 87
*Smoke Signals* 87
*Semaphores* 87
*License Plate Codes* 87
Automation Technology 88
Retail Checkout Automation 89
*Advantages* 89
*Other Advantages* 89
*Customer Acceptance* 90
*Technology Development* 90
*Societal Impact* 91
*Technology Components* 91
Universal Product Code (UPC) 91
*Reading the Code* 92
Reflections 92
Cryptology 93
*Ancient Codes* 93
Information as a Resource 94

**CHAPTER 8: INFORMATION AND CODING IN LIVING SYSTEMS** 99
The Living Cell 99
Information Coding in the Nervous System 100
Information Coding in DNA 103
Human Genome Project 105
Single Nucleotide Polymorphisms (SNPs) and the HapMap Project 106
Commercial Applications 106

**CHAPTER 9: ENERGY** 109
Standards and Units 109
Force 110
Energy 112
*Work* 112
*Power* 113
*Kinetic Energy* 114
*Potential Energy* 114
Energy Forms 114
Origins of Energy 115
First Law of Thermodynamics 115
*A Star Is Born* 116
Energy Sources 116
*Energy Conversions and Efficiencies* 117
Second Law of Thermodynamics 118
Loss of Opportunity 119
Boundary Conditions 120
*Barry Commoner's View* 120

**CHAPTER 10: A PRIMER: ENERGY SOURCES, THEIR NATURE AND USE** 123
Thermodynamics 123
Nuclear Energy 124
*Fission* 125
*Fusion* 128
Solar Energy 129
*Direct Radiation* 130
*Hydropower* 130
*Wind Energy* 131
*Biomass Energy* 131
*Photovoltaics* 132
Tidal Energy 132
Geothermal Energy 133
Fossil Fuels 134
*Coal* 134
*Natural Gas and Petroleum* 134

**CHAPTER 11: ENERGY RESOURCES AND THEIR ENVIRONMENTAL EFFECTS: THE CARBON CYCLE AND MORE** 139
Photovoltaics 139
Systems Approach 140
Electromagnetic Spectrum 142
Greenhouse Effect 143
*Natural Gas* 144
*Fossil Fuel and Auto Emissions* 146
*Coal* 147
*Plastics* 150

**CHAPTER 12: METABOLISM AND NUTRITION** 155
Entropy 155
*Energy Extraction through Metabolic Breakdown* 156
Nutrition 157
*Carbohydrates* 157
*Fats* 158
*Protein* 159
*Vitamins* 160
*Minerals* 160
*Water* 160
Calories 160
Nutritional Energy Balance 161
Weight Management 162

**CHAPTER 13: POPULATION AND GROWTH** 167
Population Growth 167
Exponential Growth 168
*Mathematical Review* 169
*Exponential Growth and Doubling Time* 170
*The Magical Formula* 170
The Exponential Function and Population 171
*Limits to Growth* 171
World3 Model 171
*Report to the Club of Rome (1972)* 174
*Population* 174
*Carrying Capacity Calculation* 174
*Population Growth and Resources Impact* 175
*Conclusions: The 1992 Report* 176

**CHAPTER 14: MEDICAL SCREENING TESTS** 181
Medical Screening Tests 181
How Effective Are Medical Screening Tests? 185
Genetic Screening Technologies 186
Fetal Screening 186
*Amniocentesis* 186
*Fetoscopy* 187
Newborn Screening 187
Risk-Benefit Analysis in Genetic Screening 188

**CHAPTER 15: TECHNOLOGY AND RISK** 193
Technology and Risk 193
Chlorinating Water Supplies 194
Cooking Steaks 196
Irradiation of Food 196
Sources of Technological Risk 197
*Hardware Factors* 197
*Human Factors* 197
*Organizational Factors* 198
*External Social Factors* 198
Quantifying Risk 198
*The Association Coefficient* 198
*Relative Risk* 199
Performing Cost-Benefit and Risk-Benefit Analysis 199
Performing Quality-Adjusted Life Year Analysis 200
Controlling Risk 200
*Automobiles* 200
*Cell Phone Regulation* 201

**APPENDIX A: REVIEW OF BASIC MATHEMATICS** 205

**APPENDIX B: SCIENTIFIC NOTATION AND SIGNIFICANT FIGURES** 213

**APPENDIX C: FUNCTIONS** 215

# Preface

The textbook *Integrated Science: The Energy Code* is written with the goal of developing a measure of scientific and technological literacy in students who are not following a path leading to a career in science and/or engineering. All citizens will participate in a society where they will be required to cast informed votes as citizens on issues containing scientific and technological content.

We have attempted to follow the recommendations of Project 2061 as outlined in the AAAS publication *Science for All Americans,* which defines science literacy and lays out some principles for effective learning and teaching. With expert panels of scientists, mathematicians, and technologists, Project 2061 set out to identify what was most important for the next generation to know and be able to do in science, mathematics, and technology—what would make them science literate.

In *Integrated Science* we shall learn how scientists and engineers solve problems. One of the most important things to keep in mind is that engineers are trained to anticipate failure in their creations and to design against it. At first we shall learn to recognize many of the blocks to problem solving we have acquired from our environment and culture.

Once we are cognizant of these we can then proceed and make use of George Polya's classic description of the problem-solving process. He devised a four-step method for solving problems, which loosely stated consists of: defining and understanding the problem, devising a plan to solve it, executing the plan, and checking the result to see if it works.

We shall examine the role of pattern recognition in problem solving and how modeling can help us. The systems approach, using the concepts in input, output, feedback, and control, which encompass the views of engineers and businesspersons alike, will be used in analyzing models and understanding them.

The role of energy and its flow through living and nonliving systems will be studied, and by applying the principles of biology, chemistry and physics we shall gain a better understanding of the world around us. Applications of science and technology will be studied so as to give us practice in improving our skills to make informed decisions on scientific issues in light of the risks and benefits they present.

At the college level the text is designed for a three-credit course given over one semester, with two hours of lecture and a two-hour "hands-on" laboratory session per week. A laboratory manual containing over 50 activities, prepared by Victor A. Stanionis and members of the Iona College Science Faculty for Iona's course in integrated science, is available. The laboratory guide is entitled "Laboratory Manual to Accompany *The Sciences*" (a text written by Trefil and Hazen) and is published by John Wiley and Sons (ISBN: 0-471-25497-5). Fourteen laboratory activities taken from the laboratory manual are scheduled in which students work in pairs. They are:

1. Orientation
2. Floating versus Sinking
3. Measuring Cell Size
4. The Small and the Large
5. Feedback and Control Models
6. Collecting Solar Energy
7. Energy Conversions
8. Problem Solving: Measuring Temperature with Only Limited Means
9. Patterns of Problem Solving
10. Solar Constant and the Earth's Carrying Capacity
11. Measuring Metabolic Rate by Indirect Means
12. Electromagnetic Spectra
13. Medical Screening Tests
14. Schneider Cardio-Vascular Test

All the illustrations and tables in *Integrated Science* are available to instructors on a CD-ROM, which may be obtained upon request. A PowerPoint presentation for each chapter has been prepared by the authors and will also be found on the instructor's CD.

A series of companion texts published by Kendall-Hunt are available for a second-semester course in thematic areas. (Levkov: *As the Earth Turns;* Rosenberg: *Exercise Science;* Stanionis: *The Art and Science of Computer Music* [to be published 2005].

# Acknowledgments

The authors would like to thank all the members of the Iona College science faculty for their helpful suggestions, criticisms, and contributions. We look forward to future comments and suggestions and invite all who use the text to help us in revising it.

The cover of the text is a reproduction of a painting by Br. Kenneth Chapman, artist-in-residence at Iona College, and we are thankful for his permission to use it for our textbook. We are especially grateful to Dr. James J. Murphy, Professor Emeritus of Physics, for his past contributions, and to our student Amy Farrell for her assistance in locating illustrations and photographs for the text. We would also like to thank Jeannie DiBuono, science secretary, for her help and assistance.

The gentle and persuasive persistence of Sue Saad, associate editor at Kendall-Hunt, encouraged us to complete the text and we are thankful. We would also like to acknowledge the help we received from Billie Jo Hefel, Colleen Zelinsky, and Angela Shaffer, editors at Kendall-Hunt, and thank them for their guidance and suggestions.

Louis S. Campisi, Ph.D.
*Professor of Chemistry*

Warren Rosenberg, Ph.D.
*Provost, Vice President of Academic Affairs*
*Professor of Biology*

Victor A. Stanionis, Ph.D.
*Professor of Physics*

# Introduction to Scientific and Technological Literacy

*"Live long and prosper."*

**Mr. Spock**

In this chapter we introduce you to the big picture. It is a summary of our goals, as follows. Our hope is that by the end of this course you will be able to:

1. Demonstrate an understanding of the nature of scientific knowledge and inquiry.
2. Apply scientific concepts, principles, laws, and theories to generate multiple solutions to contemporary issues.
3. Apply scientific and engineering principles to solve problems, make decisions, and further understand nature and technology.
4. Demonstrate an awareness of the interrelationship of science, technology, and society.
5. Cultivate the confidence to confront scientific and technological issues in areas such as human health, energy, and the environment.

## Goals

After studying this chapter you should be able to:

- Define scientific and technological literacy.
- Understand why science and technology are worthy of study.
- Distinguish between science and technology.
- Recognize the classical scientific method.
- Follow the progression of science from research to applications as in the development and use of the laser and DNA.

## What Is Science Literacy?

Project 2061 began its work in 1985—the year Halley's Comet passed near Earth. With expert panels of scientists, mathematicians, and technologists, Project 2061 set out to identify what was most important for the next generation to know and be able to do in science, mathematics, and technology—what would make them science literate. The panels' recommendations were integrated into Project 2061's 1989 publication, *Science for All Americans.* The goal of Project 2061 is for all Americans to be scientifically literate by the time Halley's comet makes its next appearance in the year 2061.

*Science for All Americans* defines science literacy and lays out some principles for effective learning and teaching. Project 2061 defines science literacy broadly, emphasizing the connections among ideas in the natural and social sciences, mathematics, and technology. *Science for All Americans* includes specific recommendations for learning in the following areas:

- *The Nature of Science* includes the scientific worldview, scientific methods of inquiry, and the nature of the scientific enterprise.
- *The Nature of Mathematics* describes the creative processes involved in both theoretical and applied mathematics.
- *The Nature of Technology* examines how technology extends our abilities to change the world and the tradeoffs necessarily involved.
- *The Physical Setting* lays out basic ideas about the content and structure of the universe (on astronomical, terrestrial, and submicroscopic levels) and the physical principles on which it seems to run.
- *The Living Environment* delineates basic facts and ideas about how living things function and how they interact with one another and their environment.
- *The Human Organism* discusses human biology as exemplary of biological systems.
- *Human Society* considers individual and group behavior, social organizations, and the process of social change.
- *The Designed World* reviews principles of how people shape and control the world through some key areas of technology.
- *The Mathematical World* gives basic mathematical ideas, especially those with practical application, that together play a key role in almost all human endeavors.
- *Historical Perspectives* illustrates the science enterprise with ten examples of exceptional significance in the development of science.
- *Common Themes* presents general concepts, such as systems and models, that cut across science, mathematics, and technology.
- *Habits of Mind* sketches the attitudes, skills, and ways of thinking that are essential to science literacy.

*Science for All Americans* also includes chapters on effective learning and teaching, reforming education, and next steps toward reform.[1]

Our definition of scientific and technological literacy rests on the premise that, even though all cannot be expected to solve complex technical problems, everyone can and should be prepared to participate in the public debate, evaluate the available information, and advance the solution through intelligent questions and informed voting.

## "Live Long and Prosper"

Mr. Spock, an alien being in a popular TV show a number of years ago, uttered these words. The greeting, meant to convey good wishes of happiness, implies that prosperity and long life are somehow related to happiness. If true, we Earthlings should be pretty happy. Our life expectancy has increased significantly. About two thousand years ago the average life span was about 22 years. (While the implications of this statement might lead one to think that teenagers would be considered middle-aged, realize that average really means that very many individuals did not make it to 22 because of diseases and accidents that were untreatable at the time). By the beginning of the 20th century the average life expectancy about doubled, to 44 years, as medical science continued to improve.

Currently the average life expectancy is about 76 years for men and 80 years for women. The increase in almost 60 years of life, on average, and the greatly enhanced quality of those years, has been an occasion of improved and generally less burdened lives. Much has happened to improve both our life expectancy and our material prosperity. Science and technology are at the heart of these changes.

We tend to accept innovative technological change as almost routine; we hold the expectation that, with time, things will only get better, be it medicine, transportation, communication, entertainment, etc. The truth is, until about one hundred years ago, technology and its impact on daily life was minimal and things changed very slowly. This is no longer the case. If we were to reflect upon the degree to which science and technology have impacted and altered our lives and society these past one hundred years, we would be astonished. In order to illustrate just how much our lives have been transformed by scientific and technological innovations, we can compare the technological innovations and changes they have brought about in our lives with those of our parents' and grandparents' generations. For example, we could divide the last century in thirds ("generational eras") and list a few of the scientific discoveries and technological innovations which came into being. See Table 1.1.

The point is clear. Life in the past was harder in a variety of ways (transportation, communication, health, creature comforts, and so on). We are the beneficiaries of technological changes, which

[1]Reference: American Association for the Advancement of Science (AAAS) Project 2061. www.project2061.org/

**TABLE 1.1 Generational Divisions**

| 1900–1933 Grandparents' Generation | 1933–1967 Parents' Generation | 1967–2004 Current Generation |
|---|---|---|
| Airplane | Transistor | Personal computers |
| Radio | Atomic energy | Video cassette recorders |
| X-Ray tube | Antibiotics | Cell phones |
| Transatlantic boat travel | Polio vaccine | Laser surgery |
| Stainless steel | Structure of DNA | Internet, DVDs, CDs |

**Figure 1.1** 1933 Atwater Kent radio (www.radioage.com).

**Figure 1.2** 1950s Transistor radio.

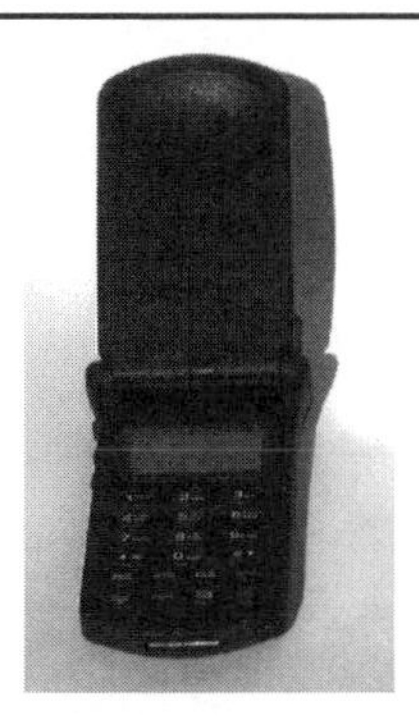

**Figure 1.3** 2000 StarTac cell phone.

have advanced society and improved our lives. In addition, we have experienced such progress in the past one hundred years at a much greater rate than in the previous millennia.

## Why Should We Study Science and Technology?

Scientific and technological literacy is important for a variety of reasons. On a personal level we make decisions dealing with topics and situations that are scientifically/technologically based. They may be related to a problem in the operation of our car, or a health problem, for which a corrective option is sought, or installing a driver to run a new printer or simply selecting a new computer.

On a community level we need to be technologically literate in order to deal with community health, environmental, or energy issues. It is important to make our opinions known, based on our understanding of the issues (rather than because someone says so). If we are scientifically and technologically informed citizens, we can make better and more informed choices and decisions concerning the election of political candidates and their policies as they relate to science and technology.

The pace of change in our world is, indeed, rapid. Whereas 20 years ago we may have listened to our music mostly on records or cassette tapes, we listen today on compact discs or digital MP3 players. Although music recording and playback technologies have reduced the size of our playback devices and improved the quality of the recordings they hold, it has introduced the previously minimal problem of copyright infringement and illegal trading of songs.

Technology, although intended to improve our lives, has always introduced new hazards and peripheral impacts. While in some ways technology has made our lives easier, in others it makes our world more complex. Technology is inescapable; it pervades everything we do, and brings with it both the good and the bad. It is always better to use technology with a full understanding of all of its impacts.

## Science as a Way of Knowing: Technology as an Act of Doing

### Science

Science is a systematic understanding of natural phenomena ascertained via experimental evidence, laws, observations, and analysis. The physical sciences include chemistry and physics, but there are others such as geology, and the life sciences, which involve biology. Sciences are organized bodies of knowledge with laws or theories to provide explanations concerning natural phenomena.

## TECHNOLOGY

Technology is the application of scientific knowledge to the solution of problems. Technology involves a skill or process, a way of doing things. For example, some of the earliest technologies used by humans involved making fire, cooking meals, and fermenting grapes to make wine. While the scientific basis for these processes is now well known, originally they were just processes that worked. We can, of course, use science to develop a technology. The production of computer chips or light bulbs are some examples of technologies based upon scientific knowledge of electron flow in conductors. (Think of a few others!)

The reason for the very rapid growth of science and technology is related to the nature of the science and technology as an enterprise that relies on the free flow of information in society. Central to the nature of how science operates is the scientific method.

# What Is the Scientific Method?

The scientific method is essentially an approach or way of trying to make sense of observations of the real world. There is no one "scientific method." However, just about all modern scientific investigation includes elements of observation, formulating explanations, and then testing them.

For more than fourteen hundred years, it was thought that the earth was the center of the universe with the planets and stars circling around it. A Greek, Ptolemy, had come up with this theory of the universe based on careful thought and reason (Fig. 1.4).

In the 16th century Galileo, who having built a telescope using spectacle lenses available at the time, observed that Io, a moon of Jupiter, did not circle the earth but rather orbited around Jupiter. It was such an observation that provided understanding based on experimental evidence as opposed to using only careful thought and reasoning methods of the ancient Greeks. It was Galileo who changed the way we study nature. It is experimental evidence that provides the foundation of our present understanding of the world around us. We will examine the scientific method in greater detail in the next chapter.

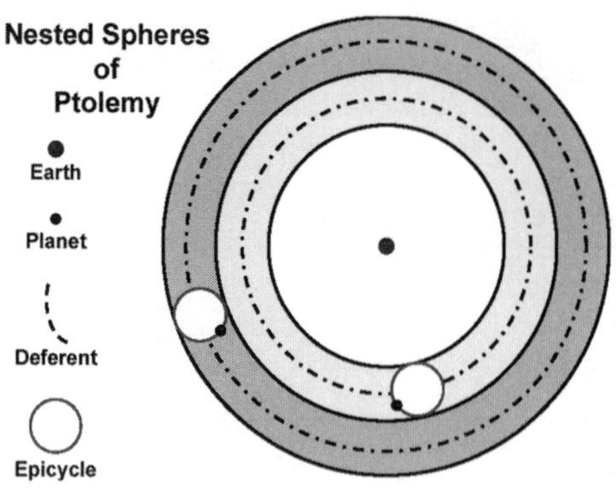


**Figure 1.4** Ptolemaic geocentric model of the universe. *http://home.fnal.gov/~rocky/Graham03/lecture.pdf*

Another key element in rapid growth of science and technology also had its beginning with Galileo and the French and English philosophers of the 17th century, and perhaps even in the American Revolution. The key element is simply the freedom of the individual to pursue science free from constraints of authoritarian dictates and preconceived notions and to publish the results in peer-reviewed journals for the world to judge and build upon.

Technology can best be described as the application of scientific knowledge, our understanding about the natural world, to the solution of societal problems. The progression from the unknown through scientific hypothesis, basic scientific research, and confirmed knowledge to technological applications can be illustrated through several examples, one being the development and use of the laser.

In the early 20th century not much was known about the nature of the atom or its interaction with energy and many prominent scientists of the day were busy working to understand it. In 1912, German physicist Albert Einstein first formulated the law of photochemical equivalence. One year later, in 1913, physicist Niels Bohr proposed his now-well-known model of atomic structure.

In 1917 Einstein, in an action that is easily considered pure science, hypothesized that when the atoms of certain elements are bombarded with light energy, the atoms' electrons would absorb the energy forcing the electrons into higher, more energetic orbit levels. Einstein further speculated that the atoms would not remain in this unnatural, excited state for long but would return their electrons to the original, or ground state, giving off the energy as wavelengths of light. In certain specific atoms, Einstein proposed, the emitted light, a form of radiant energy or radiation, would be of uniform wavelength and that each particle of light

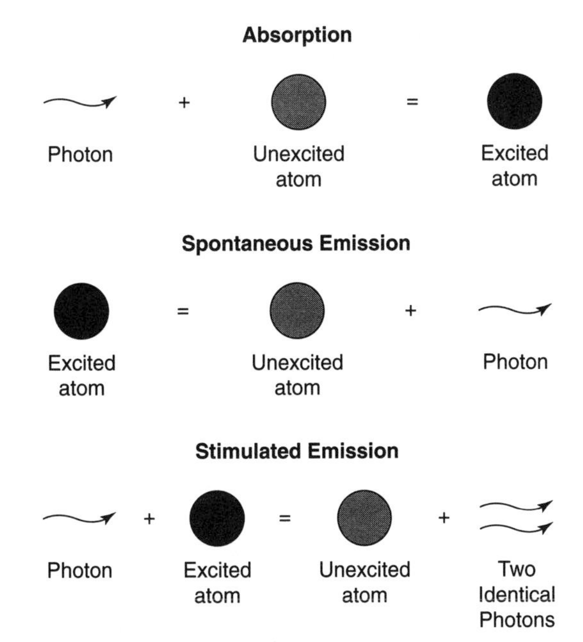


**Figure 1.5** Stimulated emission.

(photon) would be characterized as being in-phase with all of the others. Thus, random light falling upon such atoms would emerge as focused light of a specific and characteristic wavelength (Fig. 1.5).

Finally, Einstein proposed, that in a molecular arrangement of such atoms, the light emitted by one atom would induce the release of additional amounts of light in other atoms. Thus, a small amount of light striking such a material would result in a greater amount of light being emitted with the emitted light waves highly collimated and in-phase. This principle is known as light amplification through the stimulated emission of radiation. The term laser is derived from the first letters of the words **l**ight **a**mplification by **s**timulated **e**mission of **r**adiation.

The concept of harnessing the power of light energy by utilizing the principles proposed by Einstein intrigued many applied scientists who saw a broad range of applications for concentrated, collimated beams of light. To appreciate the shortcomings of typical light sources consider the flashlight. The light emitted by a flashlight bulb, even though focused by a lens, rapidly scatters after leaving the flashlight, becoming more and more diffuse as the distance from the flashlight increases. In addition, the light emitted contains radiant energy of a variety of wavelengths and not all of the waves are in phase. The result is a rather weak beam of light that is useful for illuminating a small area.

If it were possible to focus this beam more intensely, to have a more powerful beam emitted, and to create a light beam of uniform wavelength and with all waves in-phase, it might be possible to send a beam of light to the moon, reflect it off a surface mirror, and return it to the earth, a round-trip distance of approximately a half million miles. Given that the speed of light is a constant and that we know it to be 186,000 miles per second (or $3 \times 10^8$ m/s), we would be able to determine the exact distance between the earth and moon. This was accomplished in the early 1970s using a reflector placed on the surface of the moon by the Apollo 11 astronauts in 1969.

In 1960, 43 years after Einstein first proposed the principle, Theodore Maiman developed the first prototype laser. These finely focused beams of light produced by laser technology have since been used for the recording and retrieval of information. Philips Electronics introduced the first laser videodisc in 1979. The first audio compact disc players (music CDs) went on sale in 1982.

In 1985, lasers were first used to remove fat deposits from blocked arteries by focusing their intense beams of energy on fatty plaques and reducing the risk of heart attack. In 1986, lasers were first used to perform corneal surgery to improve human vision. Today's surgical lasers utilize finely focused beams of ultraviolet light to physically break carbon–carbon bonds and separate tissue without burning or bleeding.

Greater detail on the basic scientific principles of atomic structure and energy transformations will be reviewed in a later chapter of this text.

A similar progression from basic science to technological application can be seen in the timeline in Table 1.2. For a period of more than 100 years our basic scientific understanding of the nature of heredity progressed slowly, culminating in the historic discovery of the structure of DNA by Watson (Fig. 1.6) and Crick and the structural sequence of the insulin protein by Fredrick Sanger.

Technological applications accelerated rapidly after that with Kornberg's laboratory synthesis of DNA in 1957, the development of genetic engineering techniques in the early 1970s, and the rapid progress of the human genome project.

# How to "Live Long and Prosper"

In this book we shall learn how scientists and engineers solve problems. One of the most important things to keep in mind is that engineers are trained to anticipate failure in their creations and to design against it. At first we shall learn to recognize many of the blocks to problem solving we have acquired from our environment and culture.

**TABLE 1.2 Progression from Basic Science to Applications**

| | |
|---|---|
| 1830 | Discovery of proteins. |
| 1863 | G. Mendel proposes the concept of hereditary transmission of individual traits. |
| 1879 | W. Fleming discovers chromatin, the substance that composes chromosomes. |
| 1902 | W. Sutton determines that cellular chromosomes hold genetic information. |
| 1909 | DNA and RNA discovered by P. J. Levene. |
| 1925 | T. H. Morgan publishes *The Theory of the Gene.* |
| 1944 | O. Avery determines that genetic information is carried by nucleic acids. |
| 1953 | J. Watson and F. Crick determine the structure of DNA. |
| 1955 | F. Sanger determines the amino acid sequence of insulin. |
| 1957 | A. Kornberg synthesizes DNA in the laboratory. |
| 1966 | Genetic code is determined. |
| 1967 | Enzyme DNA ligase is isolated. |
| 1970 | First restriction enzyme is isolated. |
| 1973 | H. Boyer develops the DNA splicing technique, using restriction enzymes to cut, and DNA ligase to splice, DNA segments. |
| 1978 | W. Gibson develops genetically engineered bacteria to produce human insulin from synthetic DNA. |
| 1980 | U.S Supreme Court determines new life forms created by technology are patentable. Exxon patented an "oil eating" bacterium. |
| 1981 | FDA approves Eli Lilly's marketing of synthetic insulin from genetically engineered bacteria. |
| 1985 | Genetic fingerprinting is first used in U.S. courts. |
| 1987 | "Frostban," the first genetically engineered bacterium that prevents frost damage on crops, is approved for testing. |
| 1988 | Human Genome Project is announced. |
| 1994 | First human breast cancer gene is discovered. |
| 1999 | First full human chromosome is decoded. |
| 2001 | Tumor suppressor gene is discovered. |
| 2003 | Human genome is decoded. |

**Figure 1.6** Iona College Science Faculty with Dr. James Watson (center): Dr. James Murphy, Dr. Louis Campisi, Dr. George Pappas, Dr. Victor Stanionis, Dr. Frank Fazio, Dr. Jerome Levkov, Dr. Warren Rosenberg, 1991.

Courtesy Photo Bureau, Inc.

Once we are cognizant of these we can then proceed and make use of George Polya's classic description of the problem-solving process. George Polya was a mathematician who wrote the book *How to Solve It: A New Aspect of the Mathematical Method* in 1945 for solving mathematical problems. He devised a four-step method for solving problems, which loosely stated consists of: defining and understanding the problem, devising a plan to solve it, executing the plan, checking the result to see if it works.

We shall examine the role of pattern recognition in problem solving and how modeling can help us. The systems approach, using the concepts of input, output, feedback, and control, which encompass the views of engineers and businesspeople alike, will be used in analyzing models and understanding them.

The *role of energy* and its flow in living and nonliving systems will be studied and by applying the principles of biology, chemistry, and physics we shall gain a better understanding of the world around us.

Applications of science and technology will be studied so as to give us practice in improving our skills to make informed decisions on scientific issues in light of the risks and benefits they present.

# Review Questions

1. How does science differ from technology?
2. What is meant by scientific and technological literacy?
3. Why are science and technology important?
4. Is there only one "scientific method?"
5. What does the term *science* for all Americans imply?
6. What does the nature of science include?
7. Distinguish between fact, law, theory, and hypothesis. (See Internet references.)
8. Why do scientists prefer to base their work on research published in peer-reviewed journals?
9. Consider any innovation that has occurred over the last century—refrigeration, transistor, antibiotics, cell phones, etc.—and indicate several ways that it has both improved life and detracted from it.
10. How do you suppose the work of Galileo and the English and French philosophers influenced the nature of scientific inquiry?

# Multiple Choice Questions

1. Spock is a(n) _____.
   a. immigrant
   b. discoverer of the laser
   c. discoverer of DNA
   d. alien
2. Long life and prosperity are _____.
   a. a right of a U.S. citizen
   b. hereditary
   c. a product of science and technology
   d. unattainable
3. One hundred years ago the impact of science and technology on one's daily life was _____.
   a. minimal
   b. the same as today
   c. greater than today
   d. very rapid
4. Rapid growth in science and technology depends on _____.
   a. free flow of information
   b. funding
   c. authoritarian dictates
   d. philosophers
5. The word laser is an acronym for _____.
   a. light amplification system for energy radiation
   b. light amplification by stimulated emission of radiation
   c. low amplitude system for emitting radiation
   d. longitudinal analysis of sunlight emitted radiation
6. James Watson _____.
   a. worked with Sherlock Holmes
   b. discovered the laser
   c. discovered the photoelectric effect
   d. discovered the structure of DNA
7. An approach that tries to make sense of data or information is _____.
   a. the scientific method
   b. Polya's four-step method
   c. trial and error
   d. brainstorming
8. Polya's first step in problem solving is _____.
   a. devise a plan
   b. understand the problem
   c. execute the plan
   d. check the solution

9. Polya's last step in his method is _____.
   a. devise a plan
   b. understand the problem
   c. execute the plan
   d. check the solution

10. The first compact disk player was built during _____.
   a. our grandparents' generation
   b. our parents' generation
   c. the current generation
   d. None of the above.

11. A hypothesis is a _____.
   a. guess
   b. law
   c. theory
   d. large animal that stays in the water

12. Science relies on _____.
   a. hearsay
   b. heresy
   c. honesty
   d. emotion

13. The physical sciences include _____.
   a. chemistry
   b. physics
   c. biology
   d. a. and b.

14. Which is not an element of the scientific method?
   a. observation
   b. law
   c. hypothesis
   d. tarot cards

15. Which of the following is not a technology?
   a. winemaking
   b. fire
   c. cheese making
   d. the melting point of ice

16. An electron that is excited can lose energy by _____.
   a. emitting light
   b. losing weight
   c. gaining thermal energy
   d. none of the above

17. The first prototype laser was developed by _____.
   a. Einstein
   b. Bohr
   c. Maiman
   d. G. W. Bush

18. Lasers have been used _____.
   a. in surgery
   b. recording and retrieval of information
   c. to make measurements
   d. All of the above.

19. A laser is characterized by the fact that _____.
   a. many light waves constitute the light beam
   b. all the waves are peaks
   c. the beam consists of light waves all in phase
   d. it is invisible

20. The structure of DNA was determined by _____.
   a. Watson and Holmes
   b. Watson and Crick
   c. Wellington and Crick
   d. Watson and Doyle

## Bibliography and Web Resources

*Science for All Americans,* Oxford University Press, New York, 1994.

Battelle Technology Forecasts contains predictions from 2005 to 2020 as to what will be available.
**www.battelle.org/forecasts/default.stm**

Science literacy definition.
**www.summitscience.org/literacy/**

"The Importance of Science Literacy in a Computing World."
**www.npaci.edu/enVision/v15.2/director.html**

Science and Creationism: view from the National Academy of Sciences.
**www.nap.edu/html/creationism/introduction.html**

"What Is the Scientific Method?" Sir Karl Raimund Popper developed the theory of the scientific method in 1934 in his book *The Logic of Scientific Discovery.*
**www.epa.gov/maia/html/scientific.html**

# Pattern Recognition

*"Things should be as simple as possible, but not simpler."*

**Albert Einstein**

Among the many ways that scientists attempt to decipher the unknown is to attempt to find patterns in what seems to be randomness. This is, in fact, a prominent characteristic human trait. Among the many complexities of the human brain is the desire to seek out patterns that exist around us or to create patterns from randomness. Pattern recognition is one of the ways we are biologically programmed to understand the world around us.

In addition, scientists are also interested in discerning irregularities and deviations in patterns. Patterns set up expectations and sometimes observing anomalies in patterns have led to major discoveries. Galileo observed that Io, a moon of Jupiter, circled Jupiter and not the Earth and therefore contradicted the 1,400-year-old Ptolemaic theory, which asserted that the Earth was the center of the universe and all heavenly bodies orbited the Earth.

## Goals

After studying this chapter, you should be able to:

- Observe patterns in nature.
- Look for patterns in physical phenomena.
- Appreciate the value scientists place on pattern recognition.
- Realize that deviations from a pattern are important.
- Understand how important scientific discoveries were made by observing patterns.

## Periodic Waveform Patterns

Patterns exist all around us and in many different forms. Perhaps the simplest pattern that we recognize is the waveform, or periodic, pattern (not to be confused with the Periodic Chart). If we were to select a point on earth and prepare a table that plotted sunlight intensity versus time, we would quickly notice a regular pattern of cycling light and darkness. As we progressed, hour-by-hour, from midnight through noon we would notice the sunlight becoming increasingly stronger. As we continued to plot light intensity versus time we would notice, as we moved from noon back to midnight, that there was an hourly decrease in sunlight intensity. If we graphed this, it might look something like Figure 2.1.

This pattern, comparable to a surface of rough ocean waves, is a characteristic waveform pattern. Since it contains a repetitive pattern, or one in which it periodically repeats itself, it may also be referred to as a periodic pattern. A pattern such as

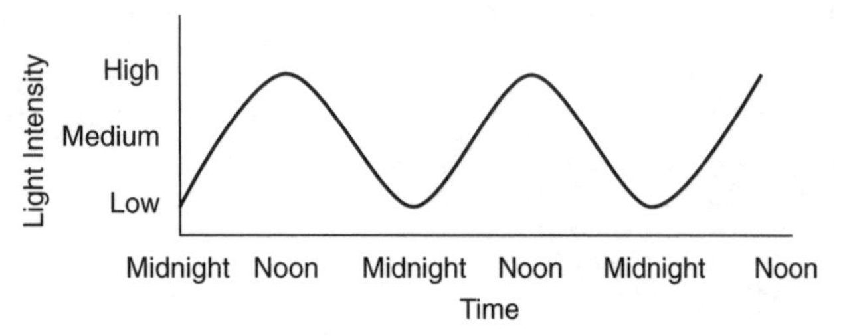


**Figure 2.1** Light intensity variation with time.

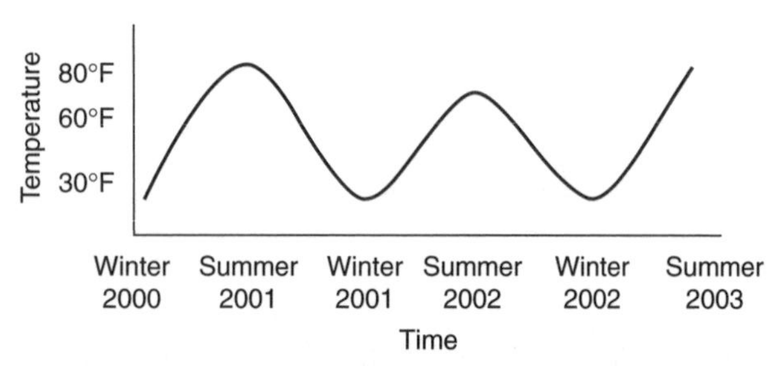


**Figure 2.2** Temperature variations with time.

this, in which some parameter is repeated on a 24-hour, or daily, cycle is referred to as **diurnal cycle.** Many biological cycles, body temperature, blood pressure, and hormone levels, also occur on a cycle that is close to diurnal. These cycles are referred to as **circadian cycles** (*circa*—about, *dia*—day). If we were to form a similar plot of the air temperature in New York over a 10-year period, we would see a similar, periodic waveform pattern. Each winter the temperatures would reach their minimum, and each summer they would reach a maximum. The plot would appear something like the pattern shown in Figure 2.2.

Waveform patterns can exist within other waveform patterns. For example, if we go back to the illustration of the daily light and dark cycle, we would notice that it was not uniform throughout the year. During the winter, when the sun's position relative to the Earth has slipped below the equator, the light intensity striking New York is less than it is in the summer when the sun's relative position places it above the equator. Therefore, the waveform pattern of daily light intensity will itself fluctuate in an annual waveform pattern with maximum amounts in the summers and minimum amounts in the winters. Such a waveform pattern within a waveform pattern would look like the pattern shown in Figure 2.3.

We can only appreciate the cycle within a cycle pattern if we look at a long enough timeframe. Consider the above pattern for example. If we viewed only a short period, perhaps the six month period from June to January, we would see the light intensity continuously decrease and would not notice the longer, annually repeating

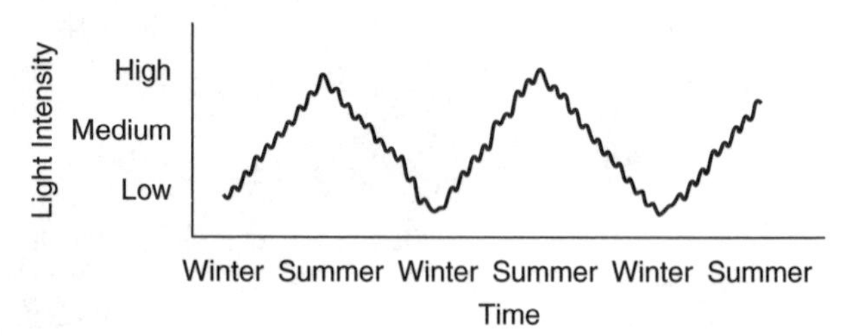


**Figure 2.3** Illustration of a waveform pattern within a waveform pattern.

cycle. If we were to draw a conclusion from such incomplete data, we might be tempted to believe that the sun had finally burned itself out and sunlight was gradually disappearing—forever. We'd have no idea that it once again would increase since that knowledge would come only from a longer observation period.

If we looked at the annual temperature waveform patterns over a million-year period, we would again see a waveform pattern within a waveform pattern. Realizing that we are now in a relatively warm period having come out of our last global ice age about ten thousand years ago, we would notice what looked like an ever-increasing annual temperature pattern.

Perhaps our fear of global warming is based on incomplete, short-term observations. Perhaps not. This is an area scientists are currently debating. Nature is full of such periodic waveform cycles. Nature is filled with simple patterns because they require the least amount of energy to generate. Can you identify any others?

# Mathematics

## Fourier Analysis

A single note sounded on a tuning fork would have a very regular and periodic waveform. As multiples of the note are added in varying amplitudes and frequencies a more complicated and fuller sound may be generated. It was Joseph Fourier, a mathematician, who discovered in 1822 that any complex wave could be analyzed in terms of simpler forms as seen in Figure 2.4. It was the observation of such patterns that motivated Fourier to develop his mathematics, which have extended the frontiers of science and engineering.

Patterns observed in nature by mathematicians and scientists in the geometry of the tetrahedral bonds (like a pyramid made up of four equilateral triangles, one at the bottom and three for the sides) of the carbon atom were employed by Buckminster Fuller in his development of geodesic structures and buildings. Patterns formed by

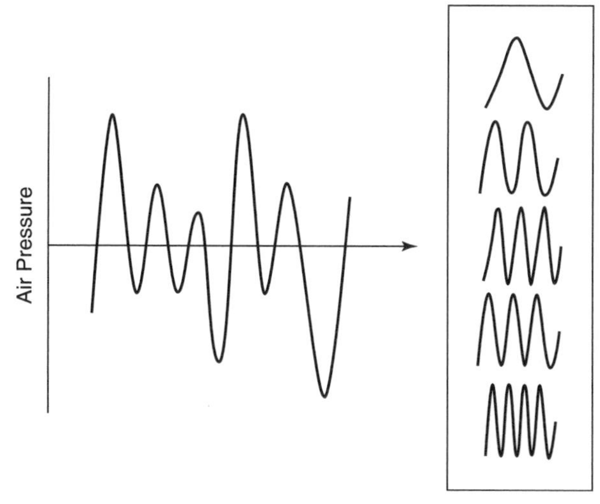


**Figure 2.4** Complex wave pattern created from summing simpler waves.

**Figure 2.5** Spiral spring with mass attached.

coalescing spherical bubbles are studied for use in designing space stations. A sphere is known to contain the largest volume using the least amount of surface area, which would seem ideal in trying to determine a shape that will maximize the volume of a space station module.

## The Spiral Spring

In studying the behavior of a spiral spring we would normally suspend it vertically and then add weights to stretch it as shown in Figure 2.5. If we were to tabulate the length of the spring as we add weights in increasing but equal increments and then look for a pattern in the data, it might be difficult to determine. However, if we draw a graph of the weight plotted against a corresponding length, the relationship between the two is revealed in that the graph turns out to be a straight line as shown in Figure 2.6. The pattern we observe is that there is a direct proportionality or linear relationship between them.

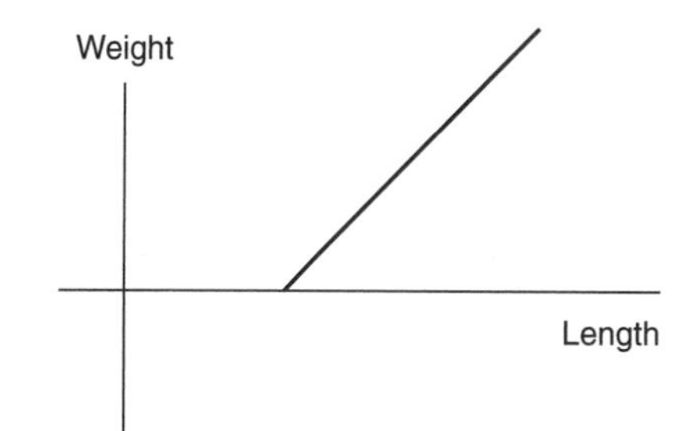


**Figure 2.6** Weight plotted vs. length of spring.

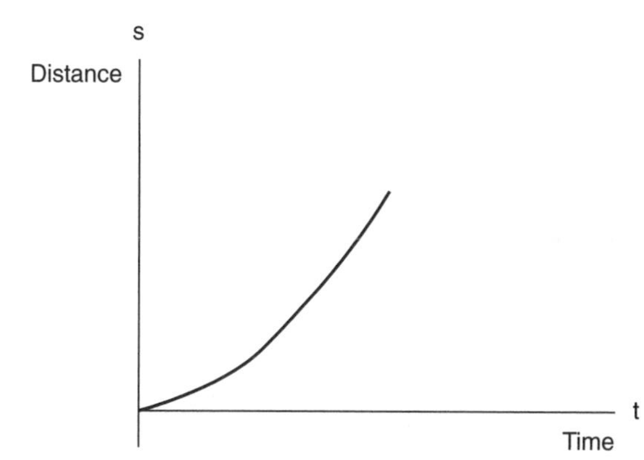


**Figure 2.7** Distance vs. time.

## The Falling Body

Suppose, like Galileo, we were to drop a sphere from the top of the Leaning Tower of Pisa and we recorded the distance it fell at equal time intervals using appropriate measuring devices. Again the table of data we would obtain might look undecipherable with no apparent pattern. However, once more when we plot the graph of displacement vs. time we see graphically the parabolic relationship between the two. The graph (Fig. 2.7) tells us that the displacement (s) varies as the square of the time (t) or s is proportional to $t^2$.

## The Fibonacci Sequence

Leonardo Fibonacci was an Italian mathematician who lived at the turn of the 13th century. Largely known for bringing us the Hindu-Arabic system of numerals (1, 2, 3, 4, 5, etc.), which replaced the Roman numerals (I, II, III, IV, V, etc.), Fibonacci is also credited with observing a specific pattern of numbers now called the Fibonacci sequence. The Fibonacci sequence is:

0, 1, 1, 2, 3, 5, 8, 13, 21, 34, 55, 89, etc.

The sequence is determined by starting with two numerals, 0 and 1, and adding them together. The sum of 0 and 1 is 1, and 1 then becomes the third numeral in the sequence.

**Figure 2.8** Spirals in nature.

0, 1

0 + 1 = 1

Therefore the sequence grows to 0, 1, 1.

To determine the next number in the sequence, add the last two numbers. Adding 1 and 1 gives us 2, so the sequence grows:

0, 1, 1

1 + 1 = 2

Therefore the sequence grows to 0, 1, 1, 2.

What number, in the Fibonacci sequence, would follow 89?

Some scientists say that nature adheres quite firmly to the Fibonacci sequence. For example, if we were to count the number of petals on a flower we would find examples of flowers with 1 petal (white calla lily), 2 petals (euphorbia), 3 petals (trillium), 5 petals (columbine), 13 petals (black-eyed susan), 21 petals (shasta daisy), and 34 petals (field daisies). Flowers with 4, 6, 9, or 15 petals would be very rare or nonexistent. The same would be found with leaf numbers.

The Fibonacci sequence forms the basis for many patterns seen in nature. The pattern seen (Fig. 2.8) in snail shells, pineapples, pinecones, and sunflower seeds within the flower all follow the Fibonacci pattern.

## The Fibonacci Constant

If we were to select any of the Fibonacci numbers and divide it by the number that came immediately before it in the sequence, starting with one we would find the following results (rounded off to the nearest tenth):

$1 \div 1 = 1$

$2 \div 1 = 2$

$3 \div 2 = 1.5$

$5 \div 3 = 1.7$

$8 \div 5 = 1.6$

$13 \div 8 = 1.6$

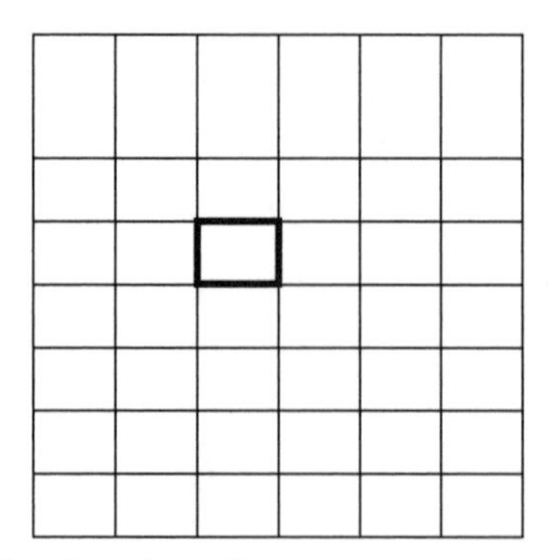

**Figure 2.9** One box 1.

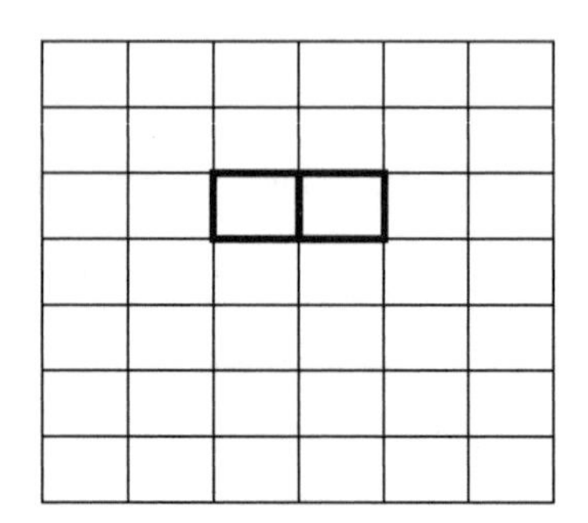

**Figure 2.10** Two boxes 1, 1.

$21 \div 13 = 1.6$

$34 \div 21 = 1.6$

If we were to continue, we'd find that all of the subsequent divisions would ultimately reduce to 1.6 (actually they would move increasingly closer to an ultimate 1.618). This is the **Fibonacci constant** and, again, many believe that it reflects, or is reflected in, many patterns in nature. The Fibonacci constant actually appears to approximate the ratio that is seen in many natural systems; for example the ratio of the length of the distance from your fingertip to your elbow relative to the length of your arm; the length of the distance between your hairline to the bottom of your nose relative to the distance between your hairline and your chin. Artists when drawing the human body use the Fibonacci constant quite frequently.

## The Fibonacci Spiral

If we were to outline a single square on a piece of graph paper and then to its immediate right, outline another square we would have Figure 2.9 and Figure 2.10. These two boxes, each with one outlined square, represent the first two whole numbers of the Fibonacci sequence, 1, 1.

Next, outline a box above them that is two squares tall by two squares wide. This represents the next number in the sequence, 1, 1, 2.

The right side of these outlined boxes is now three squares tall (Fig. 2.11), representing the next

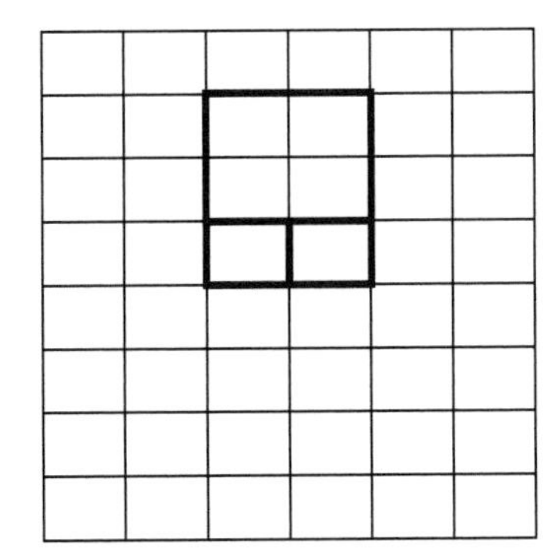

**Figure 2.11** 1, 1, 2.

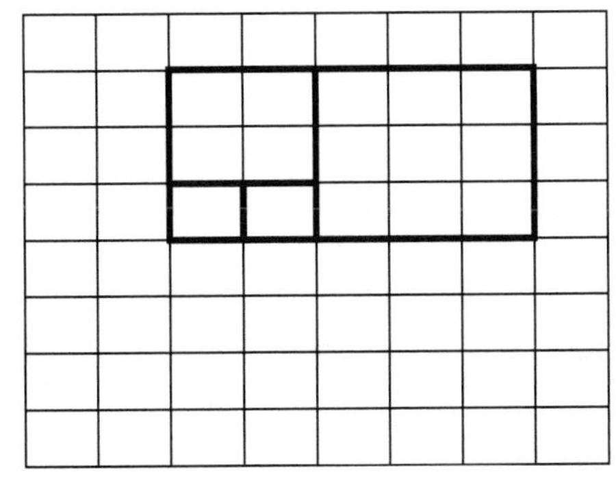

**Figure 2.12** 1, 1, 2, 3, 5.

number in the Fibonacci sequence: 1, 1, 2, 3. Complete the box that is now three squares tall by three squares wide.

We now have an outlined section that is five boxes wide on its bottom (Fig. 2.12), the next number in the sequence. If we continued on, we'd note that each successive box reflected the next number in the sequence.

If we were to draw an arc within each of the boxes, running from corner to corner, they would connect in a spiral that is reflected quite frequently in nature (Fig. 2.13).

## Music

Very closely related to mathematics is music, which is simply an expression of an audible pattern in time. It is the regularity of the pattern of the complicated sound waves striking our ears that most people find pleasing. Noise, on the other hand, is characterized by irregularity and randomness. The composer many times leads us to expect a certain regularity in his music and then deviates to surprise and create interest. If we look at the score as created by the composer, we can visibly see the regularity in the pattern of the coded symbols (notes) as they progress from bar to bar on the staff going from left to right (Fig. 2.14).

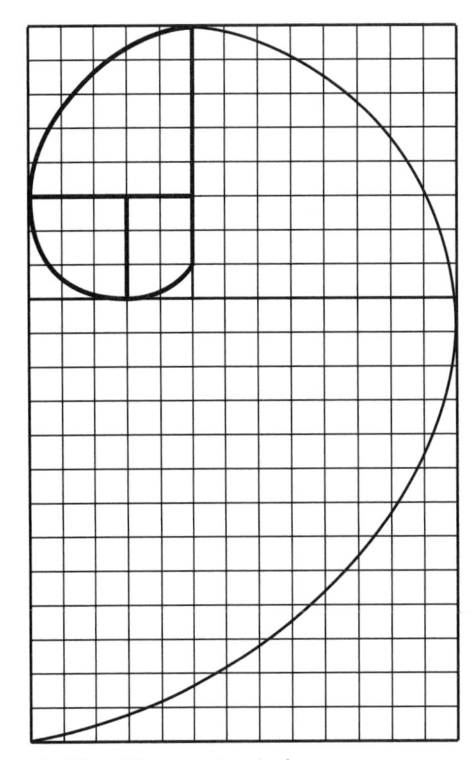

**Figure 2.13** Fibonacci spiral.

**Figure 2.14** Musical notes from computer sequencer. *http://www.niehs.nih.gov/kids/lyrics/america.htm*

## Crystallography

Crystals are solid substances that have regular shapes; that is, there are plane surfaces—the face or facet of the crystal that intersects other faces at regular angles (Fig. 2.15).

This crystallinity really reflects the fact that the solid is composed of atoms or molecules, which themselves are arranged in a regular pattern. In a diamond the arrangement is illustrated in Figure 2.16.

Understanding the use of symmetry and patterns of crystals to solve problems is a worthwhile consideration. It turns out that if one shines X-rays of a single wavelength on a single crystal of a substance, a certain regular pattern of the X-ray radiation is emitted (Fig. 2.17).

**Figure 2.15** Crystals. *http://www-structure.llnl.gov/crystal_lab/Crys_lab.html*

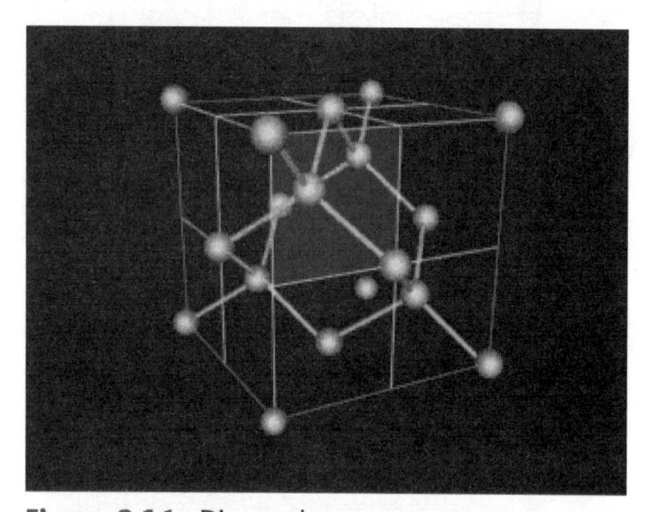

**Figure 2.16** Diamond arrangement.

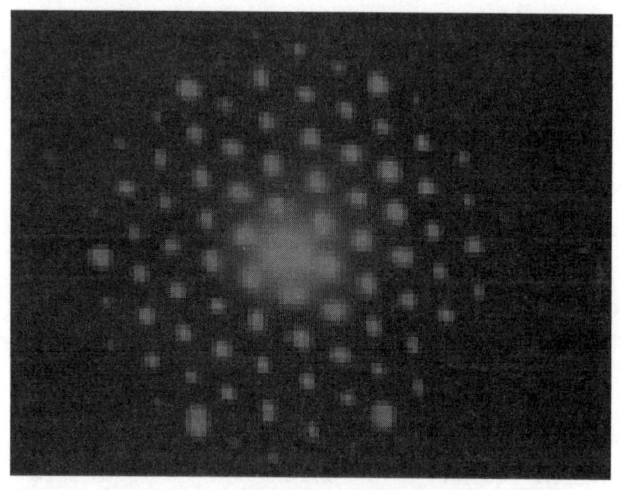

**Figure 2.17** X-ray of crystal. *http://www.nsls.bnl.gov/about/everyday/PDFs/crystals.pdf*

This pattern is related to the symmetry of the molecules within the crystal, the location of the atoms, the nature (kind) of atoms present, the structure of the molecule, and how the atoms are put together. X-ray patterns enable one to also determine how a molecule is put together. X-ray structure determination is an unequivocal means of determining molecular structure. Knowledge of molecular structure is of paramount importance if one hopes to synthesize a compound or to determine a cause of malfunction of a chemically or biologically based process.

# DNA

The technique of X-ray crystallography utilizes a two-dimensional photograph of a crystal. The photography consists of a pattern of spots and their intensities, which reflects the spacing and position of the atoms in a crystal. Working backward from this two-dimensional picture (Fig. 2.18) the three-dimensional representation of the crystal can be reconstructed. It was Francis Crick who created mathematical models that allowed him to decipher such two-dimensional patterns. Using DNA and the X-ray method he and James Watson discovered the

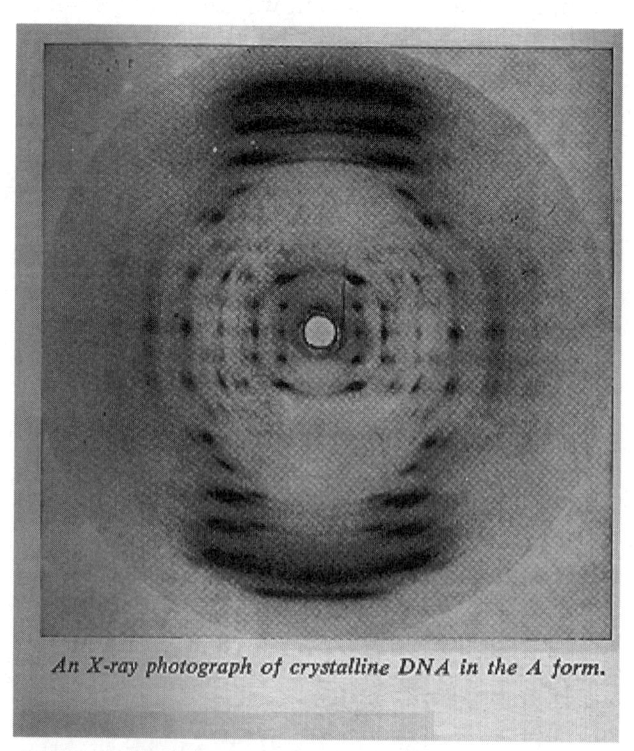



**Figure 2.18** X-ray photograph of crystalline DNA in the A form.

*The Double Helix,* by James Watson, Published by Mentor, a division of Penguin Group, New York, 1968.

double-helix structure of DNA, possibly the greatest scientific discovery of the 20th century.

It is also interesting to note this technique of studying crystal structure allowed crystallographers to identify 32 different types of symmetry that can be seen in crystals. The 13th-century Alhambra in Spain contains all the crystal symmetries identified by the crystallographers in the mosaics of the walls and floors. Muslim artists, restricted from representing the human form in their art, used geometrical (See fig. 2.19) designs and apparently exhausted all possible symmetries anticipating discoveries in X-ray crystallography in their designs for the Alhambra. We See that pattern recognition is not only useful in solving problems of science, but decorative and pleasing to the eye. In societies where human and animal figures are not used in art for religious reasons, such geometric patterns are a welcomed and inventive route of expression.

**Figure 2.19** Mosaic pattern, Moracco. Photo by C. Campisi.

## Pattern Recognition

Pattern recognition is a tool that is used by different segments of society to solve problems, deduce trends, or in general achieve a greater understanding about things. In police work for example the M.O. or modus operandi, method of operation, that a criminal uses is a pattern, which can be used to identify a perpetrator.

In an analogous vein, fingerprint patterns (Fig. 2.20) or patterns derived from DNA, DNA "fingerprints," are also used to either exonerate or accuse suspects. The ridges on the fingers create a pattern, which characterizes each individual uniquely.

The fingerprint patterns are genetically determined and, being so intricate and complicated, as manifested in their genetic variability, are believed to be unique to each individual. Analysis of these patterns, by either the expert eye or complicated computer algorithms, can be used to identify individuals with a high degree of certainty.

Every individual's DNA is essentially unique (except for twins). DNA fingerprinting consists of processing an individual's DNA and obtaining a pattern of the DNA components. The pattern is used for identification purposes just as fingerprints

**Figure 2.20** Fingerprints. *http://www.fbi.gov/kids/k5th/whatwedo3.htm*

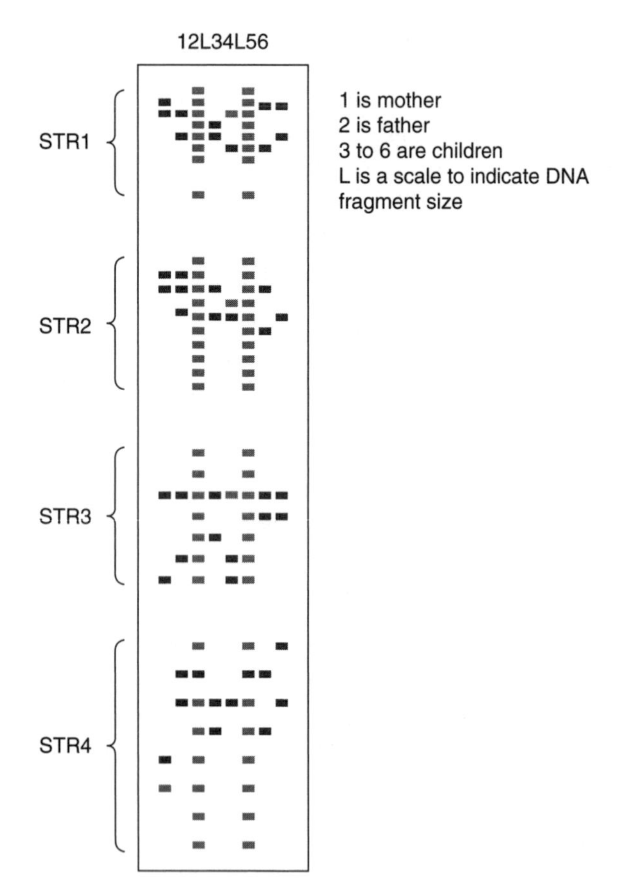


**Figure 2.21** DNA fingerprinting. *www.biotechnology.gov.au/*

are. Unlike fingerprints, which are physical ridges on the fingertips, DNA "fingerprints" are actually patterns of nucleotide sequences with the DNA of every cell. While fingerprints can be removed (acid, abrasion, scarring), and you must have one of the ten fingers available for analysis, DNA fingerprinting can be done on microscopic remnants of anything that contains cells, for example: hair, blood, saliva, skin cells left on towels and clothing, etc. See Figure 2.21.

Other patterns that characterize individuals include speech and written patterns. The fact that a check for a cash payment can be safely used by anyone is based on the uniqueness of one's pattern of handwriting. Pattern recognition involving speech patterns, including pronunciation and accents, can be used to help establish an individual's background.

## The Unabomber

In one famous case a deranged bomber was captured because of his unique mode of expression. The criminal, known as the Unabomber, was at one time a professor of mathematics. The individual, Theodore Kaczynski, at some point in his life lost contact with reality. He became angry at the direction industry and science and technology were going. Sadly, he acted on his conviction in a violent manner. Over a span of 20 years he periodically had mail bombs delivered to various places or individuals whom he felt were contributing to make this "bad" society.

About 26 of these bombs were delivered. Many resulted in injury and even death to the recipients. Over the 20-year period in which he acted, intensive searches and activity by the authorities had yielded very little. He lived alone, in the woods, far from others. His day-to-day actions were fairly normal as well as his encounters with others when they occurred. He probably would still be free today had not his mad mission caused him to act in other ways. So how was he caught?

He was so impassioned about his cause that he communicated with several newspapers, including the *New York Times,* in order to "get the word out" and "promulgate his ideas." The paper published his manifesto entitled "Industrial Society and Its Future."

Many, including his own brother who recognized some unique written expressions, which were used fairly exclusively by Ted Kaczynski, read his manifesto. This discovery set the wheels of justice in motion. Heartbroken, Kaczynski's brother notified authorities and the Unabomber was apprehended and ultimately confined. Pattern recognition was the key to solving a 20-year-old case.

## Periodic Table

Perhaps the most elegant and important application of pattern recognition applied to the sciences was the creation of the Periodic Chart of the elements. As complex as chemistry seems to us today, particularly when we are first trying to learn it, imagine how complex and puzzling the material world must have seemed when we first discovered that matter was composed of multiple different elements which we knew little about. As we began to learn more about the nature of individual elements, patterns in their individual collective structures and behaviors became apparent.

The periodic table represents an important resource where scientific relationships are summarized in a tabular and succinct manner. It is an extremely useful tool for chemists. The one individual who did more than any other individual to recognize the pattern and to organize the chemical elements into a useful tabular array was Dmitri I. Mendeleev.

The facts were these: There are a number of different chemical elements, which behave simi-

## TABLE 2.1 Dmitri Mendeleev's Periodic Table

**Tabelle II**

| REHIEN | GRUPPE I.<br>—<br>$R^2O$ | GRUPPE II.<br>—<br>RO | GRUPPE III.<br>—<br>$R^2O^3$ | GRUPPE IV.<br>$RH^4$<br>$RO^2$ | GRUPPE V.<br>$RH^3$<br>$R^2O^5$ | GRUPPE VI.<br>$RH^2$<br>$RO^3$ | GRUPPE VII.<br>RH<br>$R^2O^7$ | GRUPPE VIII.<br>—<br>$RO^4$ |
|---|---|---|---|---|---|---|---|---|
| 1 | H = 1 | | | | | | | |
| 2 | Li = 7 | Be = 9, 4 | B = 11 | C = 12 | N = 14 | O = 16 | F = 19 | |
| 3 | Na = 23 | Mg = 24 | Al = 27, 3 | SI = 28 | P = 31 | S = 32 | Cl = 35, 5 | |
| 4 | K = 39 | Cd = 40 | – = 44 | Ti = 51 | Cr = 52 | Mn = 55 | | Fe = 58, Co = 59, Ni = 59, Cu = 63. |
| 5 | (Cu = 63) | ZN = 65 | – = 68 | – = 72 | As = 75 | Se = 78 | Br = 80 | |
| 6 | Rb = 85 | Sr = 87 | ?Yt = 88 | Zr = 90 | Nb = 94 | Mo = 96 | – = 100 | Ru = 104, Rh = 104, Pd = 106, Ag = 108 |
| 7 | (Ag = 108) | Cd = 112 | In = 113 | Sn = 118 | Sb = 122 | Te = 125 | J = 127 | |
| 8 | Cs = 133 | Ba = 137 | ?Di = 138 | Ce = 140 | — | — | — | — — — — |
| 9 | (—) | — | — | — | — | — | — | |
| 10 | — | — | ?Er = 178 | ?Lo = 180 | Ta = 182 | W = 184 | — | Os = 195, Tr = 197, Pt = 198, Au = 199 |
| 11 | (Au = 199) | Hg = 200 | Tl = 204 | Pb = 207 | Bi = 208 | — | — | |
| 12 | — | — | — | Th = 231 | — | U=240 | — | — — — — |

Figure 2.5 Dmitri Mendeleev's 1872 periodic table. The spaces marked with blank lines represent elements the Mendeleev deduced existed but were unknown at the time, so he left places for them in the table. The symbols at the top of the columns (e.g., R2O and RH4) are moleculer formulas written in the style of the 19th century.

*http://pearl1.lanl.gov/periodic/mendeleev.htm*

larly. For example, sodium and potassium are metals; they are both silvery and soft. They can each be cut with a knife. They both react with water to yield hydrogen gas and a basic solution. In short, their physical and chemical properties are quite similar.

There are other examples of several elements behaving in similar ways. The properties of helium and neon for example are quite similar. They are both gases that are quite stable and nonreactive and, in fact, were once known as inert gases. Many scientists of the day were aware of the fact the groups of certain atoms had similar properties and some had attempted to represent this information in a coherent way. How did Mendeleev succeed in recognizing and organizing the vague patterns the elements' properties exhibited? Medeleev's solution to uncovering the pattern was to write the names of the elements on a card; one card for one element. The card contained the name of the element and its atomic weight.

He played a kind of solitaire, distributing the cards in different ways and looking for arrangements of the cards where elements with similar properties might be together. Mendeleev's work was made more difficult because many of the elements that are now known (helium, for example) were not known at the time of his work. In addition, there were questions concerning the atomic weight of some of the elements. Despite these impediments Mendeleev was successful. Ultimately, he organized the elements according to increasing atomic weight, listing the lightest, then the next heaviest and so on. His genius was in knowing when to start a new row and in leaving spaces for yet undiscovered elements.

(He did an amazing job predicting the properties of the missing elements, which were discovered later.) For example, in his 1871 periodic table (Table 2.1) he left a space for an element he believed would be discovered in the future. The element was to be located in the space between Galium and Arsenic and below Silicon. He called it Eka-Silicon—beyond silicon. Examine his prediction and what was observed in Table 2.2. A modern periodic table is listed in Table 2.3.

The law that governs this arrangement is called the periodic law. It states that the physical and chemical properties of the elements are periodic or recurring functions of their atomic weight. As it turns out, a better relationship follows if atomic number is used instead of atomic weight. Examine the pattern that is created when one plots atomic radius in picometers vs. atomic number (Figure 2.22).

Can you observe the periodicity and the pattern?

## EPIDEMIOLOGY

Cholera is a gut-wrenching, acute, diarrheal illness caused by infection of the intestine by the bacterium *Vibrio cholerae*. A person can get cholera by drinking water or eating food that was contaminated by the feces of an infected person. It is characterized by a rapid loss of bodily fluids, causing dehydration and shock. Death can result in as little as a week.

In 1854, there was a severe cholera epidemic in Europe and at that time its cause was unknown and the epidemic spread. It was widely thought at the time that disease was spread by the inhalation of vapors. In England, Dr. James Snow in his publication *On the Mode of Communication of Cholera* published in 1849, hypothesized that cholera was spread through contaminated food or water. At the time of the epidemic he walked through the streets of London documenting every death and plotting their locations on a map.

He discovered that at the time, two water companies supplied London. One of them pumped water from the Thames River upstream of the main city while the other pumped its water downstream of the main city. The part of the city supplied with water pumped by the Southwark and Vauxhall Company from downstream was found to have a high concentration of cholera victims, leading Snow to believe that the water might be contaminated by the city's sewage.

At the intersection of Broad Street and Cambridge he found that there were more than five hundred deaths in 10 days. Curiously, he also detected deviations from this pattern. A nearby brewery (Fig. 2.23) at Broad Street had no cases. It

**TABLE 2.2 Eka-Silicon**

| Property | Eka-silicon Predicted 1871 | Germanium Observed 1886 |
|---|---|---|
| Atomic Mass | 72 | 72.3 |
| Atomic volume | 13 ml | 13.2 ml |
| Density | 5.5 gm/ml | 5.47 gm/ml |
| Specific heat | 0.073 cal/gm C° | 0.076 ca/gm C° |

**TABLE 2.3 Modern Periodic Table**

**Los Alamos National Laboratory's Chemistry Division Presents a Periodic Table of the Elements**

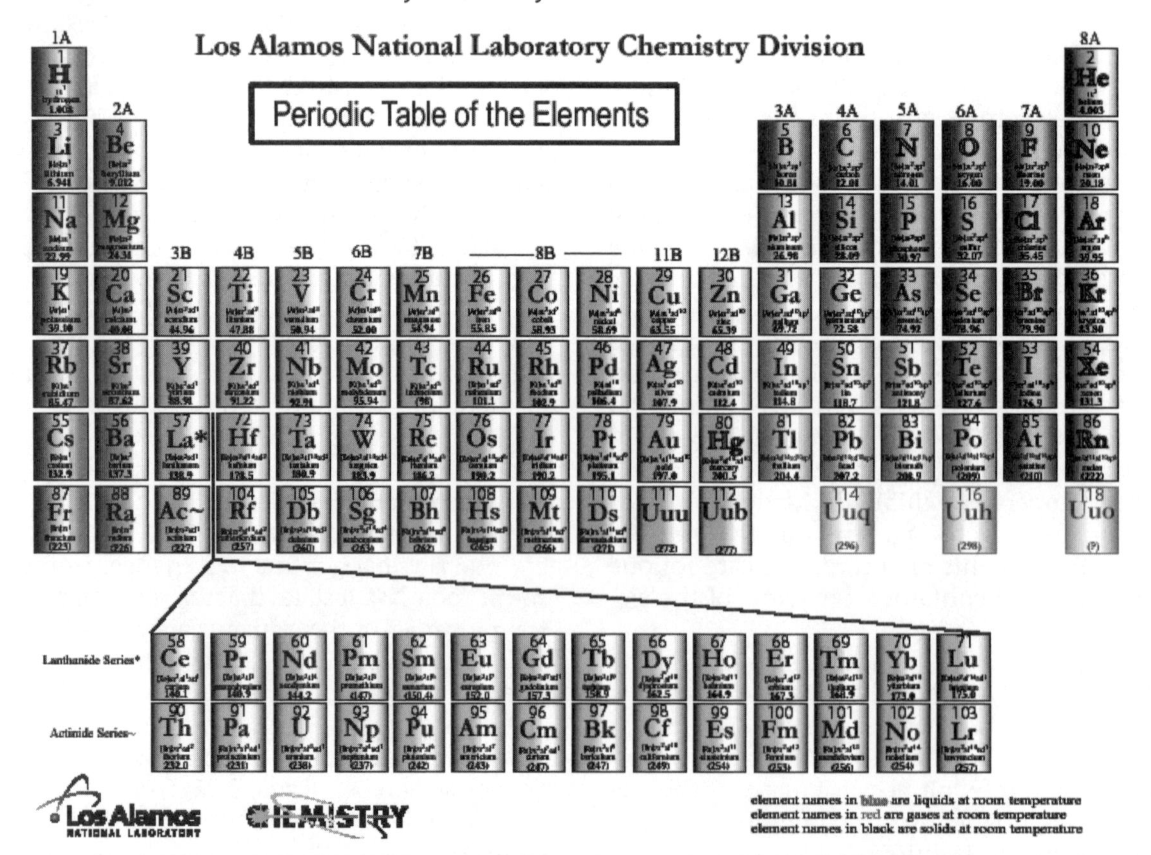

turned out upon further investigation they did not use the water from that pump, saving its 70 workers from the epidemic.

He urged the city officials to remove the handle from the Broad Street pump, and the epidemic subsided within three days. It was Snow who carefully mapped the location of cholera victims, which allowed him to determine the major causes of the disease verifying his hypothesis.

His work earned him the title of "the father of modern epidemiology."

## Evolution

Bishop James Ussher, Anglican Archbishop of Armagh, Primate of all Ireland, published a work in which he wrote that the earth was created in 4004 B.C. His arguments were based on biblical evidence. In the 18th century a great network of canals was being constructed in England. The cuts they made in the earth revealed layers of geological formations containing fossils. These strata seemed to have been laid down in layers by sediments. Some were later folded and twisted by great temperature and pressures. Certainly a pattern was evident.

James Hutton's theory of uniformitarianism, which, briefly stated, says the same processes acting today geologically, operated in the geologic past, seemed to provide the interpretation for the pattern. It wasn't catastrophic events that produced these layers and other features of the earth's landscape as the supporters of Ussher might claim, but slow, weak geologic processes, acting over long periods of time. Darwin in promulgating his theory of evolution acknowledged the work of the geologists.

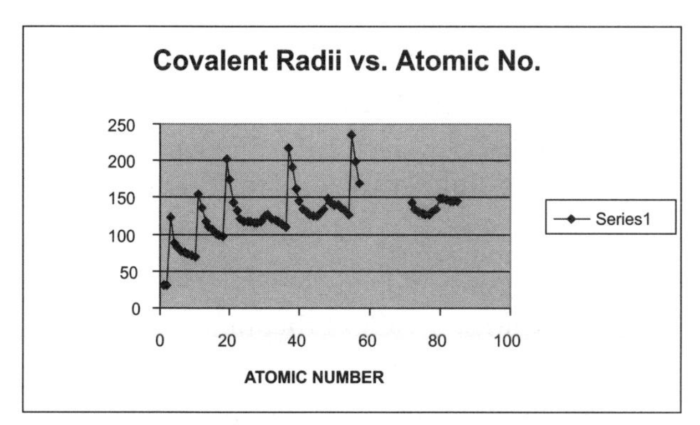


**Figure 2.22** Covalent radii vs. atomic number.

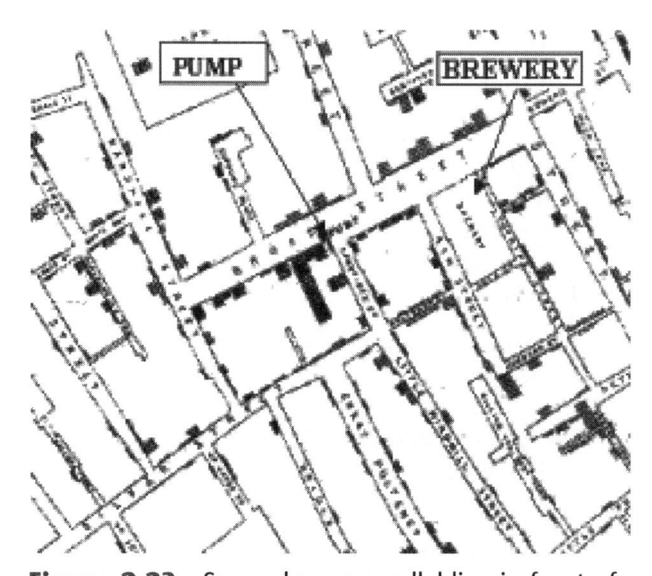


**Figure 2.23** Snow drew a parallel line in front of each building for each person who died there.

*On the Mode of Communication of Cholera* by John Snow, M.D., London: John Churchill, England, 1855 *http://www.ph.ucla.edu/epi/snow/snowmap1.pdf*

## Differential Medical Diagnoses

Pattern recognition is a critical tool in diagnosis of medical conditions. One would suspect an individual had measles if he displayed the following symptoms.

**Rubella Symptoms & Signs**

- Low-grade fever (102° F or lower)
- Headache
- General discomfort or uneasiness (malaise)
- Runny nose
- Inflammation of the eyes (bloodshot eyes)
- Rash with skin redness or inflammation
- Muscle or joint pain
- Encephalitis (rare)
- Bruising (from low platelet count, rare)

That is, particular health patterns (symptoms) are indicative of specific illnesses.

In some primitive societies the elderly were held in high esteem. There are still certain parts of

**Figure 2.24** Broad Street pump *http://www.ph.ucla.edu/epi/snow/removal.htmlhttp://www.ph*

the world where this is true today. (In the West, demographics seem to focus mainly on the young.) So why do you suppose older people were treasured? Certainly love, devotion, and religion played a role in such veneration, but perhaps also there were practical reasons as well.

In primitive societies where the ability to write and permanently record information was unknown, the elderly represented rare repositories of information critical to a group's existence and survival. *They* were the reference books. Through their lifetime the elderly had learned to recognize the patterns of various illnesses and their treatment, the patterns associated with planting and crop growing, the patterns of the seasons, and of people and their behavior. Their ability to recognize patterns and the implications associated with patterns made these individuals significant and active participants in their society. Since those early times, pattern recognition has been a key component in problem solving dealing with scientific and technological issues.

There are many more examples we can give illustrating the importance of recognizing patterns in the world around us. The skill of recognizing patterns in science and engineering cannot be overemphasized. It is one of the most powerful tools available to the research scientist.

We see that pattern recognition is a skill that every scientist and engineer develops. It is a tool that allows the scientist and engineer to see order in randomness and understand the happenings in the world around us. It is also deviations from regular patterns that provoke the curiosity of scientists leading them to make new discoveries. It is a theme that runs through the rest of this book.

## Review Questions

1. Why are the patterns we find in nature so simple?
2. What is the relationship between nature and the Fibonacci sequence?
3. Who was Joseph Fourier and why is his work important?
4. What is X-ray diffraction?
5. Discuss some practical applications of pattern recognition.
6. Who was Dmitri Mendeleev and how did his discovery contribute to the advancement of science?
7. What is epidemiology? What was Dr. James Snow's contribution to epidemiology?
8. How would Bishop Ussher explain the formation of the earth's landscape which is made up of mountains and rivers? What was his thesis?
9. How is pattern recognition used in medicine?
10. What do senior citizens have to contribute as far as pattern recognition is concerned?

## Multiple Choice Questions

1. Nature is filled with simple patterns because they require the least amount of _____ to generate.
   a. entropy
   b. energy
   c. time
   d. elements
2. The gyrations of the stock market may be characterized as a _____.
   a. cyclic pattern
   b. diurnal cycle
   c. uranal cycle
   d. circadian cycle
3. A complicated waveform may be generated by adding simpler waves. This is known as _____.
   a. Fourier synthesis
   b. Pythagorean addition
   c. Buckminster Fuller's theorem
   d. Ussher's thesis
4. The next number in the Fibonacci sequence 0, 1, 1, 2, is _____.
   a. 1
   b. 2
   c. 3
   d. 4
5. Sounds that have random and irregular waveforms are called _____.
   a. music
   b. harmony
   c. melody
   d. noise
6. X-ray crystallography allows us to view _____.
   a. electrons
   b. bones
   c. crystal structures
   d. 3-d pictures of atoms
7. The Unabomber was _____.
   a. a Mediterreanean terrorist
   b. a member of al Quaeda
   c. a deranged mathematician
   d. George Metsky

8. The periodic table contains _____.
   a. the diurnal cycles
   b. the circadian cycles
   c. chemical elements
   d. the solitaires
9. The theory of evolution was promulgated by _____.
   a. Bishop Ussher
   b. D. Mendeleev
   c. Dr. James Snow
   d. Charles Darwin
10. Uniformitarianism is _____.
    a. a religion
    b. a theory of geological processes
    c. a theory in epidemiology
    d. based on biblical evidence
11. What comes next in the sequence 2, 4, 6, 8, . . . ? _____.
    a. 3
    b. 5
    c. 7
    d. 10
12. What comes next in the sequence A, C, E, G, . . . ? _____.
    a. B
    b. D
    c. H
    d. I
13. What comes next in the sequence do, re, . . . ? _____.
    a. mus
    b. money
    c. mio
    d. mi
14. What comes next in the sequence 0, 1, 1, 2, 3, . . . ? _____.
    a. 5
    b. 8
    c. 13
    d. 4
15. What comes next in the sequence ⍋ 2Ƨ 3Ɛ Ⴤ ...?
    a. X
    b. Y
    c. ŏ
    d. ж

# Bibliography and Web Resources

*The Search for Solutions,* by Horace Freeland Judson, Holt, Rinehart, Winston, 1980.

## Cholera

DBMD—Cholera—General information on cholera from the Centers for Disease Control.
**www.cdc.gov/ncidod/dbmd/diseaseinfo/cholera_g.htm**

*On the Mode of Communication of Cholera,* by John Snow, M.D., London: John Churchill, New Burlington Street, England, 1855.
**http://www.ph.ucla.edu/epi/snow/snowbook4.html**

**http://www.ph.ucla.edu/epi/snow.html**

Snow's Map.
**http://www.ph.ucla.edu/epi/snow/snowmap1.pdf**

"Who We Are—History of John Snow." With this first known example of epidemiological research, Dr. Snow became the "father of modern epidemiology."
**www.jsi.com/aboutjsi/history.htm**

"John Snow: The London Cholera Epidemic of 1854," by Scott Crosier. This is a portion of the original map created by Dr. John Snow.
**www.csiss.org/classics/content/8**

## Crystals

Plasterwork wall decoration at the Alhambra, Granada, Spain.
**www.fandm.edu/Departments/Mathematics/a.crannell/hm/Spain/Granada/alhambra_other.html**

The main point group symmetries of interest to defect physics by operation (reflection, rotations).
**newton.ex.ac.uk/people/goss/symmetry/**

Introduction to Space Groups.
**www-structure.llnl.gov/xray/tutorial/spcgrp_tut.htm-8k-Cached**

Linus Pauling, Sir William Lawrence Bragg, *From Proteins to DNA*. Astbury's X-Rays. Ronwin Structure. Watson and Crick. First attempts.
**osulibrary.orst.edu/specialcollections/coll/pauling/dna/narrative/page6.html**

7 crystal systems.
**www.keele.ac.uk/depts/ch/resources/xtal/classes.html**

Fingerprints.
**onin.com/fp/fphistory.html**

Unabomber attacks.
**www.unabombertrial.com/**

Symptoms and illness.
**www.merck.com/mrkshared/mmanual/sections.jsp**

# Problem Solving

## Blocks and Strategies

*"Genius is one percent inspiration and ninety-nine percent perspiration."*

**Thomas Edison**

This chapter focuses on problems and the various strategies used to solve them. You will become familiar with different methods such as the gap method, Polya's method, contingency tables, and most importantly scientific methods that may be useful in solving problems. Problem-solving ability is a skill and therefore can be acquired through hard work and practice. A basketball coach can show you how to shoot "free throws" at the basket but the player has to practice to gain this skill. The instructor will demonstrate methods and strategies for solving problems, and the student will acquire the skills through imitation and practice.

In solving problems you also have to be aware of boundary conditions associated with each problem. Sometimes these occur in the form of blocks, which hinder solution. Being aware of such blocks turns out to be a first step in becoming a good problem solver. In addition, two additional ingredients are required: persistence and critical thinking skills, which will be explored in this chapter.

## Goals

After studying this chapter, you should be able to:

- Be aware of strategies for solving problems.
- Define what is meant by the "scientific method."
- Understand the common elements of all scientific methods.
- Distinguish between observations, hypotheses, testing, theory, and law.
- Be aware of blocks that hinder problem solving.
- Be aware of boundary conditions for problems.
- Be able to identify critical thinking skills.

## Problem Solving

Have you ever been driving your car alone, somewhere far from your home when suddenly it breaks down? You're towed to a repair shop where the problem is "diagnosed" and a solution is proposed along with a bill for doing a repair. What action should you take? How do you work through this problem?

Here's a second dilemma. You are given a screening test for a rare disease, which occurs in one percent of the population. The sensitivity (and specificity) of the test are both 90%. You test positive. What are the chances that you are a false positive? Do you consent to go ahead with a costly and risky treatment?

Suppose you live in a city where a debate is raging concerning the quality of the municipality's drinking water. One of the politicians proposes adding yet another chemical to the water supply

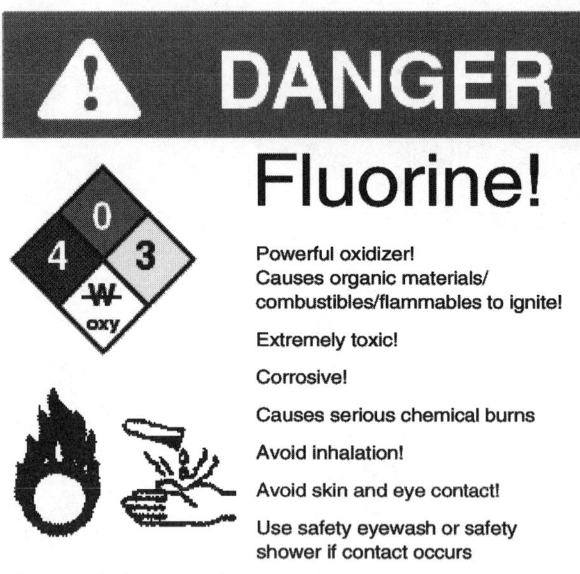


**Figure 3.1** Fluorine warning sign. *http://www.llnl.gov/es_and_h/hsm/doc_14.06/doc14-06.html#2.1*

that would further reduce the water's "natural composition." The suggestion is made to add trace amounts of fluoride to water in order to reduce the incidence of tooth decay among children. Fluoride however is a deadly poison (Fig. 3.1). The issue of adding fluoride to the city's water supply is to be put to a vote. What must you do to vote as a responsible citizen?

The examples just described represent situations involving science and technology that confront us in our daily lives. There are a variety of approaches to achieving solutions to those problems that arise, which are affiliated with scientific and techonological issues. Certainly, being extremely knowledgeable about the scientific aspects of the areas in question is very helpful. (There are many other issues that are multifaceted and have, for example, scientific and moral or ethical dimensions; these are more complex and challenging to resolve.)

In the above examples expertise in auto mechanics, in screening tests and medicine as well as chemistry, biology, and dentistry would make a solution easier to work out. While such skills and expertise are desirable, the reality is that the vast majority of the population is not so technically and scientifically well-versed. How then does the "nonexpert" arrive at solutions to problems of a scientific and/or technological nature?

Sometimes we can rely on experts. Then other questions arise. How trustworthy are the experts? How often are they wrong? It turns out there are some generalized tactics and approaches to getting answers for ourselves. In achieving answers to situations as described earlier, of course some information is needed. The more we have, the better off we are. However time constraints usually limit the amount of time devoted to learning efforts and so seeking solutions requires problem-solving tactics.

## Problem-Solving Methods

In brief, problem-solving methods consist of problem recognition, application of the scientific method, and application of critical thinking skills, which are free of blocks. Problem-solving tactics also make use of pattern recognition, models and modeling, analysis of information, and codified data, and sometimes require approaching problems in terms of a large general unit or a system. When a systems approach is used, analysis of problems is organized in terms of input, processing, feedback, control, and output.

The first step in solving a problem is to recognize that a problem does in fact exist. Once a problem has been properly identified and articulated, efforts can be directed toward its resolution. How do we know we have a problem? According to the dictionary a problem is something that requires a solution. Using this definition, problems will usually be apparent but not always.

## Polya's Method

In his 1945 book *How to Solve It: A New Aspect of the Mathematical Method,* George Polya, a Hungarian mathematician born in 1887, came up with what has become a classic four-step description of the problem-solving process (see Fig. 3.2).

1. Define and understand the problem.
2. Devise a plan for solving the problem.
3. Carry out the plan.
4. Examine the solution to see if it really solves the problem.

## Gap Model

Another problem identification mode relies on a graphical representation of a problem and is known as the "gap model" of a problem. The representation is pictured in Figure 3.3.

The actual state represents the current condition, where we are at, or the present state. The goal state or desired state is where we would like to be. If both conditions coincide, there is no problem. If

## HOW TO SOLVE IT

UNDERSTANDING THE PROBLEM

**First.**
You have to *understand* the problem.

*What is the unknown? What are the data? What is the condition?*
Is it possible to satisfy the condition? Is the condition sufficient to determine the unknown? OR is it insufficient? Or redundant? Or contradictory?
Draw a figure. Introduce suitable notation.
Separate the various parts of the condition. Can you write them down?

DEVISING A PLAN

**Second.**
Find the connection between the data and the unknown. You may be obliged to consider auxillary problems if an immediate connection cannot be found. You should obtain eventually a *plan* of the solution.

Have you seen it before? Or have you seen the same problem in a slightly different form?
*Do you know a related problem?* Do you know a theorem that could be useful?
*Look at the unknown?* And try to think of a familiar problem having the same or a similar unknown.
*Here is a problem related to yours and solved before. Could you use it?* Could you use its result? Could you use its method? Should you introduce some auxilliary element in order to make its use possible?
Could you restate the problem? Could you restate it still differently? Go back to definitions.
If you cannot solve the proposed problem, try to solve first some related problem. Could you imagine a more accessible related problem? A more general problem? A more special problem? An analogous problem? Could you solve a part of the problem? Keep only a part of the condition, drop the other part; how far is the unknown then determined; how can it vary? Could you derive something useful from the data? Could you think of other data appropriate to determine the unknown? Could you change the unknown or the data, or both if necessary, so that the new unknown and the new data are nearer to each other? Did you use all the data? Did you use the whole condition? Have you taken into account all essential notions involved in the problem?

CARRYING OUT THE PLAN

**Third.**
*Carry out* your plan.

Carrying out your plan of the solution, *check each step.* Can you see clearly that the step is correct? Can you prove that it is correct?

LOOKING BACK

**Fourth.**
*Examine* the solution obtained.

Can you check the *result?* Can you check the argument? Can you derive the result differently? Can you see it at a glance? Can you use the result, or the method, for some other problem?

**Figure 3.2** Polya's how to solve it list.
George Polya, *How to Solve It,* Fig. 6.1 1973, Princeton University Press.

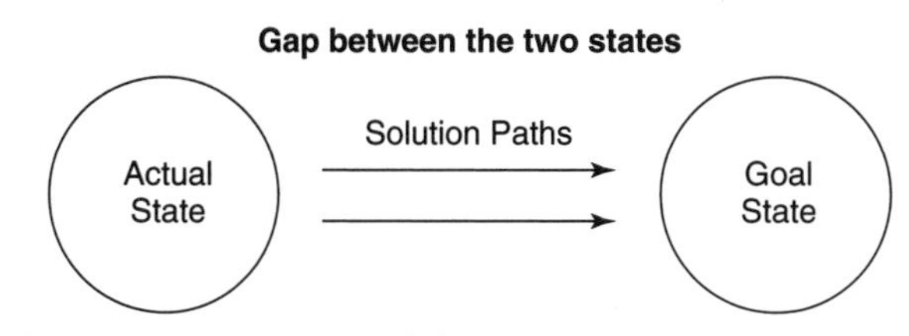


**Figure 3.3** Gap model.

the actual or present state and the desired or goal state differ, then we have a problem. If a problem does exist, then the solution paths represent different routes to resolution of the problem (solution). The model derives its name from the fact that our solution is designed to "bridge the gap."

A simple example can serve as illustration to this point. Suppose our actual state is that we are constantly late for our first class, which begins at

8:00 A.M.. Suppose further that we would like to be on time for the course. This would be our goal. Since the actual state and the goal state are different, we have a problem. The solution paths represent our options in problem resolution. Our options in this example might include waking earlier and leaving our residence earlier, traveling via a less congested route, taking a later class, having a quick breakfast. You can suggest a number of other solution paths.

# Scientific Method

One of the key methods involved in solving problems utilizes an approach called the scientific method. It is in fact one of the reasons for the very rapid growth in the area of science and technology. (Another is related to the nature of the science and technology enterprise and the free flow of information in society.)

## What Is the Scientific Method?

The scientific method is essentially an approach or way of trying to make sense of data or information. The basic premise is that events in nature occur because of a reason or cause. Things don't happen by magic, witchcraft, enchanted crystals, or pyramids. The scientific method usually begins with some observation. We see (sense, perceive) something happen or in some way we gain information about an event. If the event is an isolated occurrence, and not unusual, we probably will forget about it. However, if the event is unusual or has occurred before, we tend to remember it.

On the basis of our observations we to try to explain our observations using a tentative description called a hypothesis. The hypothesis is used to make predictions, which are then tested by further observations. As a result the hypothesis may be modified from these additional experiments.

The testing procedure is repeated many times until there is confidence in the regularity of the observations and measurements. It is then that the hypothesis may be regarded as a theory. A theory is the means we have for explaining our observations and may be used for making predictions.

It becomes known as a law after the hypothesis has been confirmed by repeated experimental tests. However, our attempt to codify the process is not set in concrete. One must view the "method" as a set of common activities including observation, formulating explanations, and then testing them (Fig. 3.4). The order is not fixed and the activities may be divided among different scientists who are called theoreticians or experimentalists.

For events, that occur repeatedly or for "uniform behavior of nature" we try to describe the

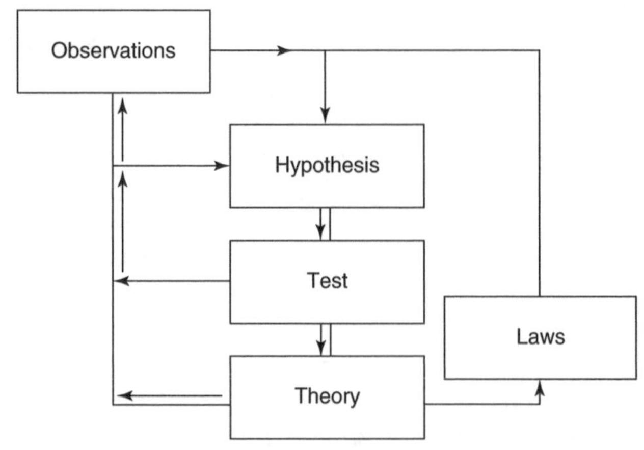


**Figure 3.4** Scientific method flowchart.

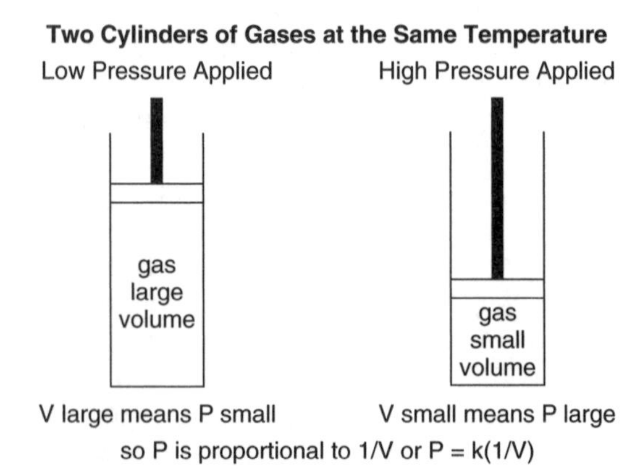


**Figure 3.5** Gas cylinders.

occurrence in a verbal or sometimes mathematical statement, which we call a law. Whether we are aware of it or not, in our daily lives, we are following laws on a personal level that have arisen and we recognize through experience.

For example, if we have a friend who is periodically late, we may regard appointments with that individual differently, a "law of lateness." Another example, the law of gravity (somewhat paraphrased) states that an object when released from some elevated position, if not supported, will fall to the earth. This law, as all laws, is a statement of the uniform behavior of nature. (The law of gravity works almost all the time under the conditions described. Can you suggest when the law appears to be violated?) For laws pertaining to the sciences, it is careful observation, under controlled conditions, that provides the supporting evidence that gives rise to laws.

There is a law pertaining to gases enclosed in containers (see Fig. 3.5) at a constant temperature. The law, known as Boyle's Law (the discoverer was James Boyle) states that the volume of the gas decreases as the pressure on the gas increases, and the harder you push on a gas the smaller the volume of the gas will be. The law seems intuitively reasonable. In mathematical terms:

$$V \propto 1/P$$

The volume of the gas is inversely proportional to the pressure on the gas as long as the temperature is constant.

The behavior of this law has been explained by assuming that gases are composed of molecules that occupy only a tiny fraction of the gaseous volume and are in constant motion. The gaseous pressure is related to the frequency with which the molecules strike the walls of the container. Actually this explanation has been verified and is part of a theory describing the behavior of gases. It is known as the kinetic molecular theory. In point of fact, the explanation just offered to account for Boyle's Law did not come about all at once.

In order to explain the basis for scientific laws, what's behind them, and why they work, some sort of working model is created. Assumptions are put forth in an attempt to account for the phenomenon the law describes. In more precise terms, hypotheses are advanced and then tested. If the hypothesis is inconsistent with the law, the hypothesis is modified and reworked and retested. Ultimately, if the hypothesis is verified, it then assumes the status of a theory. A theory is a verified and accepted explanation.

Finally, it should be noted, theories are not "set in stone." Sometimes new data or evidence arises that cast doubt on the current explanation or theory regarding the explanation of a phenomenon. It is not at all uncommon for theories to be further modified "down the road." It is within the nature of the scientific enterprise to be faithful to the facts as they are uncovered. Sometimes it takes years for conflicting theories to become resolved.

There are some Internet references at the end of the chapter regarding some theories, which were controversial. One of them was ultimately rejected.

(A brief note: Words used in a scientific context sometimes have different meanings than when used in ordinary conversation. For example, the word "theory" is used in everyday language and has a different meaning than when used in a scientific context. If we read a mystery novel, there may be a sleuth who is hot on the trail of a culprit. The sleuth has a theory; actually in this case the detective really has a hypothesis or an educated guess as to the identity of the evildoer. In science, a theory is much more than a guess; it is composed of verified [tested] hypotheses.)

As noted earlier the very rapid growth of science and technology can be ascribed to the scientific method and the rapid distribution of knowledge through the free flow of information via journals and other media.

Scientific methods are used to account for scientific behavior and phenomena. We can and do use scientific methods in our daily lives in a variety of ways, including problem solving.

**Example**

Read the following and explain what is being stated:

1. Serendipity of Neophytes.
2. Comeliness exists only within the epidermis.
3. Mini sun scintillate, scintillate.
4. Identical plumage attracts related members of a given avian species.
5. It is unprofitable to be lamenting the exodus of unintentionally departed lacteal fluid.

If you're a little bewildered, reread the sentences above and this time apply the scientific method. Make an assumption. "Translate" a few words that you know into ordinary English and see if you can make some sense out of the sentence.

If you don't see any connections, look at statement number 1, since it is the shortest sentence and translate it. A neophyte is a beginner, and serendipity has to do with luck; hence "neophyte's serendipity" is "beginner's luck." Use the scientific method to decipher the statements above. Put forth a guess or hypothesis.

"Beginner's luck" is a saying. Are the other sentences sayings or adages also?

In the example cited, we could test our hypothesis easily since there is a readily verifiable answer.

If, for example, statements are all old sayings or adages, you can check out the other phrases to see if they translate into different sayings. In deciphering the sayings, solving the problem, we translate as many words as we can and then think of different adages that use some of those words. We are using the scientific method; we form a hypothesis using the saying and then test the hypothesis.

Answers for the sayings listed previously follow:

1. Beginner's luck.
2. Beauty is only skin deep.
3. Twinkle, twinkle little star.
4. Birds of a feather flock together.
5. It's useless to cry over spilled milk.

Examples of other sayings are given at the end of this chapter.

Usually verification of a hypothesis requires efforts including experimental work and/or calculations and the application of logic. Actually while a hypothesis can certainly be disproved and we can verify a hypothesis, *absolute* proof of a hypothesis is another matter.

**Example**

There was a child who at five of six years of age developed an odd notion about the human voice. The child had heard elderly people speak and noted that they tended to speak in a more strained fashion. Their tones tended to be hushed and their voices raspy. Further, the child had heard relatives admonish one another "not to waste your voice on this or that person." Based on this data, the child concluded that a person was born with just so much voice. The older you got, the more of your voice would be used up. Old people were "running out of voice." You didn't want to waste your voice, because as you use it, you lose it.

As a five or six year old, the child's hypothesis is reasonable. Based on the data and the child's limited experience, the guess or hypothesis about "voice" makes sense. The next step, in order to verify the hypothesis, would involve experimental work. One approach might involve examination of the region "where voice resides" and data collection involving a correlation between age and remaining voice. Obviously none exists and the hypothesis is invalid.

# Hypothesis Testing

Certain hypotheses can be examined by use of a contingency diagram and their validity evaluated. A simple example illustrates this point. If we have a simple hypothesis statement of the type,

All "A" then "B"

and the negative of the statement,

All "non A" then "non B"

we can set up a contingency table and use the table to evaluate the degree to which we can believe our hypothesis.

More specifically, suppose a vaccine against a certain malady, say halitosis, is tested. We believe (our hypothesis) that "every one who receives the vaccine will not get halitosis" and the converse, that "all who do not receive the vaccine will get halitosis."

What we believe to be true (our hypothesis) is stated as a and d, and what we believe to be false is stated as b and c (Table 3.1).

**TABLE 3.1 Halitosis Hypothesis**

| | Vaccine | No Vaccine |
|---|---|---|
| No Halitosis | a | b |
| Halitosis | c | d |

**TABLE 3.2 Male/Female Chromosomes Hypothesis**

| | Have XY Chromosomes | No XY Chromosomes |
|---|---|---|
| Male | a | b |
| Female | c | d |

a, b, c, and d represent the experimental data we would collect in order to verify or refute the hypothesis.

Consider the statement,

All *males* have XY *chromosomes.*

and the negative of the statement,

All *females* do not have XY *chromosomes.*

The contingency table for this hypothesis is represented in Table 3.2. Using the data in a, b, c, and d we can calculate an association coefficient (R):

$$R = [(a + d) - (b + c)] /(a + b + c + d)$$

An association coefficient of +1 indicates 100% agreement between our hypothesis and the experimental data. An association coefficient of –1 indicates that the hypothesis is exactly the opposite of the hypothesis we proposed. An association coefficient of 0 (zero) indicates that there is no correlation between the data and the hypothesis. The results are purely random.

**Example**

Suppose a new vaccine is developed to prevent halitosis. Out of a test population of 160 individuals who have the same diet, 80 are given the vaccine and the remainder are not. Of the 80 who are given the vaccine, 10 develop halitosis. Of those who are not given the vaccine, 60 develop halitosis. Comment on the hypothesis that this vaccine will prevent halitosis. The contingency table is shown in Table 3.3.

**TABLE 3.3 Halitosis Hypothesis Contingency**

| | Vaccine | No Vaccine |
|---|---|---|
| No Halitosis | 70 | 20 |
| Halitosis | 10 | 60 |

$R = [(70 + 60) - (20 + 10)] / (70 + 60 + 20 + 10)$

$R = [130 - 30] / (160)$

$R = 100/160$

$R = .625$ or a 62.5%

Since a random result would yield a value of 0.0 or zero percent, it would appear that the vaccine has some effectiveness.

The examples cited thus far have been used to illustrate scientific methods, which we see as a way to gain greater understanding of natural phenomena. We see that the scientific method is also an approach to gain a greater understanding of other problems that sometimes confront us. In our approach to achieving scientific and technological literacy it turns out that there are a number of general ways of gaining greater competency and ability in dealing with scientific issues.

We could, of course as noted earlier, try to learn as much science as possible so as to become experts in many fields of scientific endeavor. This approach would consume most of our waking hours for many, many years and is really impractical for the nonprofessional scientist. Another approach alluded to earlier, involves learning about specific issues on a need-to-know basis and using various problem-solving techniques to help unravel any questions that we have.

For example, if we have a problem with our car, although we may not know precisely how a car's motor works, we can ask certain questions of a mechanic to have a sense of the reasonability of his/her proposed repair. Based on the mechanic's responses and other clues and information (perhaps a consultation with a friend), we can come to a point where a decision may be made.

Problem solving represents one approach to resolving questions that are of a scientific/technical nature. (We will encounter other approaches.) One problem-solving strategy, after we have identified that a problem exists and perhaps have articulated the problem, is to use the scientific method as we have already seen. The idea is to generate a hypothesis and then test it.

Several examples follow, which illustrate problems and different kinds of thinking and strategies needed to confront the problems.

# Exponential Growth

**Paper Folding.** Suppose you are given a piece of paper 0.001mm thick. Suppose the paper is folded-over and over 50 times, doubling the thickness each time; to what height would the folded-over paper reach? Could you estimate a value?

(Aside: To put the paper thickness in perspective we can convert this thickness into inches. Since there are 25.4 mm in one inch, the paper is pretty thin at (.001mm × 1.0 inch/25.4 mm) .00004 in. or $4 \times 10^{-5}$ in. (If you are "rusty" doing these operations, see Appendix B.)

Before we solve the problem, take a guess at the paper thickness (after folding it 50 times).

Select the closest answer.

a. the distance from the floor to the ceiling of an average room
b. the distance from the floor to your waist
c. two football fields long
d. the distance between New York City and Los Angeles
e. farther than the distance between the earth and the moon

We can solve this problem by first organizing and examining the data.

| Number of Folds | Thickness |
|---|---|
| 0 | $4 \times 10^{-5}$ in. |
| 1 | $8 \times 10^{-5}$ in. |
| 2 | $16 \times 10^{-5}$ in. |
| 3 | $32 \times 10^{-5}$ in. |
| 4 | $64 \times 10^{-5}$ in. |

We may observe that the increase in thickness is not linear. A linear increase would yield a proportional growth in thickness with each successive fold and would give a linear graph if we plotted thickness vs. folds (see Fig. 3.6).

A plot of thickness vs. folds gives a graph, which is nonlinear. The graph in fact is exponential in nature (see Fig. 3.7).

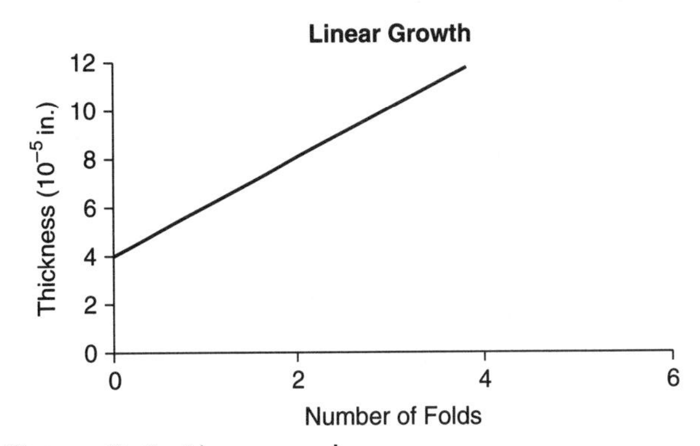


**Figure 3.6** Linear graph.

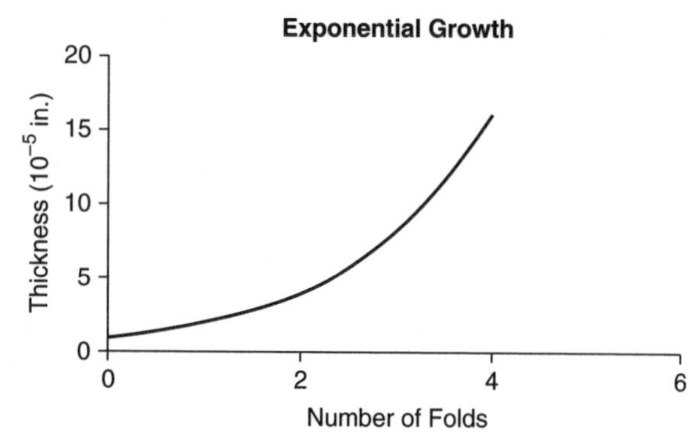


**Figure 3.7** Exponential graph.

Before we confront the problem a few words are in order concerning exponential growth. Things that change exponentially are difficult to estimate simply because they are nonlinear and we are not used to such systems.

For example, can you predict when a lake would become half-covered by algae given that a lake is totally covered by algae at the end of the month (say August), that the lake had barely any algae on the first day of the month, and each day algae growth doubled in size?

In the event you solved the problem, bravo. If you did not, the answer is that the lake would be half covered the day *before* the lake was fully covered on August 31. This surprising answer implies that it might be a good idea to try to generate some kind of relationship, which will help us to solve the problem, an equation of some sort.

We see that the effect of each fold results in a doubling of the thickness of the paper. That is,

New Thickness of paper = 2 × old thickness of paper

For the first fold, $8 \times 10^{-5} = 2 \times (4 \times 10^{-5})$

For the second fold, $16 \times 10^{-5} = 2 \times (8 \times 10^{-5})$

For the third fold, $32 \times 10^{-5} = 2 \times (16 \times 10^{-5})$ etc., a doubling in each case.

If we relate the old thickness to the original thickness, we have, using powers of 2,

For the first fold, $8 \times 10^{-5} = 2 \times (4 \times 10^{-5})$

For the second fold, $16 \times 10^{-5} = 2^2 \times (4 \times 10^{-5})$

For the third fold, $32 \times 10^{-5} = 2^3 \times (4 \times 10^{-5})$

The previous expressions may be generalized as:

Thickness of paper = $2^n \times$ (original thickness), where n is the number of folds.

We have "derived" an equation, which allows us to calculate the thickness of the paper.

To use the equation after 50 folds:

$$\begin{aligned}\text{Thickness of paper} &= 2^{50} \times (4 \times 10^{-5} \text{ in.})\\ &= 1.125 \times 10^{15}\ (4 \times 10^{-5} \text{ in.})\\ &= 4.50 \times 10^{10} \text{ in.}\end{aligned}$$

To put this value into perspective, we can calculate the number of miles thick the paper will be:

$4.50 \times 10^{10}$ in. × (1 foot/12 in.) × (1 mile/5,280 ft.)

or

$$7.1 \times 10^5 \text{ miles}$$

or almost three times the distance to the moon, all from an ultrathin piece of paper folded over 50 times.

We see that the equation, which describes how the paper thickness increases with folding, is:

$$T = 2^n T_o$$

where n is number of folds, T is thickness, and $T_o$ is original thickness.

The growth of T depends on $2^n$ or on the exponent "n." We say that T is increasing exponentially. (We will return to the concept of exponential growth later.)

For many students, the problem just examined is a turn-off. Dealing with numbers for such individuals is a "pain" and consequently those with this mind-set tend not to even attempt such a problem. This is unfortunate on a number of levels. Not only will the student be "out of the loop" by avoiding or ignoring certain problems, but initiative, confidence, and joy of solution are diminished. If dealing with numbers are turn-offs, it may well be that a mental block exists. Mental blocks are serious impediments to problem solving.

## Blocks to Problem Solving[1]

We will have difficulty in solving problems if we encounter a turn-off, or blocks. That is, there may be certain situations, conditions, etc., that prevent us from seeing a solution to a problem because of blocks, things that inhibit our thinking. Blocks to our thinking may arise because of our senses, our feelings, our background, our environment, and our "language skill."

Blocks can be classified in terms of their inhibiting or misleading action on our senses and psyche: sensory blocks (blocks that influence our senses) and emotional blocks (blocks that influence our feelings).

[1]*Conceptual Blockbusting* by James L. Adams, Basic Books, Perseus Books Group, 2001.

Blocks can also arise because of who or what we are in terms of upbringing, our culture, religion, and bias. These blocks, which come about because of our personal and societal background, include blocks associated with our physical surroundings as well.

Sometimes blocks occur because of our training or education. We still tend to solve a new problem using the same approach we successfully used in solving a similar problem in the past. Sometime this doesn't work and we're stuck. On the other hand, education provides knowledge, and specialized "language skills" such as mathematics and music so new avenues to explore and express problems are available. If we "can't do the math," we can't solve certain problems.

In gathering information in order to arrive at a judgment, we should be unbiased and not stereotype. Data or conditions may have been experienced before or seem familiar but we should not *assume* new information will be the same as in the past. We should strive to collect data and conduct experiments in an unbiased fashion.

Here are some problems, which are illustrative of the blocks outlined previously.

### Problems

1. **Given:** Six pencils of equal length. Construct four identical equilateral triangles using the pencils. The pencils ends should touch in forming the triangles.
2. **Given Figure 3.8:** Connect all the dots with four straight lines using a pencil and not taking the pencil off the paper.

Thinking "outside the box" can solve both problems. For the first problem, it will help you to first think in three *dimensions!* Four equilateral triangles attached to each other give a pyramid called a tetrahedron (Fig. 3.9).

In order to connect all the dots, thinking outside the box gives the answer (see Fig. 3.10). Sometimes we need to use all our senses in solving problems.

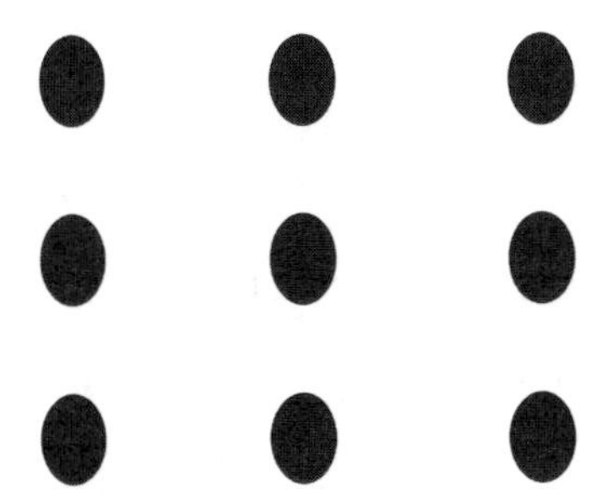

**Figure 3.8** Nine dots problem.

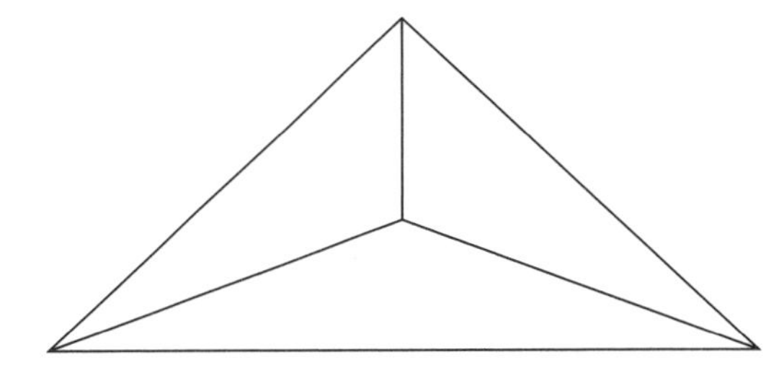

**Figure 3.9** Tetrahedron.

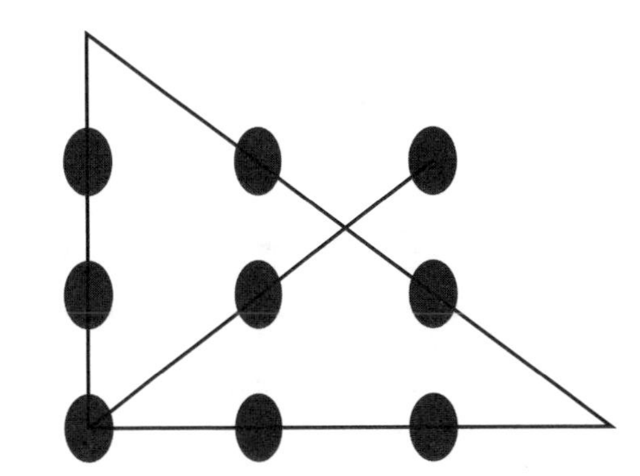

**Figure 3.10** Solution to nine dots problem.

### Problem

A white solid is on the kitchen table. We know it to be either salt or sugar (and nothing else!).

How could we resolve the identity of the unknown white solid? The sense of taste provides important information quickly, and obviates more time-consuming chemical tests.

Difficulty in perceiving the problem from different points of view is another hang-up. People sometime have a different take on a given situation. Failure to understand this can lead to difficulty in problem solving. Why do you suppose a parent has a problem with lending a new car to an adolescent new driver?

Inability to deal with "messy information"—ambiguity, disorder, and chaos—can inhibit the thinking of some individuals.

### Evian and Gecko Problem

The following problem can be solved with sound reasoning. There are no tricks or double meanings. The questions are:

1. Who owns the gecko?
2. Who drinks Evian?

Here is the information you need to solve the problem.

1. There are five houses.
2. The Democrat lives in the red house.

3. The Republican owns a dog and lives in the house to the left of the Monarchist.
4. Coffee is drunk in the green house.
5. The Conservative drinks tea.
6. The green house is to the right of the ivory house.
7. The wood carver owns snails and lives in a red house.
8. Milk is drunk in the middle house.
9. The artist lives in the yellow house, which is to the left of the blue house.
10. The Liberal lives in the first house.
11. The man who is an electric train hobbyist lives next door to the man with a fox who lives in the yellow house.
12. Miniature cars are made in the house next to the house where the horse is kept.
13. The Republican drinks orange juice.
14. The middle house is red.
15. The Monarchist plays golf.
16. The Liberal lives next door to the blue house.
17. Each man has one house, one pet, one type of hobby or recreation, and a different drink and political affiliation.

This problem can be easily solved if we are not put off by the seemingly disjointed data. If we can stand the chaos, that is organizing the information, arriving at a solution becomes a relatively easy task. See the end of the chapter for the solution.

### Swimming Problem

Our environment and our culture also influence problem solving. As an illustration, consider the following problem, dilemma might be a better word. Suppose you were in a boat along with your young child, spouse, and mother, and the boat sank far from shore. Suppose you were the only swimmer and you could save only one, who would you save?

Different cultures give different answers. In Western countries the child is usually selected as the one to be saved since the child is most helpless and dependent. In Middle Eastern nations, the mother is selected as the one to be saved. The reasoning is simple. You can remarry and have more children, but you can only have one mother.

### Cultural Block Problem

Here's an example of a cultural block, which might inhibit a solution to a problem. Suppose you are locked in a soundproof, cell-phone–proof, windowless room with six other individuals of different ages whom you have never met before. The group consists of males and females. The only exit from the room is via a heavy steel door, which is operated from inside the room by a switch attached to a pipe, which is imbedded in a concrete floor one foot thick. The switch is broken; however, you have succeeded in uncovering the pipe. It is four feet tall and half an inch wide. You can observe the terminals to the switch, which are midway in the pipe as noted in Figure 3.11.

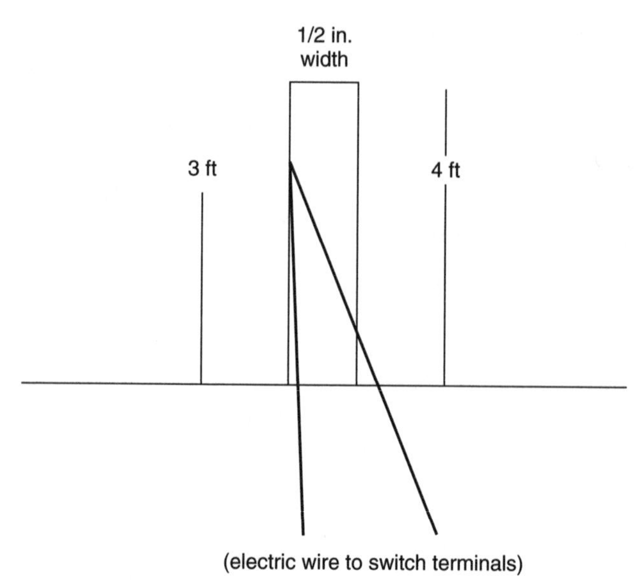

**Figure 3.11** Pipe in ground.

Assume all the individuals in the room are carrying normal items.

Indicate one solution path you and your group might use in order to exit the room. (One solution uses a taboo.)

A solution would be to fill the pipe with blood or urine to activate the switch.

### Language Block Problem

The following example is an illustration of a block, which requires the correct mode of expressing the problem. What "language" is appropriate?

A camper started at the base of a high mountain at 6:00 A.M. sharp. All day the individual hiked up the mountain, stopping at various points to rest or enjoy the scenery or snack. At sunset, base camp was reached. The camper remained at the camping site for two days, simply enjoying the beauty, reading, and relaxing.

On the third day at 6:00 A.M. sharp the camper left base camp and headed down the mountain taking the same path as before. Again there were stops along the way for various reasons. Since descending the mountain is easier than climbing the mountain, the camper reached the base of the mountain before sunset.

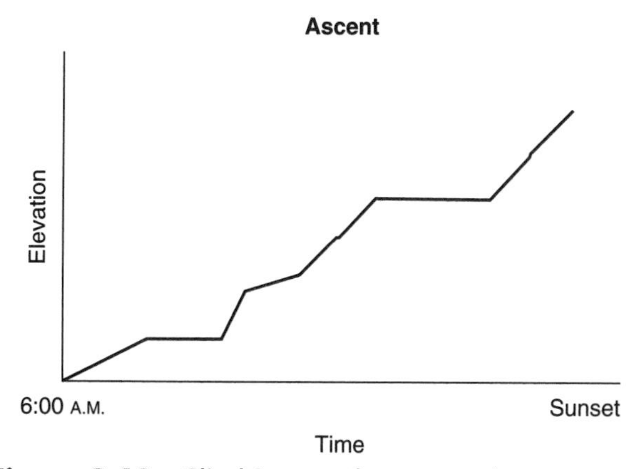

**Figure 3.12** Climbing up the mountain.

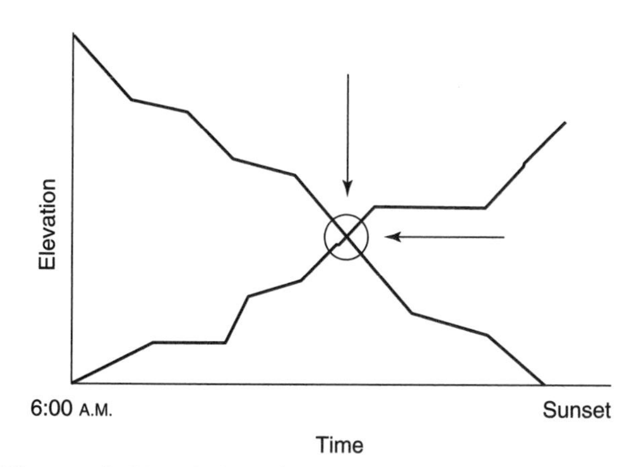

**Figure 3.13** Going down mountain.

Prove that there is one place on the mountain that the camper occupied at the *same time* both in ascending and descending

The block here is the language that is used to solve the problem. A mathematical approach leads nowhere. A graphical solution works well (see Fig. 3.12).

This graph represents the ascent of the camper, the individual's elevation with time.

Suppose we plot on the same axis the descent of the camper (the camper's elevation at different times going down) (see Fig. 3.13).

Note on this graph there is one elevation where the camper was present at the same time both in ascending and descending the mountain.

Thus far, a number of very important elements, which are vital in problem solving, have been examined. However, the process of problem solving is more than the sum of the elements needed to assist in achieving a solution. The point is, problem solving is more than a science or technology, and while one may talk of the science of problem solving, it is more an art than a science.

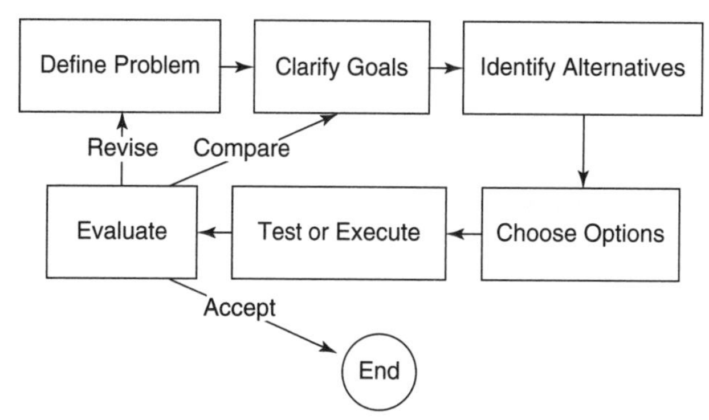

**Figure 3.14** Steps in the problem-solving process.

Figure 3.14 provides a framework for the generalized process that is involved. An explanation of this flowchart follows.

1. The first step is to define the problem. That is:
   - Recognize and clear blocks.
   - Gather needed information. These include: All pertinent facts, relevant history, and may require specialized, technical knowledge and expertise. Then we try to articulate the problem so that ultimately it may be stated in words.
2. Goals. After we define the problem, the next step is to determine what our goals and objectives are. What do we want to achieve? What do we want? What are our interests? It is useful to express these goals in a written form.

   Are we where we want to be in terms of solving the problem? Have we achieved our goals? Obviously, if the answer is "yes," the problem is solved and we are at the end of the process. If the answer is "no," we need to refine or sharpen our goals, and select alternatives consistent with our experience of having gone through the process. To achieve a satisfactory solution we need to have goals.
3. The third step is to determine the various solution paths, approaches, or alternative strategies that can be utilized in solving the articulated problem that is consistent with our goals.
4. Choose an option. Assess alternatives and select the most promising, consistent with our objectives. Here critical thought is required since judgments must be made regarding

ease of implementation, cost of achieving a solution versus benefits gained.

5. Go for it. Implement a plan for achieving objectives.
6. Examine results; evaluate the data and compare; recycle through once more.

**Problem**

In light of the information just presented, consider developing a problem-solving strategy for a successful career in a major of your choice.

While the flowchart of steps provides a general process for problem solving and serves as a reminder or checklist to insure that no essential steps have been left out, it is not a guarantee that a solution will automatically appear on the first attempt. To successfully solve problems, two other ingredients are required; these are persistence (remember Thomas Edison's quote: "Genius is one percent inspiration and ninety-nine percent perspiration.") and command of a variety of good thinking skills.

# Critical Thinking Skills

The particular kinds of thinking skills that apply to the problems we have been examining are known as critical thinking skills. Hence, in conjunction with the steps involved in problem solving, critical thinking skills must be included. What are critical thinking skills? They involve the ability to use logic and judgment in analysis of a problem or data. Critical thinking skills involve a variety of sorts of thinking. The following list is illustrative:

***Evaluative Thinking***

(Skills in which differences and alternatives are judged.)

Comparing alternatives.

Identifying inconsistencies.

Distinguishing relevant from irrelevant information.

Purchasing a stereo system represents an example of using evaluative thinking.

***Systematic Thinking***

(Skills in sequencing, in logic, and in cause and effect are paramount.)

Setting priorities.

Planning a step-by-step approach.

Identifying antecedents and causes.

The extensive renovation of a kitchen requires systematic thinking.

***Synthetic Thinking***

(Skills in making links that may not be obvious are important here.)

Seeing connections between seemingly unrelated events.

Identifying the main issues in complex situations.

Uniting different ideas into a coherent framework.

In making medical diagnoses physicians are sometimes called upon to use synthetic thinking.

***Anticipatory Thinking***

(Skills in forecasting possible consequences to actions at an early stage.)

Anticipating the consequences or implications of events.

Anticipating the long-term costs or benefits of short-term choices or actions.

Planning for obstacles that might arise later.

Part of the motivation for attendance at college is related to anticipatory thinking.

***Creative Thinking***

(Skills in thinking outside the box.)

Looking for innovative solutions.

Developing unusual ideas.

Using resources in novel ways.

The Wright brothers were creative thinkers in their belief about flying. They also exhibited all the other critical thinking skills listed above.

Critical thinking allows us to make decisions regarding the evaluation of our data, the selection of solution paths that are more likely to achieve results, and judgments regarding unique approaches and interpretations in the presence (or absence) of information. Success in application of critical thinking is related to practice in applying this skill. The more one practices, the better one gets.

During the course of the semester, we shall be confronting problems of a scientific and technological nature using the general schemes cited earlier (scientific method, problem-solving steps) in conjunction with critical thinking in different contexts (modeling, coding, etc.) involving health, energy, and the environment.

## Exercises

I. Read the following and explain what is being stated:
1. A Bic has more power than the saber.
2. One is foolish to attempt the education of a senior canine with novel activity.
3. Grimelessness is penultimate to divinity.
4. Steam and water at 100 degrees C will not be generated when the container in which they are held is constantly focused upon.
5. A prompted suture spares the future obligation by a factor of the square root of 81.
6. Reject the switch and downgrade the scion.
7. He who dwells within edifices of fused silica ought not toss rocks.
8. All resplendent matter need not contain element 79.
9. Cadavers of masculine gender cannot bear witness.
10. Supplication on the part of the needy is without choice.
11. A bevy of culinary experts working at the same time will create havoc with their broth.
12. Working a double shift as well as nights and weekends without diversion will generate a young John or Jane uninteresting.
13. A tumbling pebble will not aggregate any leafy stemmed plants.
14. Any abrupt transition involving a jump should occur only after vigilant surveillance of the surroundings.
15. Tempus fugit.

## Answers to Exercise I:

1. The pen is mightier than the sword.
2. You can't teach an old dog new tricks.
3. Cleanliness is next to Godliness.
4. A watched pot never boils.
5. A stitch in time saves nine.
6. Spare the rod and spoil the child.
7. Those who live in glass houses shouldn't throw stones.
8. All that glitters is not gold.
9. Dead men tell no tales.
10. Beggars can't be choosy.
11. Too many cooks spoil the broth.
12. All work and no play makes Jack (Jane) a dull boy (girl).
13. A rolling stone gathers no moss.
14. Look before you leap.
15. Time flies.

II. Set up a contingency table for the following and comment on the resulting association coefficient:
All New York Mets fans have red hair.
(Do your results agree with your expectations? What is the problem?)
Survey the class for data.

III. Now that you have had some experience with an unconventional problem, try to develop a problem-solving strategy to estimate the volume of water that the Mississippi River empties into the Gulf of Mexico annually. What information would you need? How would you start? Think about it. Brainstorm with one or two colleagues.

## Review Questions

1. What are Polya's four steps for solving problems?
2. What is the "scientific method?"
3. What are the common elements found in all scientific methods?
4. Define observations, hypotheses, testing, theory, law.
5. What do contingency diagrams do?
6. Name four blocks to problem solving and define them in terms of an example.
7. Name five critical thinking skills.
8. What is meant by evaluative thinking?
9. What is creative thinking?
10. Calculate the height of a piece of paper 0.001 mm thick that is folded-over twenty times.
11. List the various kinds of blocks that inhibit thinking.
12. Determine the sum of the numbers 1 to 100 as fast as possible. (Hint: Use pairs of first and last numbers.) (Ans. 5050)
13. Give an example for each of the forms of critical thinking skills listed in the chapter.

## Multiple Choice Questions

1. Fluoride is a _____.
   a. deadly poison
   b. paint dye
   c. food additive
   d. cleanser

2. The first step in Polya's problem-solving approach is to _____.
   a. devise a solution
   b. understand the problem
   c. execute the plan
   d. check the solution

3. The last step in Polya's four-step process is _____.
   a. devise a solution
   b. understand the problem
   c. execute the plan
   d. check the solution

4. The "scientific method" is _____.
   a. like witchcraft
   b. set in concrete
   c. a set of common activities
   d. a series of guesses

5. The scientific method can be used to _____.
   a. solve social problems
   b. solve scientific problems
   c. solve business problems
   d. All of the above.

6. A contingency table produces an association coefficient of –1. This means that there is _____ agreement between our hypothesis and the experimental data.
   a. 100%
   b. no agreement
   c. one negative standard deviation
   d. full correlation

7. Blocks in problem solving _____.
   a. hinder our thinking
   b. allow us to "divide and conquer" in problem solving
   c. are alleviated by concentrating on tried and true methods
   d. are conquered using tunnel vision

8. "Thinking outside the box" means _____.
   a. stereotyping
   b. saturation
   c. sublimation
   d. avoiding "tunnel vision"

9. A critical thinking skill is _____.
   a. evaluative thinking
   b. systematic thinking
   c. synthetic thinking
   d. All of the above.

10. Synthetic thinking involves _____.
    a. anticipating events
    b. looking for innovative solutions
    c. seeing connections between unrelated events
    d. comparing alternatives

11. "The volume of a gas is inversely proportional to the applied pressure at constant temperature" is an example of _____.
    a. law
    b. hypothesis
    c. a gas expansion
    d. a mathematical riddle

12. If we obtain an association coefficient of 1.00 from a contingency table we can be assured that our _____.
    a. hypothesis is absolutely correct
    b. appears to be correct
    c. is totally incorrect
    d. None of the above.

13. Given the data:

| x | y |
|---|---|
| 1 | 0 |
| 2 | 1 |
| 3 | 2 |
| 4 | 3 |
| 5 | 4 |

    The data suggests that y is increasing in what sort of fashion?
    a. linear
    b. exponential
    c. inverse
    d. sinusoidal

14. Given the data:

| x | y |
|---|---|
| 1 | 0 |
| 2 | 1 |
| 3 | 2 |
| 4 | 4 |
| 4 | 8 |

    The data suggests that y is increasing in what sort of fashion?
    a. linear
    b. exponential
    c. inverse
    d. sinusoidal

15. Given the graphs:

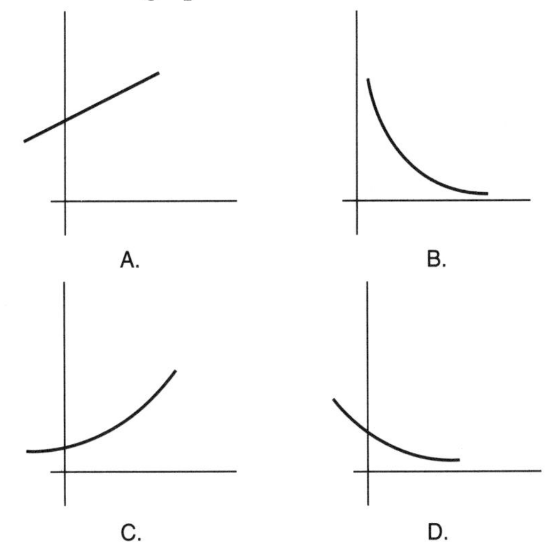


Graph A is the plot of
a. $y = mx + b$
b. $y = 1/x$
c. $y = 2^x$
d. $y = 2^{-x}$

Graph B is the plot of
a. $y = mx + b$
b. $y = 1/x$
c. $y = 2^x$
d. $y = 2^{-x}$

Graph C is the plot of
a. $y = mx + b$
b. $y = 1/x$
c. $y = 2^x$
d. $y = 2^{-x}$

Graph D is the plot of
a. $y = mx + b$
b. $y = 1/x$
c. $y = 2^x$
d. $y = 2^{-x}$

16. When Archimedes discovered the principle of buoyancy, he was employing _____.
a. creative thinking
b. synthetic thinking
c. evaluative thinking
d. All of the above.

17. The difficulty in assessing the thickness of a sheet of paper folded-over fifty times is an example of _____.
a. emotional block
b. cultural block
c. environmental block
d. None of the above.

18. Suppose data has been collected and tabulated as follows:

| | Smoker | Non-smoker |
|---|---|---|
| Cancer | 25 | 20 |
| No cancer | 40 | 15 |

The table implies that
a. smoking leads to cancer
b. smoking is unrelated to cancer
c. nonsmokers live cancer free
d. cancer is related to smoking
e. None of these.

19. Studies reveal that a students' grades are directly proportional to the number of hours they study. The mathematical relation that models this is a _____.
a. line
b. point
c. parabolic curve
d. exponential curve

The graph of grades vs. hours would look like:

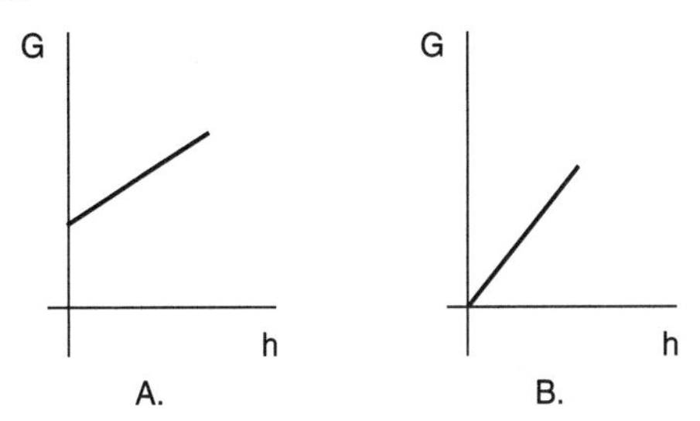


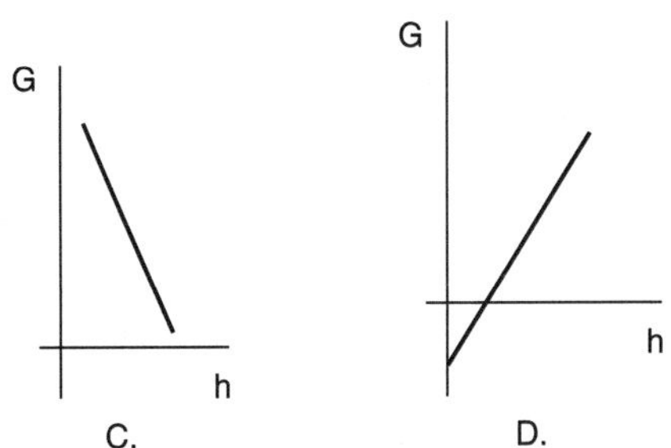


20. In manufacturing there are certain overhead costs that remain constant no matter how many units (within reason) are produced. Also there are costs associated with each unit produced. That is, for each unit produced there is a constant cost independent of the overhead costs. If T = the total cost, K = the overhead cost, c = the unit cost and x = the number of items produced, the mathematical model for this problem is:
a. $T = c + Kx$
b. $T = cx/K$
c. $T = K + cx$
d. $K = T + cx$
e. $C = T - Kx$

21. The transmission of electric power along high tension overhead lines has recently been called into question as a possible hazard to human health. The electric and magnetic fields surrounding the wires are thought to induce birth defects in mammals. A researcher interested in studying this problem subjected 100 pregnant rats to electromagnetic fields of the same strength as that encountered within 200 yards of high tension lines. Of 1,000 newborn rats (born to the 100 pregnant females), 100 were born with serious birth defects. In a control population of 2,000 newborn rats not subjected to electromagnetic fields, the number of birth defects was 150.

    Using the following contingency table, what is the association coefficient between exposure of pregnant rats to electromagnetic fields and birth defects in their offspring?

| | Outcome present | Outcome absent |
|---|---|---|
| Incidence present | a | b |
| Incidence absent | c | d |

    a. 0.03
    b. 0.3
    c. 3
    d. 300

22. The association coefficient determined in the previous question shows which of the following?
    a. no correlation between e-m fields and birth defects
    b. a high correlation between e-m fields and birth defects
    c. a moderate correlation between e-m fields and birth defects

23. An experiment is done in which a model solar collector is exposed to sunlight over a period of time. Account for the fact that the following graph was obtained when a solar collector was exposed to direct sunlight.

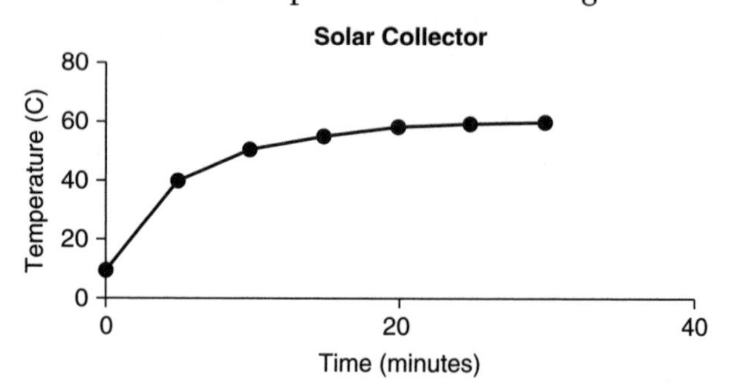


---

Answer to the Evian and Gecko problem presented earlier in chapter.

| | | | | | |
|---|---|---|---|---|---|
| **House Color** | Yellow | Blue | Red | Ivory | Green |
| **Drink** | | Tea | Milk | Orange Juice | Coffee |
| **Politics** | Liberal | Conservative | Democrat | Republican | Monarchist |
| **Hobby** | Art | Electric Trains | Wood Carving | Miniature Cars | Golf |
| **Pet** | Fox | | Snails | Dog | Horse |

The information given accounts for the politics, house color, and hobby of five individuals; however only four of the five pets and four of the five drinks are assigned. That is the drink of the liberal is not specified nor is the pet of the conservative.

Therefore the liberal drinks Evian and the conservative owns a gecko.

# Bibliography and Web Resources

*Patterns of Problem Solving,* by Moshe F. Rubinstein, Prentice-Hall, Englewood Cliffs, NJ, 1975.

Polya's four steps to solving a problem. George Polya (1887–1985), a Hungarian mathematician, wrote "How to Solve It" for high school students in 1957. Here is his four-step method.
**www.msc.uky.edu/carl/ma310/spring03/polya/Polya.htm**

George Polya.
**http://www-history.mcs.st-andrews.ac.uk/history/Mathematicians/Polya.html**
**www-gap.dcs.st-and.ac.uk/history/Mathematicians/Polya.html**

**www.mathgym.com.au/htdocs/polyab.htm**

Introduction to the scientific method. The scientific method has four steps. Description: An explanation on what the scientific method is and does.
**http://directory.google/Top/Science/Methods_and_Techniques/Scientific_Method/?il=1**

It is from these ideas that the scientific method was developed. Most of science is based on this procedure for studying nature. The scientific method explained by Jose Wudka, University of California, Riverside.
**http://phyun5.ucr/~wudka/Physics7/Notes_www/node5.html**

Internet reference to testing and false + and –.
**http://www.cnn.com/HEALTH/9802/18/alzheimers**

Internet reference on dental decay and fluoride use and toxicity.
**www.umanitoba.ca/outreach/wisdomtooth/fluoride.htm**

**Controversial Theories**
Phlogiston.
**webserver.lemoyne.edu/faculty/giunta/phlogiston.html**

Atomic theory.
**dl.clackamas.cc.or.us/ch104-04/dalton's.htm**

Galileo.
**www-gap.dcs.st-and.ac.uk/~history/Mathematicians/Galileo.html**

# Models and Matter

*"Modeling, regardless of the approach, is always a process of compromise between the tractability of simplicity and the complexity of reality."*

**Hampton N. Shirer**

Models are simplified descriptions that allow us to understand the world and allow us to make predictions. They leave out the trivial details and include only the important features. Models are sometimes employed prior to the construction of a real system.

The concept of a system as a collection of parts designed to achieve some objective is used to describe many things: transportation systems, school systems, ecosystems, computer systems, etc. Many of these systems have to be evaluated before construction (e.g., nuclear power plant). Trial and error approaches obviously cannot be used. To approximate the end result without constructing the system we use something that looks and behaves like the real thing (i.e., a model). It becomes a surrogate or substitute for the real system and allows us to determine its behavior and make predictions about its behavior in varying circumstances. This is true only after the model has been tested against known results and we have confidence in its validity. We must also be aware of its limitations imposed by boundary conditions inherent in the model and not extend its predictions beyond its limitations. Constructing models is a combination of science and art.

## Goals

After studying this chapter, you should be able to:

- Define what a model means.
- Determine how and why models are used.
- Distinguish between different types of models.
- Understand certain scientific phenomena in terms of their models.
- Gain real insights into how scientists and engineers use models in the real world.

## Models

The concept of using models is not new to us. You might not be aware of the fact that you use models every day. Sometimes when we are asked to give traveling directions to a particular location, which is rather difficult to describe in words, we make a diagram to accompany the written or oral directions in order to clarify the information we are providing. The diagram or map represents a model. Models help in our understanding and are very useful tools employed in science and engineering (and life) to describe the real world in a way that is consistent with a particular purpose or objective.

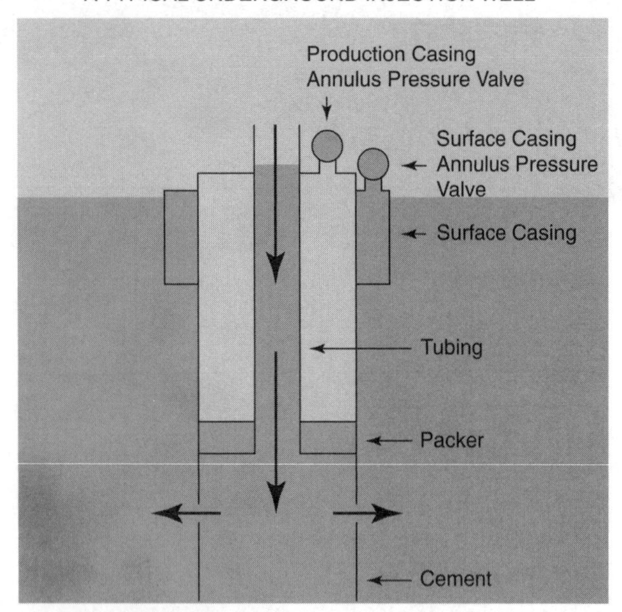


Model well. *http:www.texasep.org/html/wst/wst_4imn_injct.html*

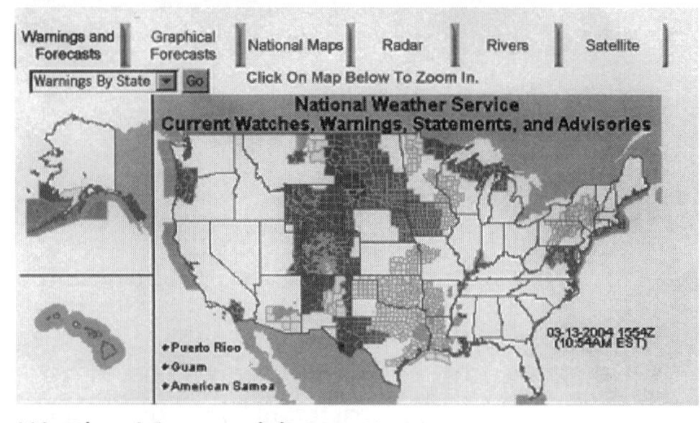


Weather Map model. *http://www.nws.noaa.gov/*

Model car.

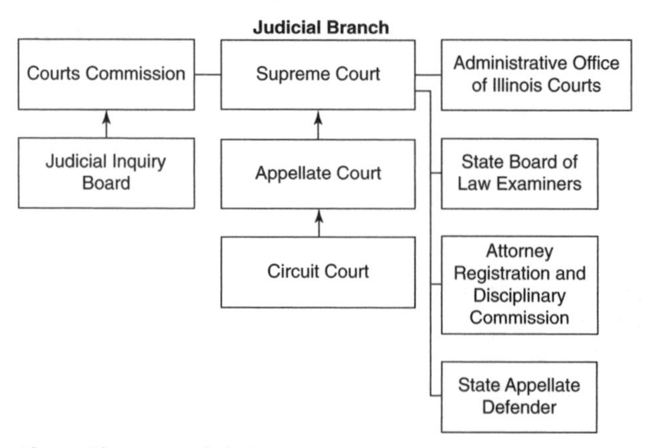


Flow Chart model. *http://www.state.il.us/kids/learn/govern/default.htm*

**Figure 4.1** Models.

You may have used modeling in solving problems on your own. For example, if you were given a mathematics problem for which the answer was not obvious, you might have represented the information on paper as a picture so as to "see" the data and the question better and from another perspective. (The act of putting down the information and integrating it in your mind so you can represent it on paper also gets your thought process going on a solution.) In this chapter we will examine the use of modeling as a pathway in problem solving. A few examples of some diverse models are illustrated in Figure 4.1. More detailed examples will be examined during the course of this chapter.

Depending on your perspective, the term "model" can conjure up a variety of images. Perhaps one's first thoughts on hearing the term might be of scaled down models of real things such as cars, dolls, or airplanes. Clearly such items are representations of "real things." They are not the real thing but capture some essence of that which they represent. But models are much more than something that is scaled down.

Models are representation of both concrete items and of ideas and concepts as well. Models can be two-dimensional representations of reality such as pictures or maps. They can be three-dimensional objects such as a sculpture or a carving or a toy. They need not be exact replicas. For example, a drawing of stick figures could serve as a useful model depending on the purpose behind the model. No model is complete in its total correspondence to reality.

## Physical Models

There are a variety of different ways a model can be expressed depending upon one's interests. There are physical models such as two- and three-dimensional miniatures that capture some aspect

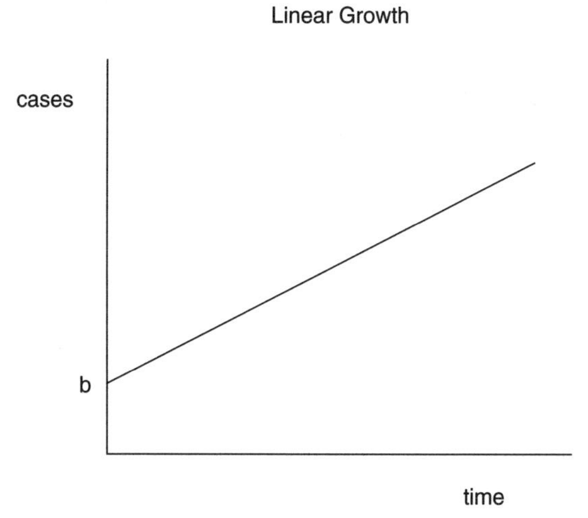


**Figure 4.2** $y = mx + b$

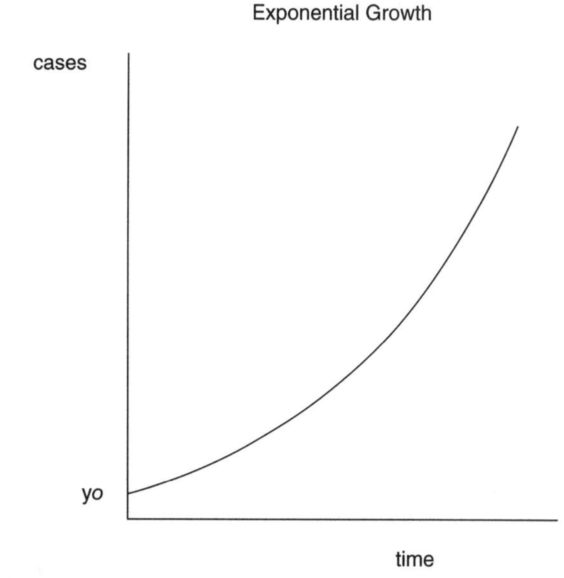


**Figure 4.3** $y = y_o 2^n$

(to varying degrees) of what we are modeling. (Note: Impressionist paintings are nonexact representations, which serve as wonderful models, and provide more than the visual picture of an object. They provide a mood.)

## Graphical Models

Other pictures, nonminiatures, can also be employed to represent or model some aspect of the *behavior* of nature. There can be graphical models, models in which data are plotted. Generally graphical models can give a quick insight concerning a process and help to make connections involving trends regarding a lot of data. For example, consider the spread of a disease in terms of two possible models. The disease might spread in a linear fashion or the disease might spread in an exponential way. How would the graphical models look?

Based on the shape of the two graphs, it would be more critical to respond to an epidemic growing exponentially than one in which there is linear growth. Why?

Graphical models sometimes have an origin in mathematics. In the two examples just cited we can express the graphical results in terms of equations. For the linear graph shown in Figure 4.2, the well-known equation

$$y = mx + b$$

can be used to describe how y (number of cases) changes with x (time) in a linear fashion.

For exponential growth, the equation

$$y = y_o 2^n,$$

where n is the number of doubling periods, may be used. To relate this equation to the graph in Figure 4.3 we can rearrange things. Let $n = x/k$, where $x$ = time and $k$ = doubling time.

Hence y (number of cases) increases as x (time) increases in an exponential way. (See Appendix C.)

$$y = y_o 2^{x/k}$$

## Mathematical Models

Equations have an advantage in modeling in that they allow for calculations and predictions so that the model can be tested. The use of computers allows us to generate large mathematical models (many equations and a lot of data) that are very complex and yield many subordinate graphical expressions.

For instance, large mathematical models are used in modeling of weather patterns so that short- and long-range predictions can be attempted. There are a variety of other similar ambitious scientific and business models, which if accurate, can have a significant societal impact. As an illustration referring to a weather model, a successful model capable of predicting spring rains, possible crop abundance, or potential flood or drought areas (weather model) would have a huge impact in terms of lives saved, societal dislocations, response, and economic costs.

## Analog Models

Another kind of model that is sometimes used is an analog model. These are models in which we use the behavior or properties of something we know or are familiar with to assist us in "modeling" something else. As a case in point, we can describe some of the properties of electricity in terms of the flow of water. We speak of the current, and think of the flow of water or electrons. In the days before transistors, devices used to regulate the flow of electricity were known as valves in certain circles.

While we hope for models to reasonably represent an aspect of reality, our aspirations may or may not be fulfilled. While some models come very close to fulfilling our objectives, others are sometimes off the mark for a variety of reasons; change, knowledge, situations, inventions, interest, etc. For example, just prior to the beginning of World War I around 1915, the use of horses represented a major mode of transportation.

No one conceived such an extensive use of the automobile as currently exists. A model that monitored the accumulation of horse manure on the streets of New York City predicted by 1950 every street in the city would be covered with three feet of the stuff. This never occurred. The technological innovation of the automobile rendered the "horse model" invalid. Yet oddly a model predicting the rate of crosstown Manhattan traffic in the horse and buggy era compares favorably to the model employing the automobile. Why?

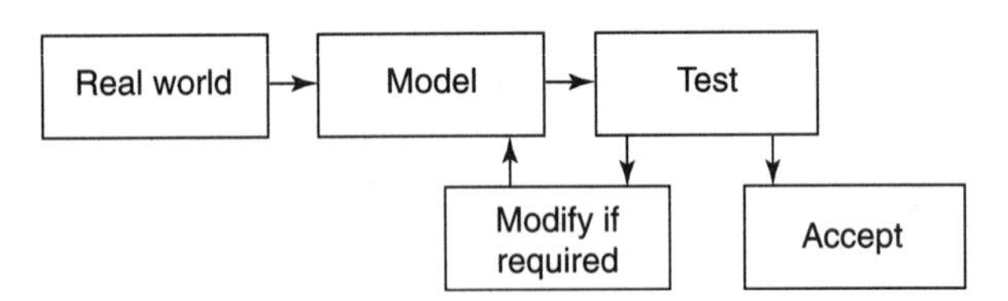


**Figure 4.4** Building a model.

We can use an analog mode to describe the behavior of a heart as a pump. Analog models are helpful way of describing the behavior of nature.

## Verbal Models

Finally, there are also verbal models. That is, the behavior, appearance, or characteristics of an object or idea might best be described in words rather than numbers, pictures or symbols, or equations. For example, how would someone express the aroma of a fruit that smells like a mixture of roses and vinegar?

In the end, a model is an external representation of a segment of reality that helps us to understand, replicate, and anticipate its behavior.

# Building Models

In order to build a model, the process involves going from the real world to some abstraction where a model is created, tested, and then modified as needed.

The model below illustrates these ideas as shown in Figure 4.4.

## Mouse Model

There are other effects which may invalidate a model and which therefore should be considered. For example, scale effect is a boundary condition. Consider the amount of food (actually food calories or energy) required to sustain a person. Proportionately one might suspect that a smaller mammal (a mouse perhaps) should require a proportionately reduced amount of food. Yet, the food calories per gram of body weight required by a mouse are roughly 10 times greater than that required by a human.

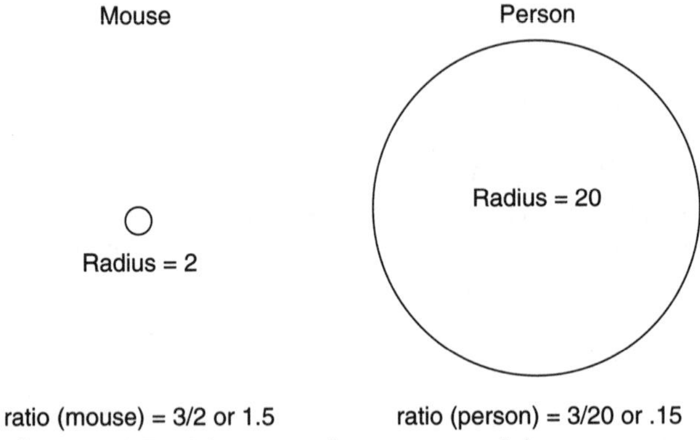


**Figure 4.5** Mouse and person models.

It is a matter of scale as related to maintaining body temperature. Both the mouse and the human are mammals and as such are warm-blooded beings. Both are required to maintain a constant body temperature, which is almost always higher than their surroundings. Maintaining body temperature requires energy. The data tell us that it is more difficult for the mouse to accomplish this task than the human. Per gram of body weight, the mouse is expending more energy to maintain body temperature than a human. Why?

It has to do with size. Relatively speaking the smaller mouse is losing body heat more readily than the larger person. Per gram of body weight the mouse has more surface area than the human. Since the mouse has a greater ratio of surface area to mass than the person, and since most body heat is lost through the skin into the environment, it is easier for a greater percentage of the mouse's body heat to escape into the surroundings than that of a person.

If we use a sphere as a model to represent both the mouse and the person (Fig. 4.5), we understand better what is going on. The mathematical expression for the volume of a sphere is:

$V = (4/3)\pi r^3$ while the expression for the surface area of a sphere is: $S = 4\pi r^2$.

$$\text{Surface Area: } S = 4\pi r^2$$

$$\text{Volume: } V = (4/3)\pi r^3$$

The ratio of surface area to volume:

$$S/V = (4\pi r^2)/(4/3)(\pi r^3) \text{ or } 3/r$$

For the mouse, when the radius is 2, the ratio is 3/2 or 1.5 For the person, the ratio is 3/20 or 0.15 when the radius is 20. Therefore, it is easier for a greater percentage of the mouse's body heat to escape into the surroundings than that of a person. You can verify for yourself.

The model we have just employed helps us to understand why the amount of food required for different mammals is not proportional to the size of the mammal but actually decreases (per gram of body weight) as the size of the different species increases.

## Model of Matter

The first model we will create and explore involves matter. We will use this model as a springboard both to look at the generation of a model and as a "back door entrance" in providing us with some of the basic concepts and language of science.

We define matter as anything that has mass and occupies space or volume. If its mass can be determined and it takes up space, it is matter.

Mass is a property of matter which is constant and does not change depending upon location. We often think of weight as being the same as mass but in fact there is a significant difference between the two. The weight of an object changes with location. Your weight is different on the Earth than on Jupiter; however your mass is constant at both those sites. When an astronaut is in a capsule orbiting the Earth, the astronaut is in a condition of weightlessness but still has mass. Mass is a quantity of matter while weight is a force. Your weight represents the force the Earth exerts on you due to the pull of gravity.

On Jupiter there is a greater pull of gravity, hence a greater force of attraction, and therefore your weight will be a bigger force. Mathematically,

$$\text{Weight (or Force)} = \text{Mass} \times \text{Pull of Gravity (or acceleration due to gravity)}$$

$$F = m \times g$$

When the pull of gravity is zero, the force or weight becomes zero while the mass is maintained.

That is, $g = 0$, so $F = m \times g$ becomes

$$F = m \times 0, \text{ hence}$$
$$F = 0$$

In expressing the mass of an object there are a variety of units we can use. In scientific studies units called SI units (International System of Units) are employed.

# Units

The metric system (see Table 4.1 and Table 4.2) had its origin after the French Revolution where one of the consequences was to reform their system of weights and measures

# Metric System

The unit of mass was originally the gram and was defined as the mass occupied by a specific volume of water at a certain temperature (the mass of one cubic centimeter of water at about 4 degrees Celsius). The system is rational in that quantities are related in a "power of ten" fashion as denoted by prefixes (see Table 4.3).

Hence one kilogram, or 1 kg, is $10^3$g where we substitute the $10^3$ for the k; one milligram, or 1 mg, is $10^{-3}$ g where we substitute the $10^{-3}$ for the m. The prefixes tell us the connection between different quantities in a rational way. In the English or American system there is no rational connection. The name of the unit gives no clue as to the relationship between quantities. For example, the connection between yards, feet, inches, and miles is unclear unless we have memorized the connection. The metric prefixes apply to all units, mass, volume, etc.

**TABLE 4.1 SI Base Units**

| Base Quantity | Unit | Symbol |
|---|---|---|
| **Length** | meter | m |
| **Mass** | kilogram | kg |
| **Time** | second | s |
| **Electric current** | ampere | A |
| **Luminous intensity** | candela | cd |
| **Temperature** | kelvin | K |
| **Amount of substance** | mole | mol |

**TABLE 4.2 Units**

| Unit of Measure | The Metric System | The American System |
|---|---|---|
| **Length** | millimeter | inch |
| | centimeter | foot |
| | meter | yard |
| | kilometer | mile |
| **Mass** | kilogram | slugs |
| **Weight** | Newtons | pounds |
| **Volume** | milliliter | ounce |
| | liter | cup |
| | | pint |
| | | quart |
| | | gallon |
| **Time** | second | second |
| | minute | minute |
| | hour | hour |
| | day | day |
| **Temperature** | Celsius | Fahrenheit |
| **Pressure** | Pascal | inches of mercury |
| | Kilopascal | pounds per square inch |

**TABLE 4.3 Prefixes**

| Prefix | Symbol | Meaning |
|---|---|---|
| tera | T | $10^{12}$ |
| giga | G | $10^{+9}$ |
| mega | M | $10^{+6}$ |
| kilo | k | $10^{+3}$ |
| deci | d | $10^{-1}$ |
| centi | c | $10^{-2}$ |
| milli | m | $10^{-3}$ |
| micro | μ | $10^{-6}$ |
| nano | n | $10^{-9}$ |

The unit of mass is the kilogram, gram, milligram, and so on. In order to give a sense as to the magnitude of a kilogram (a unit of mass), its weight is about 2.2 pounds on Earth.

Another characteristic of matter is that it has volume. In order to discuss the unit of volume, the unit of length should be examined first since volume requires length in three dimensions.

The base unit of length in SI units is the meter. (The unit was originally generated by a commission, which had its origins after the French Revolution). A meter is about 39.4 inches. There are 100 cm or 1000 mm or 1,000,000 μm in a meter.

The unit of volume can be examined in terms of a container, which is in the form of a cube one decimeter by one decimeter by one decimeter.

$$\text{Volume of cube} = \text{length} \times \text{width} \times \text{height}$$
$$= (1 \text{ dm}) \times (1 \text{ dm}) \times (1 \text{ dm}) = 1 \text{ dm}^3$$

This volume is called the liter; hence 1 liter = $1 \text{dm}^3$.

Since 1 dm = 10 cm, 1 liter = (10 cm) × (10 cm) × (10 cm) = $10^3$ $\text{cm}^3$.

So 1 liter = 1 $\text{dm}^3$ = $10^3$ $\text{cm}^3$ = 1000 $\text{cm}^3$ = 1000 cc, where cc stands for cubic centimeters.

The liter is also divided into 1000 parts so that 1 liter may be expressed as 1000 ml. As a point of reference to the more familiar English or American system, a liter is about 1.1 quarts. (Think of a 2-liter soda bottle, which contains about 2.2 quarts or about 68 liquid ounces.)

# Density

Matter is characterized by mass and volume. One of the ways of describing matter is in terms of its density. Density is the mass of an object divided by its volume or

$$\text{Density} = \text{mass/volume}$$
$$D = m/v$$

Density is expressed in units such as grams/milliliter (g/ml) or grams/cubic centimeter (g/$\text{cm}^3$).

The density of water is 1.0 g/$\text{cm}^3$. (This corresponds to one $\text{foot}^3$ of water weighing about 62.4 lbs.)

Knowledge of density is useful. It enabled Archimedes to detect counterfeit gold objects and it can help us understand the conditions that influence objects sinking or floating. If the density of an object is less than the density of water, it will float in water while those with a greater density will sink.

In general an object will float in any media (liquid or gas such as air) if the density of the object is less than or equal to the density of the medium in which they are placed. (Why does a steel ship float if the density of steel is greater than the density of water?)

We return to our model of matter. Matter can be divided into two forms, pure matter known as substances, and mixtures where mixtures implies the blending or two or more substances. As a further subdivision, mixtures may be either homogeneous (also called solutions) or heterogeneous, while substances, the pure form of matter, may be either compounds or elements. A model of this classification is noted in Figure 4.6.

# Matter

We start with substances. Substances represent a pure form of matter. This means that a given substance will be only one thing—not a mixture; these things may be either compounds or elements. Let us explore the idea of elements in terms of a pure form of matter. A list of the known elements is presented in Table 4.4.

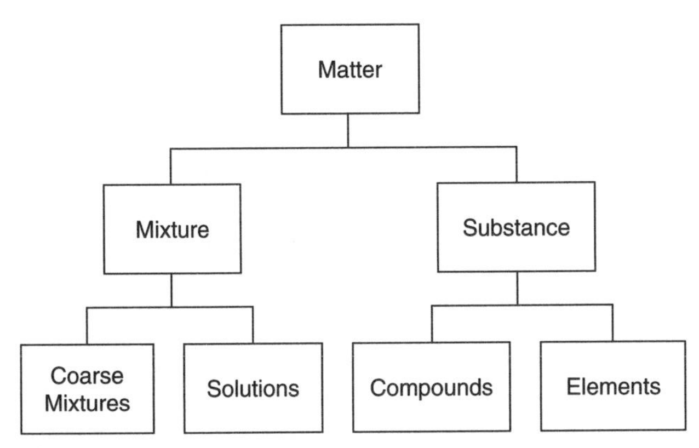


**Figure 4.6** Matter.

There are around one hundred known elements although perhaps only about 35 are commonly encountered. A given element, say gold, is called an element because it is totally one thing, which cannot be broken down further into anything simpler. (We could heat the gold and subject the metal to a variety of ordinary physical and chemical processes, but the gold cannot be broken down further.) The reason for this, it turns out, is that gold, like all elements, is composed of minute discrete and "indivisible" particles called atoms. (The mass of an atom is around $10^{-23}$ grams.)

The element is composed only of atoms of one type (well, almost; there are small mass differences between atoms in an element, which will be discussed later.) All atoms in a given element are the same; all atoms of a different element are different. The point is that atoms of a given element are composed of atoms unique for that element.

**TABLE 4.4 Modern Periodic Table**

**Los Alamos National Laboratory's Chemistry Division Presents a Periodic Table of the Elements**

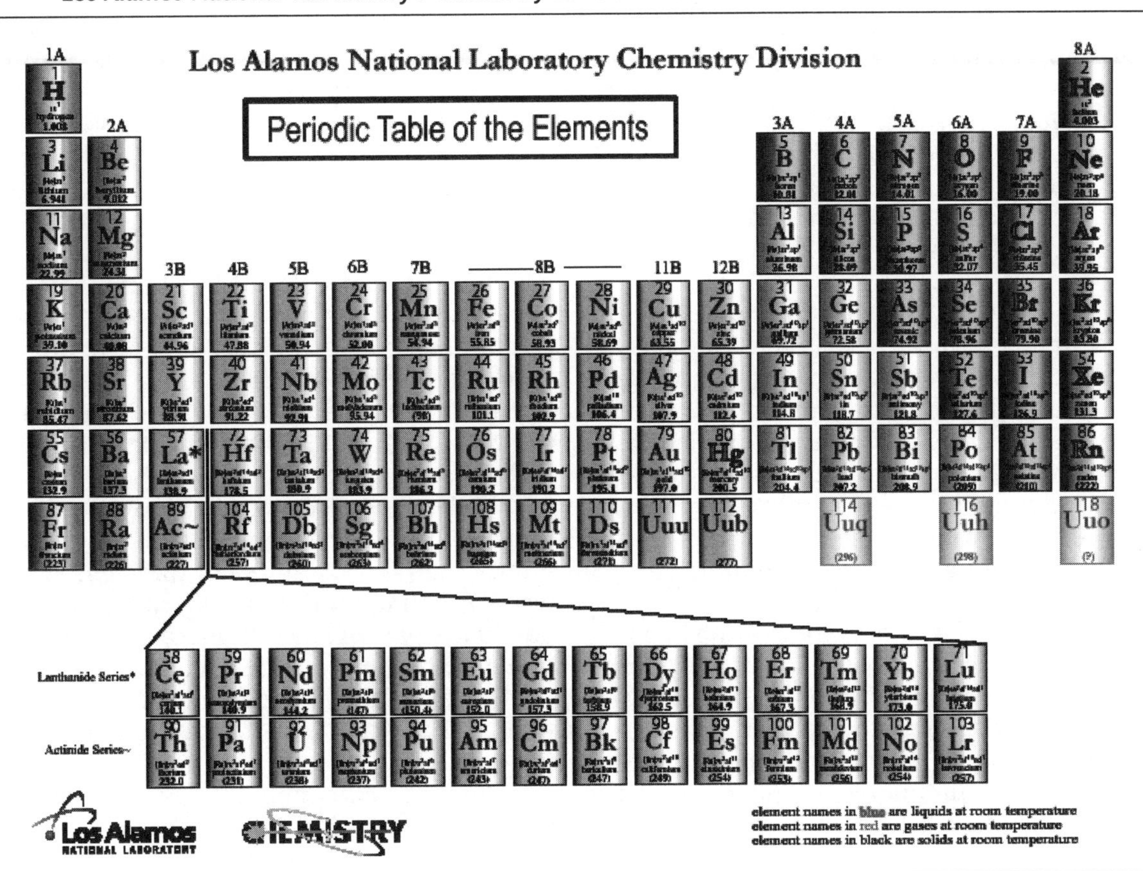


*http://www.nsls.bnl.gov/about/everyday/PDFs/crystals.pdf*

**TABLE 4.5 Some Common Chemical Symbols**

| | |
|---|---|
| He | Helium |
| O | Oxygen |
| C | Carbon |
| N | Nitrogen |
| Cu | Copper |
| Cl | Chlorine |
| Au | Gold |

There are symbols the chemist uses to represent the elements. Some symbols are intuitively simple O for oxygen, H for hydrogen, He for helium. Some require that we learn or memorize the symbol such as Pb for lead, Fe for iron. These symbols arise from the ancient Latin names used for the elements (see Table 4.5).

# Chemical Structures

Certain elements exist only in the atomic form. For example, the element helium (as well as some of the other Noble gases) exists only in the atomic form, that is to say the element is composed only of individual He atoms. Other elements, the vast majority, do not exist in the atomic form but rather exist in a form where two or more atoms are combined. When two atoms combine, the unit formed is called a molecule.

Consider the element oxygen, which can exist in three different elemental forms. These are atomic oxygen, molecular oxygen, and ozone. We can represent these forms of the element as shown in Figure 4.7.

Note that in molecular oxygen two oxygen atoms are united to form a diatomic molecule, which can be expressed by the formula $O_2$, while in ozone the molecule is triatomic and can be expressed as $O_3$. The properties of molecules change with the chemical structure. In the case of the element oxygen we see the contrast between molecular oxygen, which is needed for life, with ozone, which would kill us if we were to breathe the gas.

Another example of an element existing in different molecular forms is the element carbon. Carbon can exist in the form of graphite. The "lead" in a pencil with which one writes is really graphite (pure solid, carbon-gray in color, and a conductor of electricity) and diamond (pure solid, clear, and a nonconductor of electricity). The difference in properties depends on how the atoms are put together, that is their structural formula.

| Atomic oxygen | Molecular oxygen | Ozone |
|---|---|---|
| O | O-O | O-O-O |

**Figure 4.7** Different forms of oxygen.

Graphite Diamond

**Figure 4.8** Two different forms of carbon.

The model for the structural formula is shown in Figure 4.8 for two different forms of carbon.

# Compounds

Compounds are also a pure form of matter. They differ from elements in that compounds are composed of two or more elements and the atoms of the different elements are chemically united in a specific and particular way. As an illustration, the compound water is composed of two atoms of hydrogen and one atom of oxygen such that the two H atoms are chemically bound to the O atom, as noted:

H-O-H

and represented by the formula $H_2O$. In the case of compounds we have another example of a molecule but the molecule is not of an element. The line joining the H atoms to oxygen represents a chemical bond. In the case of water, the two H atoms are chemically bound to the O atom. (Actually there are different kinds of chemical bonds. We will examine the different forms of bonding and their properties later.)

While a given compound always contains the same elements in the same percent by weight this information is sometimes not sufficient to characterize a compound. How the atoms in a compound are put together also is important in determination of the properties of the compound.

For example, the compound with the formula $C_2H_6O$ can have two different structures (see Fig. 4.9).

The first compound will put you out if you drink too much. It's the stuff in beer, wine, and

```
 H H          H   H
 | |          |   |
H-C-C-O-H    H-C-O-C-H
 | |          |   |
 H H          H   H
```

ethyl alcohol    dimethyl ether

**Figure 4.9** $C_2H_6O$ ethyl alcohol and dimethyl ether.

whiskey that makes it alcohol. The second compound will put you out if you smell it. It is the stuff once used as anesthesia.

The chemical model (structural formula) helps us to start to learn about the properties of matter. As further information is gained through experiment the chemical model is modified so that distances between atoms (bond lengths) and angles between atoms (bond angles) can be determined. Once the structure of a compound has been determined the chemist can ultimately synthesize the substance in the laboratory. For important, but rarely found, natural products this can be lifesaving.

### SOLUTIONS

Solutions differ from substances (elements or compounds) in that solutions are homogeneous mixtures. Solutions are composed of two of more components; they are homogeneous mixtures that can have different variable compositions. Consider a solution of salt and water. Within certain limits we can have a small amount of salt or a larger amount of salt in water. While compounds have a fixed composition of chemically bound components, solutions do not.

Mixtures need not be homogeneous. We can also have coarse mixtures. Granite is an example of a mixture that is not uniform in composition, hence not a solution. (Solutions can also be solid.)

## States of Matter

Finally, a few words should be stated with regard to states of matter. The state in which matter exists basically reflects the environment around a particular molecule or molecular species. In general the same element or compound (or mixture) can be present in either the solid or liquid or gaseous state. These states exist at different temperatures (and pressures). As the temperature is lowered the (kinetic) energy of molecules is lowered and so the forces that exist between molecules become more significant. As the intermolecular forces increase the molecular environment

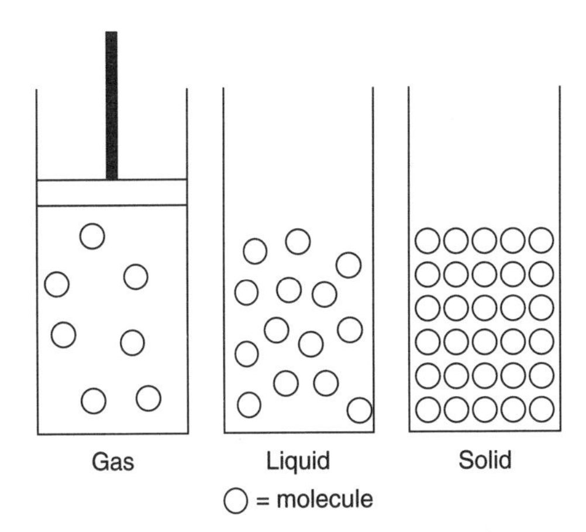


**Figure 4.10** Gas, liquid, solid states of matter.

becomes altered, leading to the transition from gas to liquid to solid as the temperature of a gaseous substance is lowered.

The molecular "picture" of gas, liquid, and solid state is represented in Figure 4.10.

### GASES

In the gaseous state molecules are generally far from each other. However they are in constant motion and so collide with each other, bouncing off each other and off the walls of the container. The arrangement of the molecules in the container is totally random. Because the forces between the molecules are relatively weak, a lid on the container is necessary to keep the molecules from flying out. The large distances between molecules give rise to the low density that gases possess. How does this model help to explain that balloons filled with hot air or helium gas float? What does it suggest about the density of helium or hot air?

### LIQUIDS

Liquids and solids possess a higher density than gas. In fact the density of liquids and solids are roughly about the same. The molecular picture or model reflects this fact. These states are characterized by molecules "squeezed" closely together, to about the same degree, so that there are many more molecules packed in a given volume than in the corresponding gaseous state. The closer packing reflects the fact that the molecules have less kinetic energy in the solid or liquid state and so the intermolecular forces become more significant.

In the liquid state, while the molecules are well-packed together, the molecular arrangement is totally random and the molecules are free to "roll over" one another. The net effect of this is that liquids are fluid or nonrigid. Like gases they have no definite shape but since the molecular forces are greater than gases, their volumes are fixed. The fact that liquids are shapeless reflects the fact that the molecules are mobile and are not in fixed positions.

## SOLIDS

The solid state differs from the other two states of matter in that solids are rigid. In the solid state as in the liquid state the molecules composing the material are packed closely; however in the solid state the molecules are in fixed positions and do not "roll over" one another. In fact a pure solid substance generally is composed of molecular species, which are closely packed together, are in fixed positions, and arranged in some sort of regular repeating pattern. On a grand scale this regular arrangement gives rise to the crystallinity manifested in solids.

## CHANGES IN STATE

In summary, changes in state represent changes occurring in the environment of molecules or molecular species, which are brought about as molecules became more or less energized by the transfer of energy, as noted in the following example:

$$\text{Water as ice } (H_2O) \xrightarrow[\textbf{(melts at 0° C)}]{\textbf{Change in state}} \text{Water as liquid } (H_2O) \xrightarrow[\textbf{(boils at 100° C)}]{\textbf{Change in state}} \text{Water as vapor } (H_2O)$$

# Review Questions

1. Define what a model means.
2. What is meant by a scale effect? Discuss the example of the mouse given in this chapter.
3. Describe three types of models and give an example of each.
4. How is matter characterized? What is meant by density?
5. What are molecules? What are compounds? How do they differ?
6. Discuss the states of matter.
7. Oleic acid is oil, which does not dissolve in water. When a drop of oleic acid is added to a large volume of water, the oleic acid spreads out to form a layer one molecule thick. Explain how one can estimate the length of the oleic acid molecule if we know the volume of the drop used, and the area to which the oleic acid spreads.

# Multiple Choice Questions

1. Models of the real world are simplified versions of the real phenomenon that enhance understanding and _____.
   a. are fun to build
   b. allow for predictions
   c. are usually not very useful
   d. None of the above.
2. A map may be considered to be a(n) _____.
   a. plan
   b. outline
   c. model
   d. picture
3. Matter is anything that has mass and _____.
   a. occupies space
   b. has velocity
   c. atoms
   d. substance
4. When two atoms combine, the unit form is called a(n) _____.
   a. element
   b. compound
   c. molecule
   d. prion
5. Compounds are formed from _____.
   a. protons
   b. electrons
   c. atoms
   d. single elements
6. A pure element may exist as a _____.
   a. compound
   b. solution
   c. mixture
   d. molecule

7. Given two objects of the same mass and density, one object is shaped as a cube and the other is spherical. Which of the following is true?
   a. The cube will have a larger volume than the sphere.
   b. The sphere will have a larger volume than the cube.
   c. Both volumes are equal.
   d. The sphere will sink in water while the cube will float.

8. A cell was found to be $12 \times 10^{-3}$ mm in diameter. The diameter of the cell in meters is _____.
   a. $12 \times 10^{3}$
   b. 12
   c. $12 \times 10^{-6}$
   d. None of the above.

9. For the four following examples, select the kind of model present from these choices:
   a. analog
   b. verbal
   c. mathematical
   d. scale model

   The flow of electrical current is like the flow of water. _____

   A model of the statue of liberty. _____

   When he laughed his belly shook like a bowl full of jelly. _____

   F = ma. _____

10. The compound $(CH_3)_2C(OH)H$ contains how many H atoms?
    a. 8
    b. 6
    c. 7
    d. 4

11. Place the following in the correct order from smaller to larger.
    a. mm ft um
    b. µm cm m
    c. µm ft mm
    d. yd km cm

12. Given a cube-shaped object that is 2 cm on an edge, the mass of the object is 8 grams. The density of the object is how many $g/cm^3$?
    a. 1
    b. 2
    c. 1/2
    d. 4

13. Liquid water and ice are examples of _____.
    a. two different compounds
    b. the same compound in different states
    c. the different compounds in the same states
    d. None of the above.

14. Salt, NaCl, dissolved in water is an example of a _____.
    a. solution
    b. element
    c. compound
    d. heterogeneous mixture

15. An object has a density of 1.5 $g/cm^3$. In which substances will it float?
    a. water (d = 1.0 $g/cm^3$)
    b. mercury (d = 13.6 $g/cm^3$)
    c. ethanol (d = 0.80 $g/cm^3$)
    d. carbon tetrachloride (d = 1.59 $g/cm^3$)

16. Use this information to answer the following two questions: An object has a mass of 9 and a volume of 10.

    The object, when placed in water, _____.
    a. floats
    b. sinks
    c. floats and then sinks
    d. None of the above.

17. The object is placed in a fluid whose density is 0.80 $g/cm^3$. It _____.
    a. floats
    b. sinks
    c. floats and then sinks
    d. None of the above.

18. A model is _____.
    a. an abstract representation of the real thing
    b. a framework for understanding
    c. a means of making predictions
    d. a means to trial and error
    e. All of the above.

19. Engineers are constructing a technologically advanced submarine for the Navy. One of the considerations is how to allow the submarine to first submerge and then to resurface. One suggestion is to develop a system to change the density of the submarine upon

command. From your experience in lab, to allow the submarine to first submerge and then resurface, which of the following would you suggest?

a. change density to 2.5 and then to .75 $g/cm^3$
b. change density to 2.5 and then to 3.5 $g/cm^3$
c. change density to 0.8 and then to 1.5 $g/cm^3$
d. change density to 3.5 and then to 2.5 $g/cm^3$

# Bibliography and Web Resources

Scientific modeling. Why is scientific modeling important? Types of scientific models.
**www.mcrel.org/epo/resources/sci_modeling.asp**

About scientific modeling.
**www.wcer.wisc.edu/ncisla/muse/philosophy/**
**(More results from www.wcer.wisc.edu)**

Floating and sinking.
**www.topscience.org/float_sink.htm**

"On Being the Right Size," essay by Haldane, J. B. S., in *Writing About Science,* edited by Mary Elizabeth Bowen and Joseph A. Mazzeo, Oxford University Press, 1979, pp. 21–27.

# Models and the Atom

*"If anything like mechanics were true then one would never understand the existence of atoms. Evidently there exists another 'quantum mechanics.' "*

**Werner Heisenberg**

Our current model of the atom has evolved significantly over the last century. During the 19th century scientific investigations had shown that there was an electrical nature associated with matter, and that there were several basic particles that made up all matter. One of these was called an electron and was defined as possessing a negative charge; the other was called a proton and was positively charged.

In the early 1900s it was believed that all atoms were made up of both the negative (electrons) and positive (protons) in order to be electrically neutral. It was believed that these charged particles were homogeneously distributed throughout the atom to create what was called the "plum-pudding" model of the atom. It was accepted for a number of years until an experiment conducted in 1910 by Ernest Rutherford changed the model and the way we think about atoms.

## Goals

After studying this chapter, you should be able to:

- Understand how modeling was used to study the atom.
- See how the study of blackbody radiation helped our understanding of the structure of the atom.
- Understand the role of the photoelectric effect and photons.
- Understand how the study of line spectra contributed to further understanding of the atom.
- Follow the development of models to explain DNA and cardiac output.

## Rutherford's Atomic Model

Ultimately in the first quarter of the 20th century the masses and absolute charges of the particles were experimentally determined. It turns out that the absolute magnitude of the charges on a proton and an electron are exactly equal to each other but opposite in sign. For convenience sake we shall say the proton's charge is +1 and the electron's charge is –1. (In actuality the charges on the particles are $+1.6022 \times 10^{-19}$ coulombs and $-1.6022 \times 10^{-19}$ coulombs respectively.) The mass of the proton is about 1,800 times that of the electron.

Rutherford and his graduate student, Hans Geiger (to become famous for his invention) and an undergraduate student Ernest Marsden conducted simple experiments in which high-energy charged particles were allowed to strike thin foils of some metallic elements. The high-energy particles were

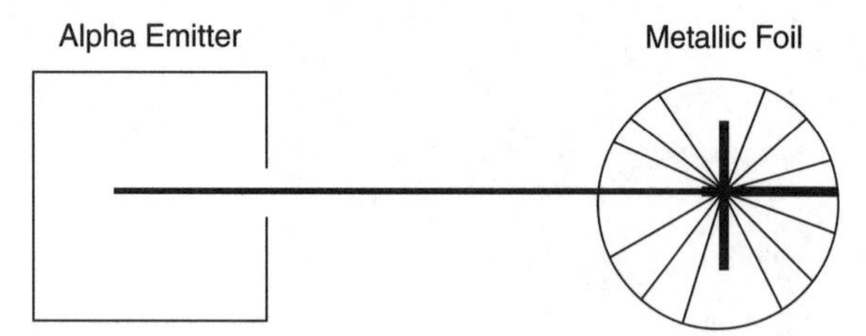


**Figure 5.1** Alpha particles striking a target.

minute ($10^{-15}$ m in diameter) but extremely dense, and possessed a very high concentration of positive charge. It was expected that these particles, called alpha particles, would readily pass through the thin foils with ease. Indeed it was observed that some alpha particles passed through the foil of metal with ease. However it was also observed that some alpha particles were deflected through various angles, some of which were quite large (see Fig. 5.1).

The fact that some of the alpha particles were deflected back to the source was quite significant to Rutherford. In a letter Rutherford sent home, he expressed the magnitude of his surprise. "It was as incredible as if you fired a 15-inch shell at a piece of tissue paper and it came back and hit you." If the model of the atom was changed, it would be possible to explain how a 15-inch shell (the alpha particle) could be deflected back to the source after hitting the tissue paper (the metal foil).

The only way the very high density, highly concentrated, positively charged alpha particle could be made to change direction by 180 degrees would be if it encountered another particle of like mass and charge density. In other words the model of the atom was not a homogeneous distribution of + and – charges. To get the alpha particle to reverse direction, an atom would have to have a nucleus where virtually all the mass and positive charge resided. Only when an alpha particle encountered such a nucleus directly would there be enough force to cause the observed phenomena. Figure 5.2 illustrates this point.

While this model explains much, there are a few questions that arise; one obvious question is related to the distribution of electrical charge. That is, if the nucleus of the atom is positively charged, and the atom has no charge, what is the role and location of the electrons in the atomic structure?

# Photons

Insight regarding the composition and structure of the atom, which as explained previously was puzzling, came about as a result of attempts to explain certain natural phenomena. Three of the most

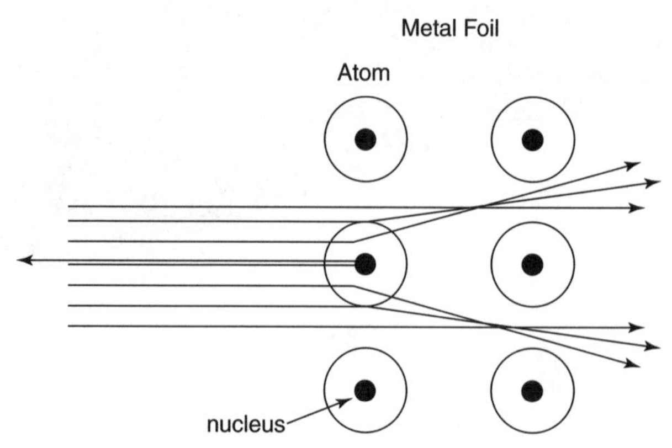


**Figure 5.2** Alpha particles colliding with atoms.

important unexplained phenomena were known by the terms blackbody radiation, photoelectric effect, and line spectra. The explanation of the first two redefined our understanding of the relationship between light and energy. This work, accomplished by the German physicist Max Planck and Albert Einstein, ultimately resulted in the idea that light, which was propagated and behaved like a wave, could simultaneously also behave like a particle. The massless particle, called a photon, possessed different amounts of energy depending upon the wavelength of light associated with the photon. That is:

Energy of light (photon) is proportional to
1/wavelength of light.

So if the wavelength of light is long (big), the energy of the photon is smaller than if the wavelength of light is short (small).

Interpretation of line spectra by the Danish scientist Niels Bohr brought us on the path toward a more modern picture of the atom. In order to gain insight into the process of building a model of the atom, it is helpful to first review some concepts involving light and spectra.

# Light and Waves

Light is a form of electromagnetic radiation. This means that there are both electric and magnetic properties associated with light. These properties need not concern us. What is important for our discussion is the mode of transmission of light. Light is propagated in the form of a wave. We characterize the transmitted light in terms of its wavelength, frequency, and velocity. The wavelength of light symbol, $\lambda$, is the length of the smallest distance between two successive points on a wave that are executing the same motion. The wavelength is expressed in distance units (actually distance per wave or distance per cycle).

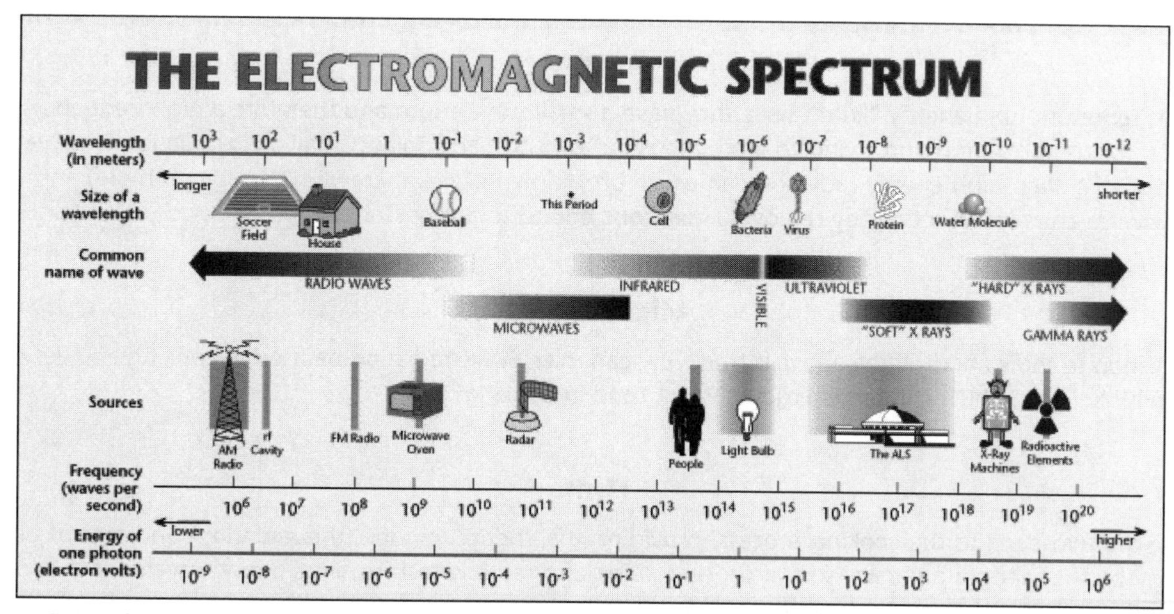


**Figure 5.3** Electromagnetic spectrum. *http://www.lbl.gov/MicroWorlds/ALSTool/EMSpec/EMSpec2.html*

For example, X-ray radiation is about 0.10 nm long, visible red light is about 650 nm long, and radio waves are around 100,000,000,000 nm. (Recall that nm = $10^{-9}$ m.) X-ray radiation is short wavelength while radio waves are long wavelength. The entire electromagnetic spectrum (a spectrum is a collection of wavelengths) is listed in Figure 5.3.

Note that various regions in the electromagnetic spectrum within a certain wavelength range are collectively described by names designating specific spectral regions. As illustrations there is the X-ray region, the radio wave region, the visible region, and so on.

Within the visible region spectrum from longer to shorter wavelengths the associated colors are:

Red Orange Yellow Green Blue Indigo Violet

The shorter wavelength region just beyond the violet is called the ultraviolet or UV. Those longer wavelengths just beyond the red region of the spectrum are called the infrared or IR. Examination of the figure locates some of the other spectral regions.

Another characteristic of a wave is its frequency. We can think of the frequency as "how often" the waves (wavelengths) are coming. If we are at a point and could observe or count the number of waves (wavelengths) passing a point during a given time, we would determine the number of waves (or cycles) per time (waves per sec or waves/sec or wave/min, etc). The symbol used for frequency is $\upsilon$. From the spectra shown, you can see that that the frequency for X-ray radiation is $10^{18}$ waves/second or $10^{18}$ cycles/second or $10^{18}$ Hz (1 wave or cycle/second is called a Hertz or Hz) . Frequency in the AM radio range is on the order of $10^6$ Hz.

A station located at 93.9 MHz on the radio dial is really at $93.9 \times 10^6$ Hz or 93.9 megahertz, 93.9 MHz. Sometimes frequency is simply expressed as 1/time; in the example $93.9 \times 10^6$ 1/sec or $93.9 \times 10^6$ sec$^{-1}$.

The wavelength and frequency are related in an inverse way. As one quantity increases, the other decreases as noted:

$$\lambda \text{ is proportional to } 1/\upsilon.$$

This is true because the product of wavelength and frequency equals the speed of the wave

$$\text{Wavelength} \times \text{Frequency} = \text{Speed of wave}$$

or

$$\lambda \times \upsilon = c$$

Another characteristic of light is its amplitude, which is related to the intensity of light.

# Planck's Theory

In 1900, Max Planck made a remarkable statement based on his study and analysis of the intensity of light associated with the electromagnetic spectrum of heated glowing matter. The radiation study (commonly called blackbody radiation) and associated theoretical conclusions led Planck to the nonintuitive but inescapable conclusion that the energy of a light wave depends

### X-rays

X-rays represent high energy "light" since they have a small wavelength and therefore a high frequency. The reason for using caution and trying to limit exposure to X-rays and solar (ultraviolet) radiation is related to the fact that such high energy radiation can cause breakdown of cell material (DNA for example) and initiate adverse chemical reactions by removing electrons bound to atoms.

### Microwaves

A question to think about: If the above is true, why can microwave radiation be used in cooking and yet ordinary visible light (which is more energetic) is not used in cooking?

### Hint

There are two parts to the cooking process, providing the energy to cook (the radiation) and receipt of the energy by the food. In a parallel sort of process, think of a radio, which receives many wavelengths of radio transmission but plays only the one to which it is tuned.

upon the frequency of the light wave. His equation, known as the Planck relationship, is:

$$E = h \upsilon$$

$$\text{Energy of light wave} = \text{h (a constant)} \times \text{frequency of light wave}$$

or the energy of light is directly proportional to the frequency of the light wave.

The relationship, when originally proposed by Planck, was rejected by the scientific community. It was not accepted until Einstein gave validity to the relationship when he used it in the solution of another problem that had baffled scientists.

The problem that Einstein solved involved the photoelectric effect. Basically the photoelectric effect was the observation that only high frequency light could dislodge electrons from matter. Using Planck's conclusions that energy and frequency were proportional, Einstein extended Planck's ideas in that he concluded that not only was the light emitted in bursts, rather than continuously, but traveled through space as localized bundles of energy called photons. It was the photons that knocked the electrons out of the matter in a particle-like collision. Some implications of this fact are noted in the box on this page.

## Emission Spectra

Heating a substance to a high enough temperature will ultimately cause the material to glow. The tungsten wire in a lightbulb is an example. If we were to allow a thin beam of that white light to be resolved into its constituent components by passing it through a prism, we would obtain a spectrum of visible light, a rainbow of colors (Fig. 5.4). This spectrum is a continuous emission spectrum (light is being given off, all wavelengths are represented, and none is missing). When light from the sun passes through raindrops, the same phenomenon of a continuous spectrum, a rainbow may be observed.

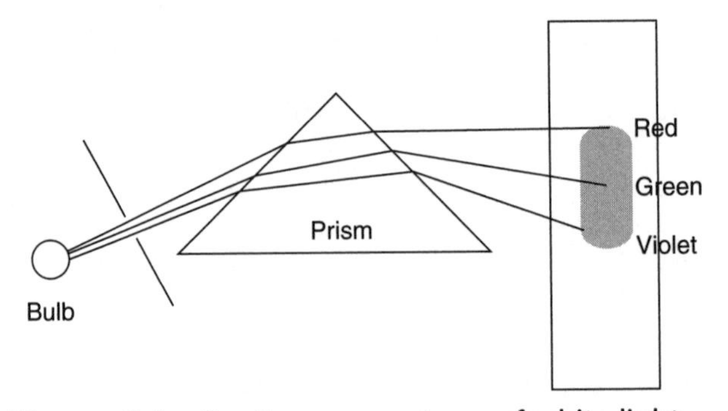


**Figure 5.4** Continuous spectrum of white light.

However, for many substances, especially those that exist as atoms or simple molecules, the emitted light resulting from excitation using heat or electricity resulted in a baffling spectrum. A beam of light from such an excited source ($H_2$ for example), when resolved by passing through a prism, gave a discontinuous spectrum or line spectrum. It was observed that only certain colors (and wavelengths) of light appeared while other wavelengths were totally absent (Fig. 5.5). The appearance of only select or discrete wavelengths was perplexing. Different lines appeared for atoms of different elements to the degree that the line spectrum could be used like a fingerprint to identify the element responsible.

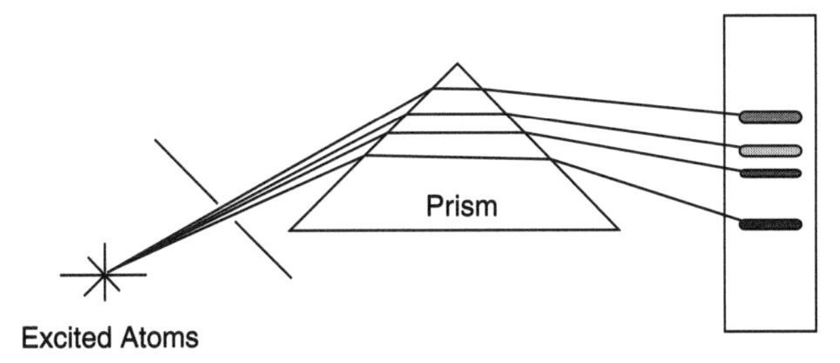


**Figure 5.5** Line spectrum of an atomic source.

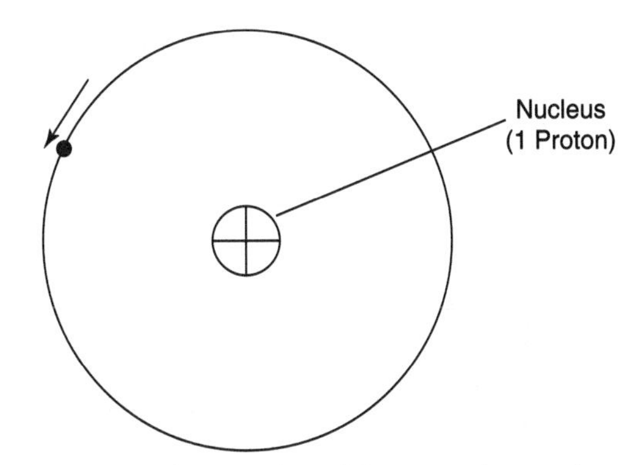


**Figure 5.6** Electron in orbit around a nucleus.

# Bohr's Atomic Model

In order to understand why excitation of certain elements gave a line spectrum, Bohr created a model. He selected the simplest system possible, the hydrogen atom. Using the hydrogen atom, which possesses one electron and one proton, and applying standard or classical physics (Fig. 5.6), he envisioned a system where the negatively charged electron rotated in a circular orbit about the much smaller positively charged nucleus. Bohr recognized the analogy between the solar system and atom although the forces were quite different.

In the case of the atom, the negative electron's attraction by the positive nucleus is "balanced" by the electron's motion about the atom. Bohr had the insight to further recognize that the electron could rotate about the nucleus of the hydrogen atom only in certain discrete, fixed orbits at specific distances from the nucleus. There were only certain allowed orbital paths.

Bohr's hypothesis was a radical departure from conventional thinking. He also contradicted classical concepts in stating that electrons traveling in a fixed orbit did not radiate energy and were continually stable in that pathway. The import of this statement is that H atoms could only have discrete or quantized energies depending on which orbit contained the electron.

When an electron was in a certain orbit (Fig. 5.7), the H atom would be in a certain energy state (possess a certain amount of energy) and a certain distance from the nucleus. If the atom was excited, the electrons could be elevated to one of a number of higher quantized energy states and located at a further distance from the nucleus.

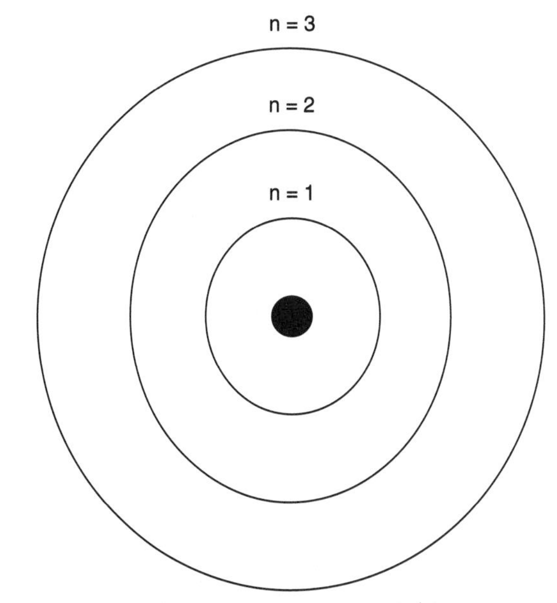


**Figure 5.7** Electrons in quantized orbits.

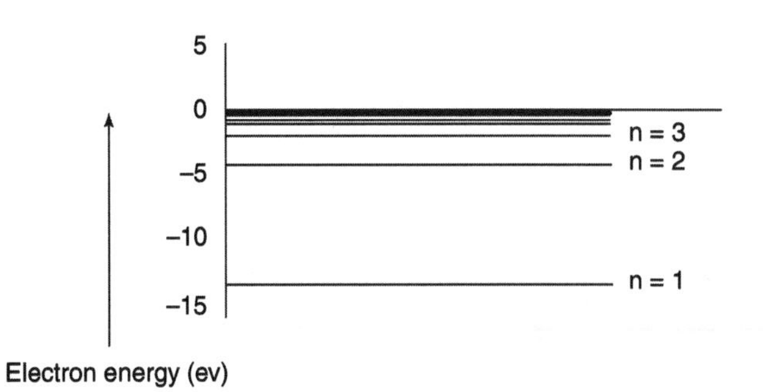


**Figure 5.8** Illustration of quantized energy levels for an electron in orbit.

In order to explain the line spectrum it is also necessary to consider energy changes associated with electrons as they undergo transitions from one circular orbit to another (Fig. 5.8); that is to consider different "states" of the atom and transitions between them.

To understand the reasoning, consider the following: As a negatively charged electron moves into orbits closer to the positive nucleus, it becomes more difficult to remove it away from the attractive influence. If we wish to move an electron from a close-in orbit to a more distant one, we must put in energy. Conversely, when an electron moves closer to the nucleus by dropping from a larger, farther away orbit to a smaller one closer to the nucleus, energy is given off.

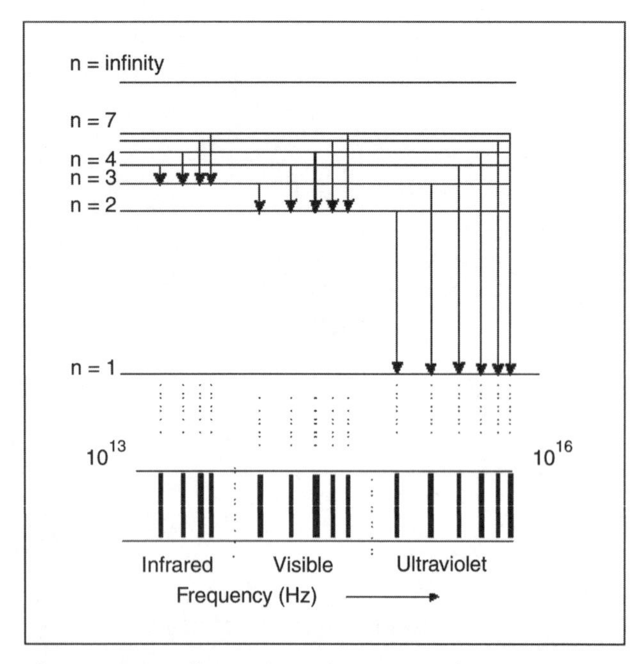


**Figure 5.9** Illustration of transition line spectra.

Since there are only discrete orbits (and free space that can not be occupied between them), the electrons in atoms can be excited further away from the nucleus to only certain specific energy states or orbits. Because the electrons in the atoms have only certain energies, we say the energy is quantized, as opposed to being continuous energy.

Eventually, electrons, which have been excited and driven further from the nucleus, are "de-excited" and drop down to lower quantized energy levels. In the process they give off a specific amount of energy by radiating light. The specific amount of energy is translated to a packet of light of a specific wavelength.

$$E = h\nu \text{ or } E = hc/\lambda$$

Each specific energy release of an orbital transition corresponds to a specific wavelength of light. The line spectrum (Fig. 5.9) obtained is a collection of the "finite" or limited number of possible electron jumps from one quantized orbit or energy level to another. This pictorial model of the H atom represents Bohr's hypothesis in an attempt to explain line spectra. In order to verify the assumptions made, Bohr created a mathematical model based on forces and energies required to sustain the pictorial model. Bohr's mathematical model was impressive. The model yielded wavelengths for hydrogen that agreed with the measured values to an exceptionally high degree of accuracy.

Bohr received the Nobel prize for his work on the H spectra. However as time progressed and additional studies were performed on a variety of different elements, weaknesses in the Bohr model began to emerge and the model began to lose its luster.

Although Bohr's theory has now been superceded and replaced by the modern theory of quantum mechanics that is capable of better explaining the behavior of the atom, the Bohr model is still used by scientists to visualize and describe atomic behavior as a first approximation. For accurate work quantum mechanics is used. Quantum mechanics applies models that are more abstract and mathematical than the Bohr version, and it remains a departure from classical scientific ideas. Apparently the world at the atomic level behaves much differently from the larger-scale world we observe with our senses.

# The Atom

Finally, a few words concerning the makeup of the atom are in order. We see as a consequence of the work of Rutherford and Bohr and others that the atom consists of a very dense core called a nucleus, which contains virtually all the mass and all the positive charge of the structure. The nucleus is surrounded (at various distances if we use the Bohr model) by the extremely light negatively charged electrons.

In the case of a hydrogen atom there is present one proton (nucleus) and one electron. Actually, there is another subatomic particle that also exists, and with only one exception is present in the nuclei of all atoms. This particle is the neutron. Neutrons are subatomic particles that bear no electrical charge and have almost exactly the same mass as that of protons. We can characterize atoms by their mass number and atomic number. The mass number is equal to the number of protons and neutrons present in the atomic nucleus, while the atomic number represents the number of protons present.

The convention is to indicate this information in the following way:

$$^{\text{mass number}}_{\text{atomic number}}\text{M}$$

For example,

$$^{1}_{1}\text{H}$$

indicates one proton and one electron.

The number of protons defines the element. That is to say every atom of the same element has the same number of protons or the same atomic number. Different atoms of different elements will have different atomic numbers. For a given atom, which is neutral, the number of electrons and the number of protons will be equal. However, while atoms do not lose or gain protons they can and do lose or gain electrons. In addition, the number of neutrons of the same element can vary. Such varieties are called isotopes and they will have different masses, although they are atoms of the same chemical element and possess essentially the same properties. Some examples of isotopes are:

$^{1}_{1}H$ (protinium) $^{2}_{1}H$ (deuterium) $^{3}_{1}H$ (tritium)

These three isotopes of hydrogen all possess one proton and one electron. They differ in that one isotope (deuterium) has one neutron while tritium has two.

Atoms can gain or lose electrons to become charged particles called ions. For example:

$^{23}_{11}Na$ (Na or sodium has 11 electrons)

In the process $Na \rightarrow Na^{+1}$ + 1 electron, the sodium atom has lost an electron to become a positive ion with only 10 electrons.

In the process 1 electron + $Cl \rightarrow Cl^{-1}$, the chloride atom has gained an electron to become a negative ion.

In general, metals tend to lose electrons and nonmetals tend to gain electrons in the formation of ions.

Compound formation either involves the attraction of + and – ions to hold the ions together or the sharing of electrons between the atoms. The compounds thus formed are called respectively, ionic or molecular compounds.

The periodic table, which is organized according to atomic number of elements, enables one to discern properties of elements, including their metallic or nonmetallic nature and tendency toward ion formation.

As noted earlier, some models can be pictorial in nature, and others may be mathematical or involve a blending of the two approaches. Some examples follow.

## DNA

The discovery and structural determination of DNA represents one of the hallmarks of scientific achievement, not only for its own sake but also because of the promise it may mean in terms of human health and well-being. But what is DNA? Over the next few chapters the structure and workings of DNA will be explained so that an appreciation of its function can be achieved. Before we know how it works, we should know something about what it is and looks like.

First the name: DNA is an abbreviated form of the common chemical name for the compound.

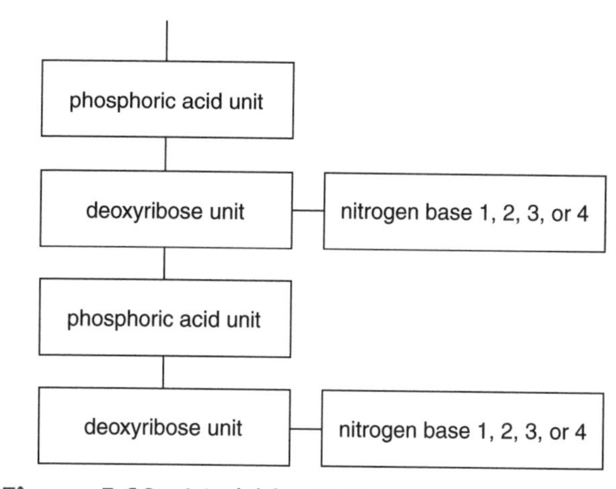


**Figure 5.10** Model for DNA.

The full name is DeoxyriboNucleicAcid. The letters, which are capitalized and underlined, are the source of the letters DNA. DNA is actually a long chain molecule; that is to say, the atoms, which make up the compound, are bound one to the other and continue in a chainlike fashion over and over to make a very long giant molecule. Long chained molecules with repeating parts connected together are called polymers. (The name polymer, poly [many] and mer [parts] is apt). The segments, which are repeated over, are the phosphoric acid and the deoxyribose units (see Figure 5.10). DNA is called a biopolymer because it is a polymer, which comes from living things.

As the common chemical name implies DNA is composed of three units, deoxyribose, (deoxyribose is a carbohydrate—carbohydrate names usually end in "ose"), a nitrogen "rich" base unit of which four possibilities exist (purines or pyrimidines), and phosphoric acid. The deoxyribose units are bound to the phosphoric units in a repeating fashion to create the "backbone" of the biopolymer. One of the possible four nitrogen base compounds is attached to the deoxyribose. A model for the structure is noted in Figure 5.10.

As noted, there are four nitrogen bases (Fig. 5.11). Their names are cytosine, guanine, thymine, and adenine. Their chemical structure, which shows how the atoms in the base unit are attached, is shown in Figure 5.11.

The segment of the DNA with all components present is represented in the model shown in Figure 5.12. This picture represents the DNA in more detail. As the figure shows, DNA is actually two strands of the biopolymer, which in nature exist as two strands in a coil or helix about each other. While everyone's DNA is exactly the same with regard to the deoxyribose phosphoric acid backbone, there are differences between individuals in terms of the sequence in which the nitrogen bases

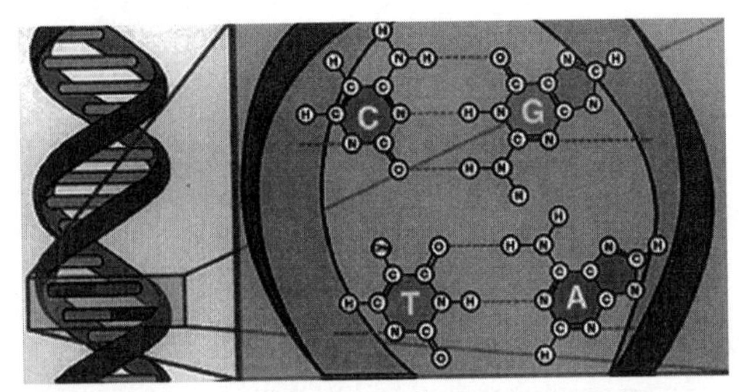


**Figure 5.11** DNA structure with nitrogen base pairs indicated: cytosine (C) and guanine(G), also thymine(T) and adenine(A). *http://genetics.nbii.gov/Basic1.html*

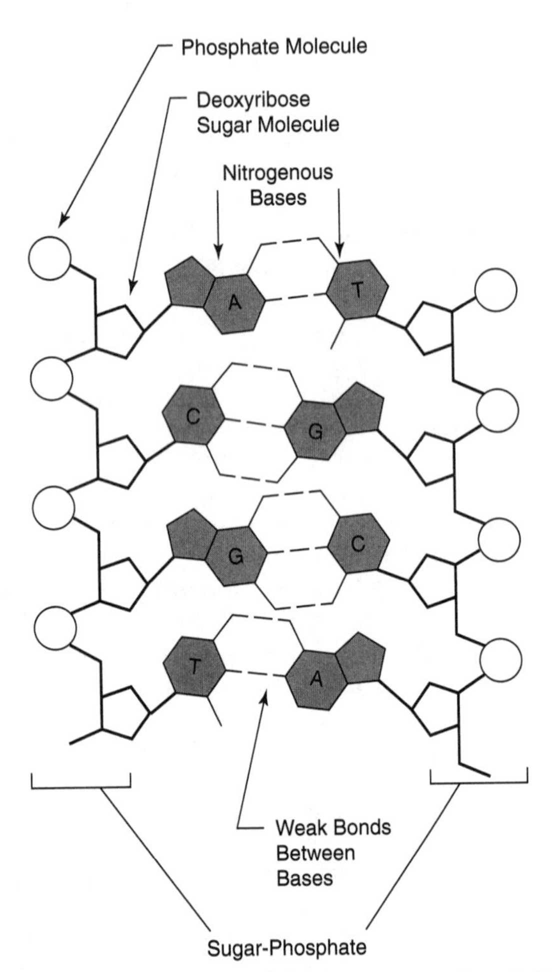


**Figure 5.12** DNA model. *http://www.ornl.gov/sci/techresources/Human_Genome/publicat/primer/fig2.html*

are placed. This difference (which is quite minor from individual to individual) is what makes us unique.

One final point should be noted. For reasons that will not be discussed here, the nitrogen base pair exists as partners when in the double helix

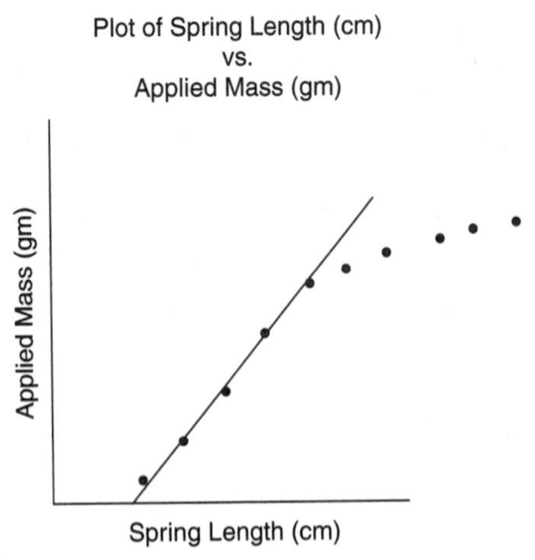


**Figure 5.13** Graph of change in spring length with change in load.

and is a critical factor in transmitting the genetic code. How protein is synthesized from the food we eat, under the direction of DNA and genetic code, will be examined in the subsequent chapters after we have examined coding.

## Mathematical Models

Mathematical models are commonly used in science, industry, and business. For example, short- and long-range predictions concerning weather and weather patterns involve mathematical models. The exactness or precision of the model is a function of the reliability and timeliness of the information provided as well as how well the model has been constructed. As the model is used and experience gained, improvements to the model may be forthcoming. In addition, the model may have limitations, conditions under which it cannot operate, because the premise upon which it was constructed has been compromised.

Consider modeling the behavior of a spring coil; that is can we mathematically express how a spring reacts when forces are applied to it. We know that when we pull on a spring it stretches. We can conduct a series of experiments where different weights are applied to a suspended spring and collect data, which relate to the length of a spring, associated with the different weights. We can then graph the data. A typical graph, which would result from this experiment, is shown in Figure 5.13.

Note that for the data collected when smaller masses are applied there is a linear relationship; that is, applied mass and spring length are directly proportional. At larger applied masses, there is deviation from linearity and the spring doesn't work in a predictable way.

In addition, the applied mass is really acting as a force.

Using our knowledge of graphing we can obtain an equation for the linear portion of the graph. We can determine the slope of the line and the y (or x-intercept). Ultimately an equation of the form

Applied Weight = slope of line (spring length – "x-intercept")

or

W = slope (L – "x-intercept")

where slope (m) and "x-intercept" (or $L_o$) are experimental.

$$W = m (L - L_o)$$

or

W = m (L) – "y-intercept"

where slope (m) and "y intercept" = $mL_o$ = b are experimental,

$$W = mL - b.$$

Hence, an equation of the type: W = 10(L – 20), if the slope were 10 and the "x-intercept" were 20, would be a model for the behavior of the spring. One could predict how the spring would react. The spring length increases upon the addition of any weight to the spring (within the range that is applicable). We could simply do the calculation rather than actually doing the measurement. Modeling allows us to forecast or predict. (See Appendix C.)

Finally, sometimes a model may cause us to examine our system in more detail so that new things are learned about the system that causal association precludes.

As shown in the equations below, the human heart and cardiovascular system can provide us with some clear examples of models and their use in understanding how systems work and how we can manipulate them. The role of the heart is to pump blood into the arteries for distribution throughout the body. The amount of blood pumped by the heart in one minute is known as the *cardiac output* (c.o.) and is usually about 5,000 milliliters/min (about 1 1/4 gallons). Cardiac output, the volume of blood pumped by the heart per minute, is determined by the volume of blood pumped with each beat, or *stroke volume,* (s.v.) and the number of beats per minute, or *heart rate* (H.R.). As shown in the equation, or mathematical model that follows, cardiac output is seen to be the product of heart rate and stroke volume. Multiplied together, they give the amount of blood pumped by the heart each minute. The following provides an example of this relationship.

$$\text{C.O.} = \text{H.R.} \times \text{S.V.}$$

$$4900 \text{ ml/min} = 70 \text{ beats/min} \times 70 \text{ ml/beat}$$

The model shows us that, on average, the heart beats at a rate of 70 beats/min (the H.R.) and pumps approximately 70 ml of blood with each beat (the S.V.). According to the model, the cardiac output would therefore be HR × SV, or 4,900 ml/min.

One of the uses of such models is their value in prediction. For example, using this model we can predict that if the heart rate were to increase to 100 beats/min, the cardiac output would increase to 7,000 ml/min (100 × 70). When we observe the actual system, do we find this prediction to be true? If so, our model may very well be a valid one.

In fact, it is true. We know that during exercise, because of the increased levels of activity, the muscles greatly increase their demand for oxygen, nutrients, and waste removal. To supply the muscles with this increased demand, the cardiac output of the heart must also increase and we therefore find that our heart rate rises, as predicted by the model.

## Review Questions

1. What is the Rutherford model of the atom?
2. What are photons? What do they tell us about electromagnetic waves?
3. What is color?
4. Discuss the photoelectric effect and Einstein's explanation.
5. Discuss the Bohr atomic model.
6. What is DNA?
7. Explain how the human heart can be expressed as a model.
8. Account for the fact that exposure to X-ray or UV radiation is generally more damaging than exposure to longer wavelength radiation.
9. Account for the fact that excitation of the H atom yields a line spectrum.
10. Each chemical element yields its own characteristic spectrum. Indicate how this can be used to identify the element.
11. The element helium was first discovered on the sun before it was found on earth. How is it possible to discover an element on the sun?

# Multiple Choice Questions

1. One of the first models of the atom was _____.
   a. Bohr's model
   b. Einstein's model
   c. plum-pudding model
   d. solar system model

2. Einstein and Planck believed that light could behave like a particle called the _____.
   a. proton
   b. electron
   c. neutrino
   d. photon

3. The energy carried by a photon is directly proportional to its _____.
   a. wavelength
   b. frequency
   c. period
   d. amplitude

4. Microwaves can cook food if the food contains _____.
   a. photons
   b. water molecules
   c. heat absorbers
   d. alpha particles

5. The Bohr model of the atom is analogous to _____.
   a. the Ptolemaic model of the universe
   b. the plum-pudding model
   c. the solar system model
   d. the nuclear model

6. A model is a tool used to describe an aspect of reality, depending on how it can be constructed; it can be classified according to the following categories:
   a. visual
   b. quantitative/mathematical
   c. verbal model
   d. analog

   Categorize each of the following:

   The Statue of Liberty. _____

   Electrical flow is like the flow of water. _____

   Matter is composed of minute discrete particles called atoms. _____

   The force of attraction by the earth on a body is equal to the product of the mass of the body and the acceleration constant due to gravity. _____

7. The fact that excitation of hydrogen gas in a gas discharge tube yields a line spectrum is indicative of the fact that _____.
   a. hydrogen is a buoyant gas
   b. hydrogen emits energy that is quantized
   c. hydrogen possess only one proton
   d. hydrogen is found in water

8. The equation that models a certain spring is W = mL – b where b = 10 g and m = 10 g/cm. Determine the length of the spring if 50 g are applied.
   a. 6 cm
   b. 10 cm
   c. 20 cm
   d. None of the above.

9. Isotopes are atoms of elements that have the same number of _____.
   a. neutrons
   b. electrons
   c. croutons
   d. protons

10. The element ${}^{17}_{8}O$ has nine _____.
    a. electrons
    b. neutrons
    c. protons
    d. grams of mass

11. DNA _____.
    a. is a biopolymer
    b. contains the elements D, N, and A
    c. was known to early man
    d contains ribosomes

12. The ion ${}^{35}_{17}Cl^{-1}$ contains _____.
    a. 17 electrons
    b. 35 protons
    c. 18 electrons
    d. 35 neutrons

13. The thing that makes an individual's DNA unique is _____.
    a. the sequence of deoxyribose units
    b. the sequence of phosphoric acid units
    c. the sequence of nitrogen bases
    d. the order of deoxyribose and phosphoric acid

14. Given ${}^{14}_{7}N$ (atomic number 7), the element has _____.
    a. 7 neutrons
    b. 8 electrons
    c. 8 protons
    e. 8 neutrons

15. Salt, NaCl, dissolved in water is an example of a(n) _____.
    a. solution
    b. element
    c. compound
    d. heterogeneous mixture

# Bibliography and Web Resources

Scientific modeling. Why is scientific modeling important? Types of scientific models. Scientists develop models in many different forms.
**www.mcrel.org/epo/resources/sci_modeling.asp**

About Scientific Modeling. If you would like to learn more about scientific modeling, please read "The Nature and Structure of Scientific Models," found on this Web site.
**www.wcer.wisc.edu/ncisla/muse/philosophy/**
**(More results from www.wcer.wisc.edu.)**

# Systems

*"Systems thinking is a discipline for seeing wholes."*

**Peter Senge**

This chapter introduces you to the systems point of view and the concepts and language needed to understand it. It provides the foundation that will be used to analyze, describe, and understand the nature of the world around us.

We shall learn a language that will be used to analyze the behavior of many living and nonliving systems in terms of their inputs, outputs, feedback, and the controlling mechanisms involved.

## Goals

After studying this chapter, you should be able to:

- Define a system.
- Define feedback and its role in systems.
- Distinguish between positive and negative feedback.
- Know what is meant by systems analysis.
- Understand the application of the concepts of systems analysis.
- Be able to identify systems, system components, and feedback pathways in the world around us.

## Systems

The world around us is indeed complex. This is even more so the case with issues regarding science and technology. When trying to understand and manipulate scientific or technological issues such as the environmental impact of human activities, or the design, operation, and failure possibilities of the Space Shuttle, the task can seem daunting. The task can be greatly simplified by recognizing that all such issues, including the two mentioned above, are not single, monolithic entities but, rather, are systems made up of relatively simpler and more easily understandable parts. By understanding the nature of a system and the way in which its parts interact, even the most complex events can be understood.

We shall use the concept of a system (Fig. 6.1) in order to analyze and understand many of the phenomena of science and technology. Most people have an intuitive understanding of what constitutes a system. For example, we are familiar with the terms: hi-fidelity audio system, computer system, burglar alarm system, economic system, and educational system. What are the common elements of all these systems?

A system is made up of a collection of parts that work together to achieve a common goal. When dealing with systems, we have to be able to distinguish the system from its surroundings and be able to identify the boundaries separating them. Being able to do this in a careful manner helps avoid one of the most common errors in systems analysis, failure to consider the entire system.

Consider the anatomy of a hi-fidelity system. There are four major parts in a hi-fi system (Fig. 6.2): the sound sources (CD, turntable, tuner, microphone, tape deck), the preamplifier, the amplifier, and the speakers.

The speakers themselves constitute a subsystem, pictured in Fig. 6.3, made up of voice coil, permanent magnet, diaphragm, flexible edge, spider and basket that holds it all together.

Another system is the computer system with its input devices, processing unit, and outputs (Fig. 6.4). The processing unit itself or system unit is a subsystem containing CPU, memory chips, motherboard, disk drives of all types, network card or modem, sound card, video graphics card, etc. In a computer system we can identify input devices, processing, and output devices. If a system has input and output devices such as in the above examples, then it can interact with its surroundings and is said to be an open system. If on the other hand, it has no inputs or outputs, it is closed.

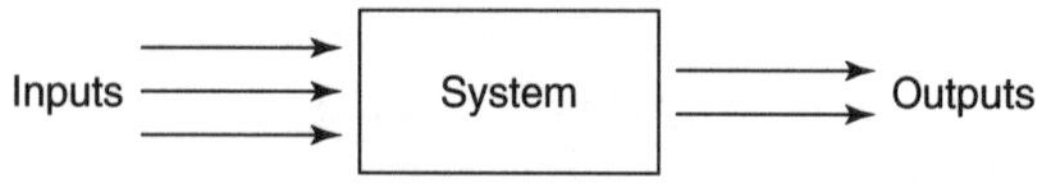

Inputs
System
Outputs

**Figure 6.1** Model of a system.

There is no system that is so isolated that it is truly closed, unless we take the universe to be our system. A well-sealed thermos bottle approximates a closed system. Almost no heat can enter or leave, but as we all know even the best insulated thermos will not retain heat for days on end.

Many biology research labs handle pathogens and dangerous life forms in sealed environments that prevent microbes or viruses from entering or leaving. Some physics experiments are conducted in specially shielded rooms that prevent radiation from leaking to the outside. They do, however, rely on outside sources of energy and external, filtered air. These would be examples of partially closed systems.

Another thing we notice about systems is that they are made up of subsystems. For example, an automobile (Fig. 6.5) has a braking system, a steering system, a suspension system, a drive-train system, lighting system, transmission system, etc. The nesting of these subsystems results in some very complex results. Usually, we confine ourselves to the smallest subsystem that contains the relationship we want to study. The brake specialist does not have to work on the engine to replace the brake shoes or pads.

Besides characterizing systems as made up of a collection of parts that also could be characterized as subsystems, we can usually identify the input and output portions of a system.

The two versions of systems diagrams (Fig. 6.6) for an automobile are shown. They illustrate the

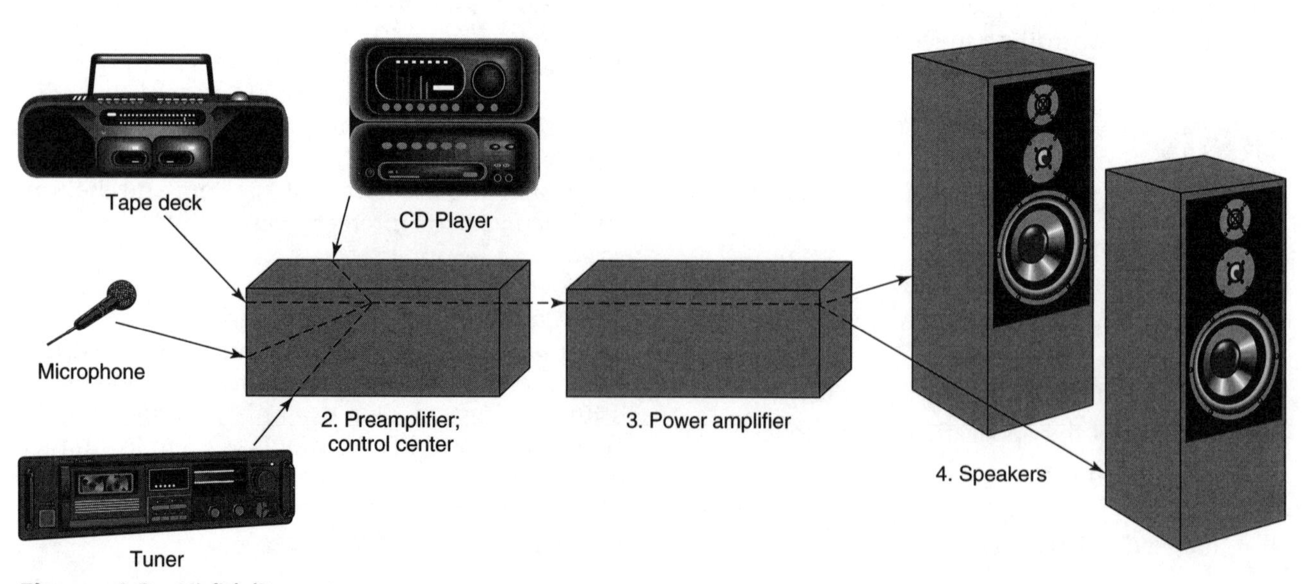


**Figure 6.2** Hi-fidelity system.

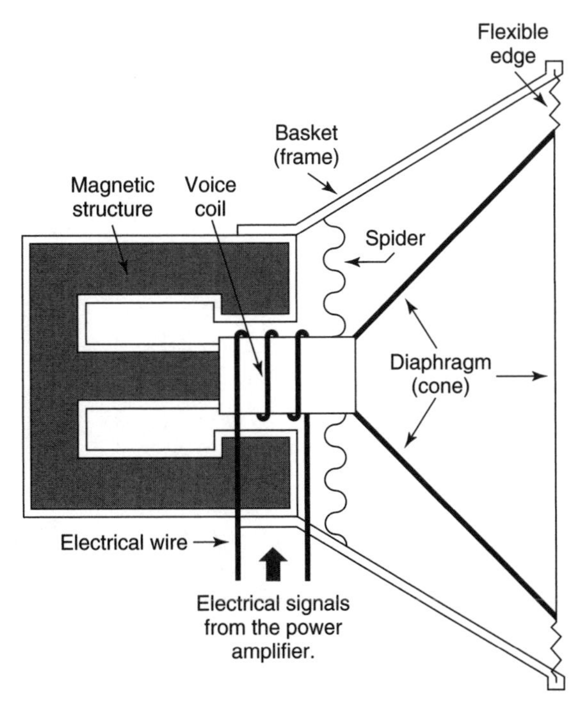


**Figure 6.3** Speaker system.

From *The Science of Hi-Fidelity* Third Edition by Johnson, Walker and Cutnell. Copyright © 1994 by Kendall/Hunt Publishing Company. Reprinted by permission.

**Figure 6.4** Computer system.

different ways in which inputs and outputs might be specified using different points of view. The first deals with the issues of fuel consumption, efficiency, and pollution, while the second treats driver-related concerns.

In most systems the purpose or function of the system can be identified from one or more of its outputs. Usually systems have a control function, which regulates or controls variables within it. Usually it is desirable to regulate or control the output within certain limits.

Suppose we wanted to increase the gasoline mileage obtained by a given automobile. To increase the number of miles per gallon we have to minimize the amount of fuel used. The control element in the automobile is the fuel computer, which meters the gas/air ratio, and adjusts the spark timing and automatic transmission to select

**Figure 6.5** Automobile. *http://www.bcarchives.gov.bc.ca/cgi-bin/www2i/.visual/img_med/dir_68/a_00485.gif*

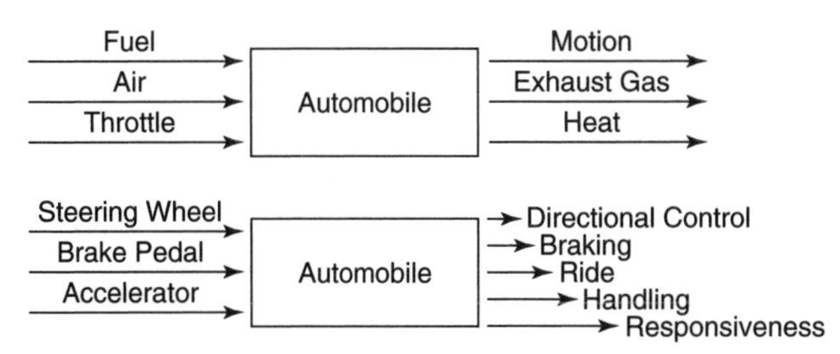


**Figure 6.6** Automobile systems.

the gear that gives the maximum fuel economy for a given speed.

# Feedback

Besides including a regulatory function, systems usually employ feedback, which basically consists of some of the output being returned or fed back into the system as input so as to modify the net output of the system. For example, if a business were to double the price of an item and consumers refused to buy it, management would have to readjust the price to a more acceptable level.

Sometimes, control is established by feedback in which some of the output is transferred back to the system and is used as input. In the mileage example, the driver would be a feedback path if the car had a standard transmission. Based on his observations of road speed and engine rpm, he would have to manually select the proper gears to achieve the highest mileage.

Feedback mechanisms are common ways that systems can exercise control over their outputs. An audio amplifier (Fig. 6.7) employs negative feedback in trying to enlarge and reproduce a signal accurately while minimizing the effects of distortion produced by the imperfect electronic components themselves.

Suppose you were blindfolded and told to walk down a street on the sidewalk. You are given a long pointer-like stick that you may use. You walk down the street moving the stick first to the right and then to the left. Information is being fed

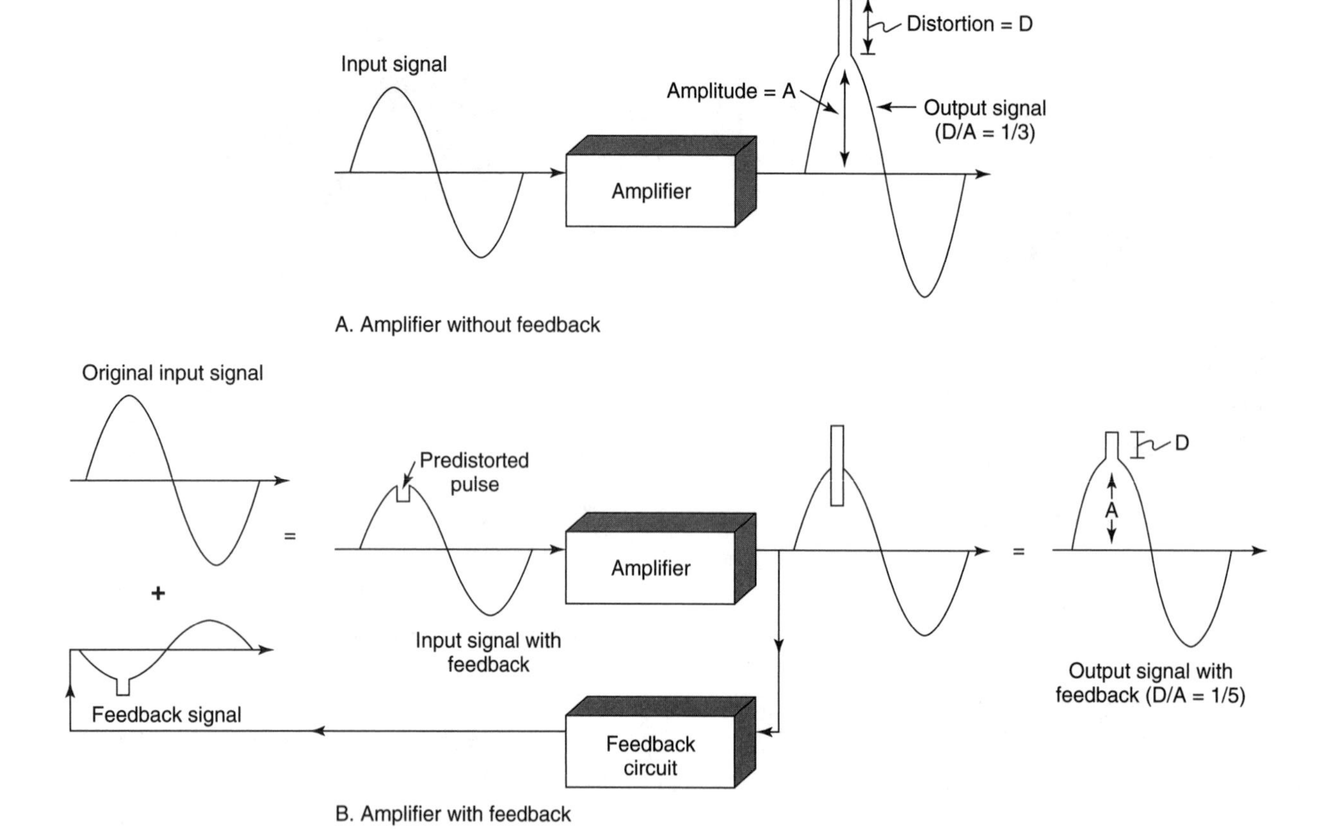


**Figure 6.7** Audio amplifier with negative feedback (K-H Johnson, *Science of Hi-Fi,* 3rd ed. p. 250).


into your brain from the tip of the stick. If you hit the side of the building, you would change direction and begin moving away from the building. Hitting the curb causes another change in direction, back the other way.

The feedback from the tip of the stick, which causes you to change direction, is called negative feedback. Adjusting the input in the opposite direction to eliminate the error in walking is negative feedback. If you were to continue in the same direction, we would have positive feedback.

With negative feedback the system is controlled in such a way to reduce the error between the desired output and actual output to zero. In other words, negative feedback allows you to remain on the sidewalk.

Sometimes the type of control illustrated in this example is referred to as "bang-bang" or "on-off" control causing it to send the person to the other side. However, running down the sidewalk might be a problem in that you would not be able to change direction quickly enough (i.e., your brain would not be able to process information fast enough to keep the system stable).

In driving on a highway separated by a median guardrail, turning the steering wheel only after hitting the guardrail on one side of the road and turning the wheel after hitting the curb on the other side would be a disaster. As you drive, information is fed to the brain in visual form, causing it to send instructions to the hands. The eyes provide the needed feedback pathway for the brain to exercise continuous control (Fig. 6.8). Such control is known as proportionate control.

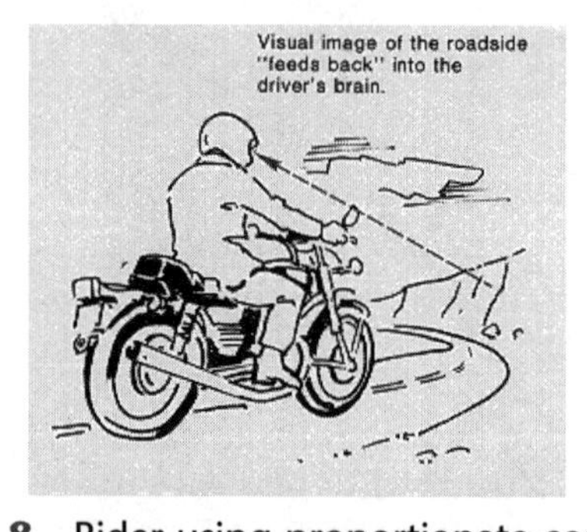


**Figure 6.8** Rider using proportionate control (K-H Johnson, *Science of Hi-Fi,* p. 249).


Systems that exhibit negative feedback tend to be stable. Stable, however, does not imply that a

Figure 6.9 Illustration of oscillations or "hunting" about equilibrium path.

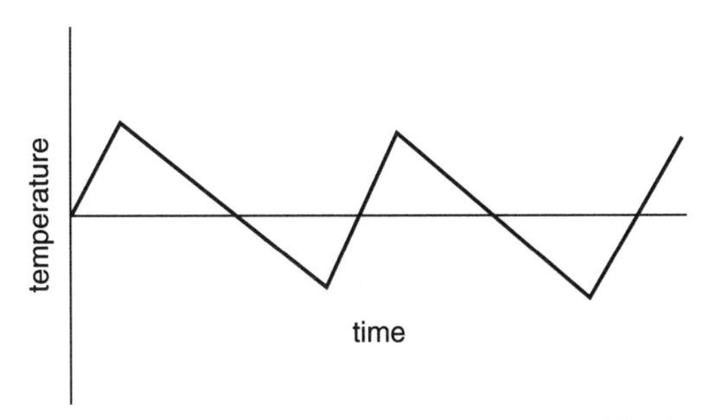


Figure 6.10 Variation of temperature with time.

system is static, or unchanging. Rather, a stable system is one that maintains its output parameter at an average value and within a reasonable range. In the case of the blindfolded person running down the street we saw the response time was not fast enough to keep the person on the sidewalk. Usually the controlling mechanism cannot hold the system at an exact value but allows it to "hunt" about the value.

When negative feedback is present, the system is able to achieve stability. If we were to look, from above, at the progress of our blindfolded person in moving along the sidewalk from left to right, we would notice a small oscillation (Fig. 6.9) from side to side as the person moved down the axis of the sidewalk.

Although the person's path would be making general progress towards the right of the page as illustrated in Fig. 6.9, it would not be in a perfectly straight, or static, direction. The feedback control mechanism would, however, keep these variations from the centerline within a reasonable range and keep our person from wandering off into the street on one side, or into the buildings on the other.

Similarly, any system regulated by negative feedback control will be characterized by an output that oscillated around an average, or mean, within a controlled range (Fig. 6.10). Take, for example, a home heating system set to 72° F. The home's temperature will actually vary from a high of 74° F to a low of 70° F thereby maintaining an average temperature of 72° F. At any point during the day, however, the actual temperature is unlikely to be exactly 72° F.

Feedback relates the output to the input and thus controls the net output. The art and science of control theory developed over a long period of time has been given the name cybernetics. The concepts of cybernetics, which deal with systems, information, feedback, and control, have extensive applications in the physical and biological world.

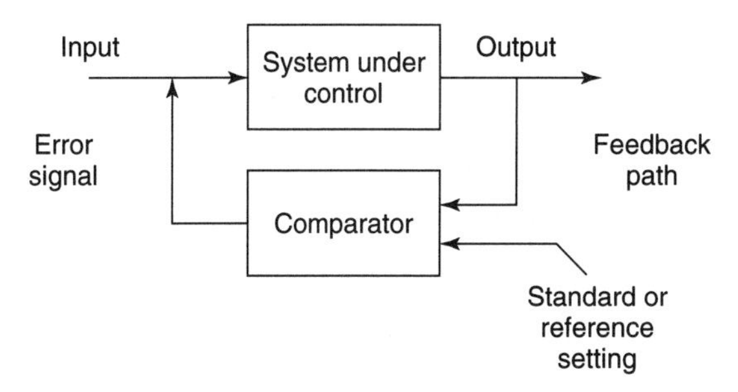


Figure 6.11 Operation of comparator.

Another example would be to consider the case of a manufacturer who finds that there are very few sales of his product because of its high price. Probably, he would lower the price to increase sales. If the increase in sales did not produce the desired profit, he might increase prices again slightly to make greater profits on slightly less sales volume.

Here negative feedback is being used to adjust the operation of a free market system to maximize profits. Each change is made to increase the output profit. On the other hand, a politician running for office tests his ideas by introducing them gently to obtain a response from the public. Those ideas that seem unpopular, he deletes. That's negative feedback.

There are special systems where a limited amount of positive feedback is valuable and necessary to the system's goals, but when control and stability are required, negative feedback is necessary.

The failure of a system to execute the input instruction is interpreted as an error and the correction is made by adjusting the input in the opposite direction to eliminate the error. In a negative feedback loop the error between the desired and actual output can be reduced to close to zero. A device that is used as an error detector in a feedback loop is called a comparator (Fig. 6.11).

The comparator itself can be thought of as a system with inputs and outputs. The comparator and the system it helps control are stabilized by negative feedback. A common example that illustrates the operation of a comparator occurs in a home heating system. The comparator that provides the feedback path is the thermostat. If the thermostat (Fig. 6.12) in a room were set to 70 degrees, it would shut off the boiler when the room achieved that temperature. This is another illustration of negative feedback. On the other hand, if it were to signal the boiler to send more heat, when the temperature reached 70 degrees, we would have an example of positive feedback.

**Figure 6.12** Thermostat. *http://www.pnl.gov/uac/images/t87f_thermostat.jpg*

# Systems Analysis

A systems analyst when examining a system focuses on how the system interacts with its surroundings and how the inputs are modified to produce outputs. The details of its design and construction are not of primary interest but rather how the outputs are produced from given inputs. It doesn't matter whether the parts of the system are mechanical, electrical, hydraulic, and biological or some combination of these.

Understanding the operation of the system is broken down into identifying the basic inputs, outputs and feedback and control mechanism. In examining the various phenomena of matter energy and life, we shall adopt the systems analyst's point of view. It will be the overriding method of analysis we shall use as we proceed to gain an understanding of science and technology.

For example, physiologists have cataloged the operation of human body systems (cardiovascular, skeletal, neuromuscular, etc.) and as a result bioengineers have been able to design prostheses and artificial organs to allow people to regain some essential functions.

In examining and interpreting physical and biological systems from this point of view, it is seen that a system is made up of parts that in turn are subsystems connected in a network.

# Equilibrium

Interestingly enough, many systems both natural and man-made when displaced from their equilibrium position will try to return to the original state. This is true of systems in stable equilibrium. Consider the ball at the bottom of a bowl (Fig. 6.13). Move it to the left or right and it attempts to return to the bottom but overshoots and goes into a periodic back-and-forth rolling motion. This is an example of stable equilibrium. If the ball were balanced on the top of an inverted bowl, then when we displaced it, it would move farther from the equilibrium position and not execute a periodic motion. Neutral equilibrium occurs when the ball sits on a flat surface and moving it to the left or right leaves the ball at its new position with no tendency to return to the original point.

**Figure 6.13** Stable, unstable, and neutral equilibrium.

**Figure 6.14** Spiral spring with hanging mass.

There was some concern among some scientists at the time the United States was landing probes on the moon, that the moon might be in an unstable orbit and hitting it with a missile or probe might cause it to move off into outer space. Think how disastrous this would be for much of the world's populations that depend on moonlight to see and travel at night. Have you ever been in an isolated area and outside on a moonless night?

## Hunting

Those stable systems when displaced from equilibrium will tend to oscillate and hunt, like the ball in the bowl, around the equilibrium position. Take a look at the variations that take place in the stock market. How about the variation in temperature and rainfall over a year's time? There are restoring influences that tend to bring the systems back toward their equilibrium position.

Consider a mass attached to a spiral spring as in Figure 6.14.

When the mass is pulled down out of the equilibrium position, it will vibrate or oscillate in a periodic or repetitious fashion. The stiffness of the spring determines how strong the force on the mass will be to vibrate or oscillate it in a periodic or repetitious fashion. A restoring force, like a negative feedback effect, provides a stabilizing influence on system behavior. Sometimes the restoring forces are insufficient to return the system to its equilibrium position.

## Tacoma Narrows Bridge

On November 7, 1940, the Tacoma Narrows Bridge in the state of Washington was destroyed by periodic oscillations. The bridge was designed by its engineers to be aesthetically pleasing with extreme flexibility and lightness (Fig. 6.15). All systems have associated with them a natural frequency. For example, when a wine glass is struck lightly at its rim with a spoon, it rings or generates sound as it vibrates at its natural frequency.

With a wind velocity of about 30 to 46 mph the bridge began to swing but the horizontal gusts seem to induce oscillations that approached the natural frequencies of the structure causing resonance. The wind blowing horizontally across the deck of the bridge produced vortices or whirlpool-like areas of wind on the other side of the span, which exerted a periodic force on the bridge as they spun away, causing a destructive oscillation. When the shedding frequency of the vortices was equal to the natural frequency of the bridge, it absorbed energy from the wind according to one explanation for its collapse. The energy delivered by the wind built up in the span so that the restoring forces could no longer cope with them and the bridge collapsed. (See Internet reference at the end of the chapter.)

It is interesting to note that the Whitestone Bridge connecting the boroughs of the Bronx and Queens in New York City had the same design as the Tacoma Narrows Bridge and was opened for traffic officially on April 29, 1939, so that motorists could cross it to attend the 1939 World's Fair, which opened the next day in Flushing Meadow Park in Queens. As a reaction to the collapse of the Tacoma Narrows Bridge in 1940, diagonal stays and braces were installed at the towers to reduce motion. In 1946, the roadways were widened to their current six-lane configuration, and stiffening trusses were added along both sides of the bridge to further reduce motion in the wind. This increased both its bulk and weight.

Most recently the steel stabilizing trusses were replaced with much lighter but just as sturdy fiberglass trusses. With the emergence of new technologies and advances in aerodynamic analysis, these heavy trusses were replaced with lightweight fiberglass aerodynamic structures, triangular in shape, that were installed along both sides of the bridge. Today, the bridge looks as modern and elegant as when it served as the gateway to the 1939 World's Fair "World of Tomorrow."

**Figure 6.15** Tacoma Narrows Bridge collapse. *http://cflhd.gov/agm/engApplications/BridgeSystemSuperStructure/31DeckStabiltiyAnalysis.htm* [Figure 1. Collapse of Tacoma Narrows Bridge, photo by Ed Elliott, The Camera Shop, Tacoma, WA].

# The Human Body

The human body is a complex structure. Systems analysis may be used when studying the human body, in health or disease, or when attempting to diagnose and treat a health problem. The human body is composed of a number of systems (Fig. 6.16), each with its own purpose and mechanisms of normal operation. It is the interplay of all these systems, functioning normally, which accounts for the healthful state. A disturbance of one or more of these systems will usually have some measurable effect upon the others. These disturbances often result in disease.

The systems of the human body include the nervous system, musculoskeletal system, cardiovascular system, respiratory system, digestive system, and excretory system among others.

When a physician or researcher is concerned with treating or studying the disease of arthritis, specific attention would be paid to the musculoskeletal system; for emphysema, the respiratory

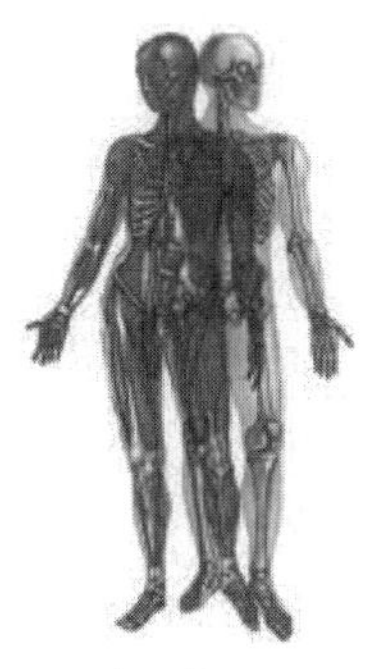

**Figure 6.16** Human body systems. *http://training.seer.cancer.gov/module_anatomy/unit1_1_body_structure.html*

system; and for hypertension, the cardiovascular system. As with the automobile engine, a subsystem of the car that was itself subdivided into electrical and fuel systems, body systems can also be further subdivided. The digestive system, for example, can be further broken down into its component organs such as the stomach, pancreas, liver, intestines, etc. The organs can be subdivided into tissues and the tissues into cells.

The structural systems for all life, in all organisms, can be subdivided down to the level of the cell. At that level, any of the subsystems of the main system (the organism) can itself be considered a living thing. In fact, one of the basic courses in the study of medicine is cell biology. So basic to health and to medicine is the subsystem the cell, that the following excerpt begins paragraph one of page one of the most popular medical school textbook of pathology (the study of disease):

> The emerging physician is told so often—be concerned with the whole patient—that he sometimes forgets that behind every organic illness there are malfunctioning cells. Indeed it is more correct to say that when a sufficient number of cells become sick, so does the patient. It is appropriate, therefore, to begin the study of pathology with the consideration of derangements at the cellular level before we turn to the diseases of the whole organs and of the organism (person). The student of pathology is understandably impatient to get to the "people diseases". But, in this day of sophisticated medical technology, one cannot understand the cause, development and clinical implications of a disease without deep penetration into the cell. (S. L. Robbins, *Pathologic Basis of Disease,* p. 1.)[1]

To roughly paraphrase this passage we might say: in order to study, understand, and treat disease, perform a systems analysis and pay careful attention to the smallest living subsystems of the main system.

## The Pacemaker

A brief discussion of one of the most extraordinary and successful biomedical devices may help to indicate how analysis that springs from a systems viewpoint can lead to an engineering artifact. Here we turn to another body subsystem, the heart, and think about the models that have been built of its substructure.

[1]Robbins, S. L., R. S. Cotran, V. Kumar, and W. B. Saunders, *Pathological Basis of Disease,* Saunders, Philadelphia, PA, 1984.

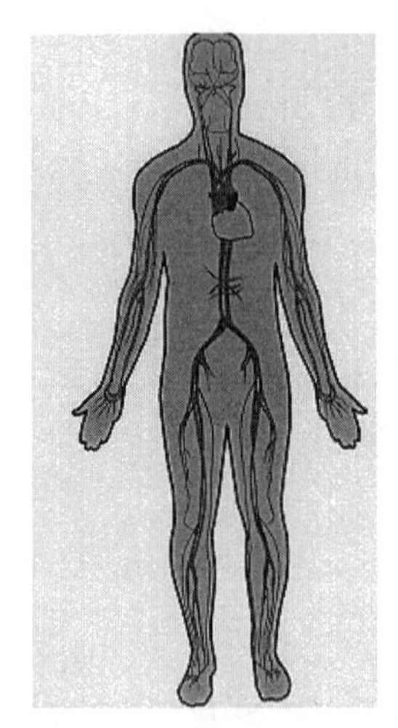

**Figure 6.17** Circulatory system. *http://www.betterhealth.vic.gov.au/bhcv2/bhcarticles.nsf/Pictures/_system?open*

Recall that the heart is made up of two pumps. One pump is concerned with supplying oxygen to the blood and the other pump forces the blood to the rest of the body. The right side of the heart squeezes blood to be oxygenated in the lungs while the left side pumps the oxygenated blood to the rest of the body (Fig. 6.17). Each of these pumps goes through a carefully synchronized pumping cycle each time the heart "beats."

The pumping mechanism of the heart is controlled and kept regular and synchronous electrically. The electrical signal does not come from batteries and wires from outside the heart but is biologically generated by cardiac muscle cells. There is a specific site called the *sino-atrial node,* or SA node, which is a collection of modified cardiac muscle cells in the right atrial wall of the heart that send out electrical impulses. It is somewhat analogous to a microprocessor. Input defined by the body's need for oxygen is fed by the senses to the brain, which then controls the output by adjusting the heart's pumping rate electrically through the SA node. Pulse rate and heart action increase as a result of external stimuli that raise oxygen demand.

After the blood fills the upper chamber of the right side of the heart *(right atrium)* (Fig. 6.18) it is squeezed into the lower chamber only after the SA node has generated the needed signal for the heart to contract. The valve connecting the right atrium to the *right ventricle,* the lower chamber, is a one-way valve (i.e., blood can flow only in one direction). The time of transfer is on the order of one or two tenths of a second. The ventricle then contracts after a short delay, squeezing the blood out to the lungs to be re-oxygenated. This delay is required so as to allow the ventricles adequate time to fill following atrial contraction.

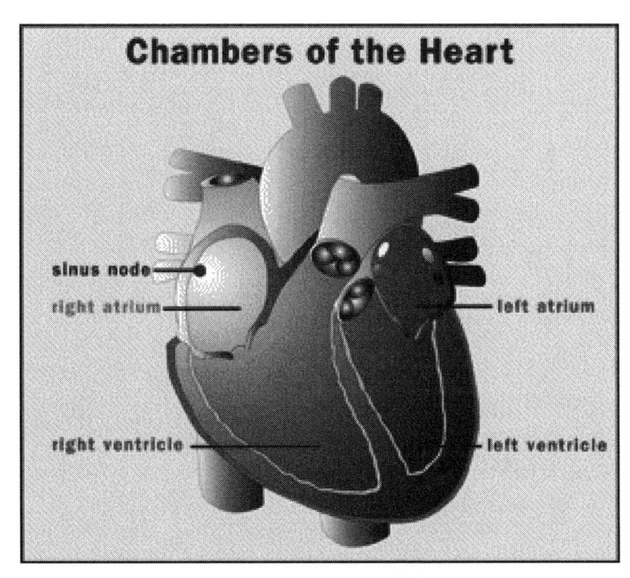


**Figure 6.18** Human heart. *http://www.nlm.nih.gov/medlineplus/ency/imagepages/19566.htm*

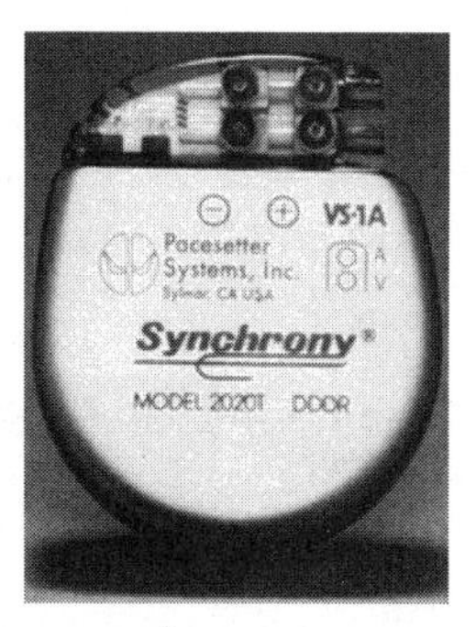


**Figure 6.19** Pacemaker. *http://www.grc.nasa.gov/WWW/PAO/images/fullsize/pacemkr.gif*

The delay is produced by another microprocessor, the AV, or *atrioventricular node.* It senses the electrical impulse generated by the SA, introduces a delay, and then sends out a signal for the ventricles to contract. The upper chambers on the right and left side of the heart contract in unison (simultaneously) because of nerve interconnections. The same is true for the lower chambers after the time delay.

Knowing something about the pumping sequence prepares us to understand a certain type of heart malfunction. In some hearts the electrical signals generated by either or both of these nodes is weak, delayed, or intermittent. The artificial pacemaker was devised to generate the electrical signals normally furnished by the SA and AV. It does so by using circuitry and battery power to augment or supplant the biogenerators. In less than one human generation, these devices have gone from large, cumbersome external instruments, which anchored the patient to one location to small, implantable units, which are virtually maintenance-free.

In most defective hearts the nodes work most of the time and therefore need help only intermittently. Early pacemakers were designed to send the electrical impulses whether they were necessary or not. This sometimes confused the heart. Present day pacemakers function on demand. They sense the electrical signals emitted by the SA and AV nodes and act only when needed. The circuit is designed so that if the signal from the heart is received over a single connecting wire the stimulating impulse is not sent and the pacemaker is reset, getting ready to look for the next signal. If the patient's heart activity stops for too long, the pacemaker takes over.

The pacemaker is usually connected in such a fashion that it supplements or replaces the SA node. However, it can also be connected in a similar fashion to generate the delay signal if the AV node is malfunctioning. Under these circumstances the right ventricle rather than the right atrium would be wired to the pacemaker. Sometimes the pacemaker will have to supplement both the SA and AV nodes so that two wires must be connected to it, one from the atrium and one from the ventricle.

A typical pacemaker (Fig. 6.19) measures about 1⅜ inches by 1/4 inch and weighs about 1 ounce. It is implanted under local anesthetic usually beneath the skin just below the left collarbone. An insulated wire is threaded through a blood vessel to the atrium or ventricle. It is battery operated using a battery similar to those used in electronic watches but having a lifetime of about five years and sufficiently larger in size to be able to supply the energy required.

Many improvements in safety, reliability, biocompatibility, and cost have been developed. Some electronic equipment can interfere with pacemakers, for example airport walk-through metal detectors and mobile phones. Phones should be kept at least six inches away from a pacemaker and should be held to the opposite ear. If it is exposed to such equipment, a pacemaker probably will not stop working but will revert to its unprogrammed state.

The pacemaker is a success story that owes its beginnings to the understanding of the heart, which emerges from a systems analysis. The design of the device itself proceeds according to systems concepts such as input and output requirements as well as the capacity of feedback methods to allow automatic control of firing rate. While the systems ideas are keys to understanding the development of the device, it should also be mentioned that pacemaker improvement has also profited greatly from technological advances, especially those in electronics, chemistry, and surgery, during the time this development has taken place.

In the same period of time, other artifacts to aid ailing hearts have also been developed. Replacement valves, left ventricular assist devices, and other implantable or attachable parts are being used with regular success. The completely artificial heart is still under development. Versions have actually been implanted in human patients in a number of highly publicized cases. The program to develop an implantable artificial heart as well as the desirability of the device itself remains a controversial issue. The NIH Artificial Heart Program in the United States has adopted a NASA-type systems management approach to its design and development problem—with limited success. Apparently, systems approaches sometimes need to be augmented by other, more holistic viewpoints.

# The Environment

An *ecosystem* (or ecological system) is a system (Fig. 6.20) that includes all the life forms that interact strongly with one another or occupy the same locale, and the nonliving resources they draw upon to sustain their lives and the continuance of their species. If the human being, a single organism, is as complex a system as described above, the environment in which we live, or any ecosystem for that matter, is most assuredly a tremendously complex system. If one were to attempt to study an ecosystem by looking at the whole, she would find herself overwhelmed by the complexity of it. Alternatively, as with the study of the automobile or the human body, it would be more efficient to break an ecosystem down into its component subsystems and then study each subsystem individually.

The subsystems of an ecosystem include the physical environment, or abiotic components, and the living organisms, the biotic parts. Again, in analogy to the car subsystem, the engine, which was subdivided into electrical and fuel systems, the ecological subsystems can also be further subdivided. The *abiotic* subsystem is composed of the energy, which has been supplied by the external input (sun), the climate, rainfall, topography, and the chemistry of the air, water, and soil.

The *biotic* subsystem can also be further subdivided into its components. These include the populations of different organisms classified as producers, consumers, and decomposers; themselves each capable of being further subdivided.

In a normally functioning ecosystem, it is the interplay of all these systems, each functioning normally, that accounts for the self-sustaining and balanced nature of the ecosystem as a whole. A disturbance of one or more of these systems results in an ecological imbalance and potential environmental disaster.

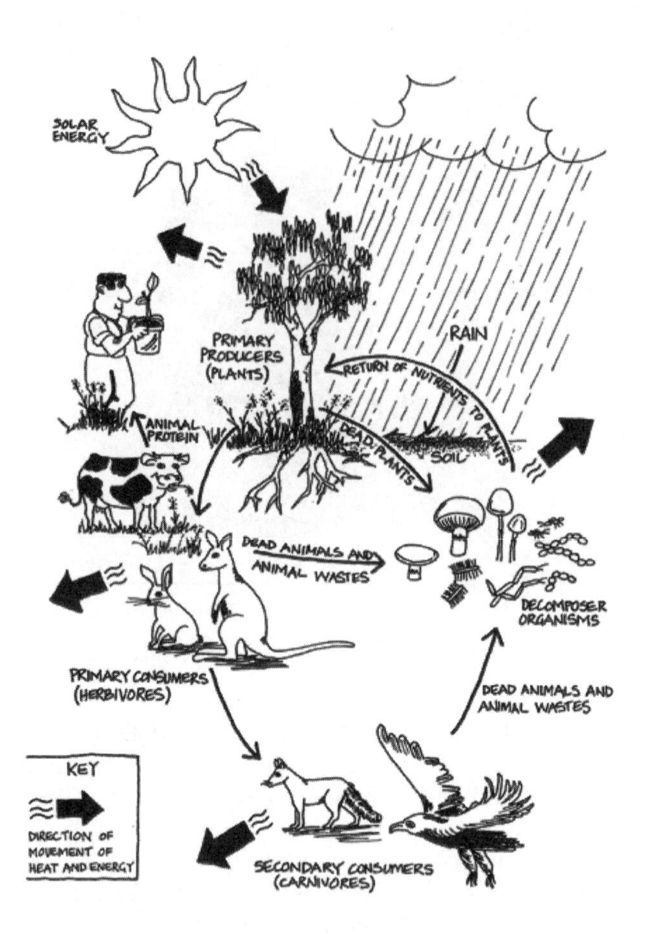


*An ecosystem showing abiotic and biotic components and the flow of energy (Adapted from Recher 1986)*

**Figure 6.20** An ecosystem.

From *Biology: Understanding Life,* 3rd edition by Sandra Alters. Copyright © 2000 Jones and Bartlett Publishers, Sudbury, MA *www.jbpub.com* Reprinted with permission.

In attempting to study the impact of a phenomenon on the environment, it is helpful and revealing to use a systems approach, first looking at the effect on each subsystem component, then determining the aggregate effect on each subsystem, and finally assessing the overall ecological impact.

Total understanding of a system and the various inputs and outputs associated with its smooth operation can sometimes be a daunting challenge. For example, one might believe that we sufficiently understand the process of photosynthesis and respiration on a global level so that we should be able to construct an enclosed self-sustaining ecosystem involving living creatures and plant matter. In fact in the last decade such an attempt was made. It was a dismal failure

In 1991 eight people were sealed in a 140,000-square-foot facility, which had miniature forests, lakes, streams, as well as an ocean that mimicked the natural systems of our planet. The $200 million facility called Biosphere II was located in Arizona. It was supposed to be a glass-covered "copy" of an efficiently running Earth designed to support eight humans who were to be isolated for two years, enclosed in the biosphere.

Self-contained systems involving plants to generate oxygen and food and recycle carbon dioxide were part of the Biosphere. For a variety of reasons, unanticipated events resulted in a decline in the atmospheric oxygen level from 21% to 14%. Almost 80% of the species introduced to the Biosphere became extinct. (The big survivors were ants and roaches.) Insects that pollinate plants perished. Obtaining pure water became a problem. The project could not be completed as planned. If anything, the Biosphere II experience indicates how fragile our ecosphere is and how destruction of natural systems can lead to precarious departures from a stable equilibrium.

## Review Questions

1. What is a system?
2. What is feedback?
3. What is the function of a comparator?
4. What is the difference between positive and negative feedback?
5. What is systems analysis?
6. What is a pacemaker and how does it work?

## Multiple Choice Questions

1. A system is _____.
   a. a bunch of parts
   b. a way of operating
   c. a measure of probabilities to predict outcomes
   d. a collection of parts that work together to achieve a goal

2. Feedback is _____.
   a. when some input is returned as output
   b. when some output is returned as input
   c. a loud howling sound
   d. when a system hunts

3. A characteristic of negative feedback is that the system tends to be _____.
   a. unstable
   b. stable
   c. neutral
   d. exponential

4. The direction a system is headed changes when there is _____.
   a. positive feedback
   b. negative feedback
   c. no feedback
   d. no control

5. _____ mechanisms are common ways that systems exercise control over their outputs.
   a. Control
   b. Response time
   c. Feedback
   d. Processing

6. Systems analysts focus on how the system interacts with its surroundings and how inputs are modified to produce _____.
   a. control
   b. feedback
   c. outputs
   d. equilibrium

7. Stable systems when displaced from equilibrium tend to oscillate and _____.
   a. hunt
   b. resonate
   c. generate feedback
   d. become unstable

8. The Tacoma Narrows Bridge collapsed primarily because of _____.
   a. hunting
   b. resonance
   c. frequency oscillations
   d. restoring forces

9. The structural systems for all life can be subdivided down to the level of the _____.
   a. atom
   b. molecule
   c. cell
   d. organ

10. The sino-atrial node is a _____.
    a. piece of heart tissue
    b. an acupuncture site
    c. part of the brain
    d. part of the human skeletal system

11. In the system involving photosynthesis, which of the following is an output?
    a. oxygen
    b. carbon dioxide
    c. sunlight
    d. water

12. In the system involving combustion, which of the following is an output?
    a. oxygen
    b. carbon dioxide
    c. sunlight
    d. water
13. In the system involving photosynthesis followed by combustion, which of the following is a feedback?
    a. oxygen
    b. carbon dioxide
    c. sunlight
    d. water
14. The water reservoir used in some toilets for flushing purposes consists of a tank that fills with water. The filling stops when a float is raised to a certain level by the water filling the reservoir. This simple system is an example that uses _____.
    a. negative input
    b. positive output
    c. negative feedback
    d. positive feedback
15. If a basement were flooding as a result of a broken pipe and the drain was plugged up, we could say this was an example of _____.
    a. negative input
    b. positive output
    c. negative feedback
    d. positive feedback
16. What kind of feedback is important to the workings of a pacemaker?
    a. negative input
    b. positive output
    c. negative feedback
    d. positive feedback
17. If a person were pushing a swing such that each push occurred when the swing was on its downward motion, this would be an example of _____.
    a. negative input
    b. positive output
    c. negative feedback
    d. positive feedback
18. Positive feedback tends to be _____.
    a. stabilizing
    b. destabilizing
    c. fatiguing
    d. satisfying
19. Systems requiring control tend to employ _____.
    a. negative input
    b. positive output
    c. negative feedback
    d. positive feedback
20. The fact that eating causes our appetite to diminish is probably the result of _____.
    a. negative input
    b. positive output
    c. negative feedback
    d. positive feedback
21. The return of output or part of the output to a system as input is called _____.
    a. information
    b. feedback
    c. control
    d. modeling
22. When the temperature falls, negative feedback causes the thermostat to _____.
    a. let more heat flow
    b. let less heat flow
    c. shut down the system
    d. cause instability in the system
23. After leaving the heart, blood passes through _____.
    a. veins, venules, capillaries, arterioles, arteries
    b. arteries, veins, capillaries, venules, arterioles
    c. arteries, arterioles, capillaries, venules, veins
    d. None of these sequences is correct.
24. A system _____.
    a. has inputs and outputs
    b. is usually made up of parts
    c. is usually cyclic because it employs feedback
    d. All of these.

# Exercises

1. Consider the systems listed below:
   i) a flashlight
   ii) an aquarium for tropical fish
   iii) an electric coffeemaker of the "Mr. Coffee" type
   iv) College
   v) "Dining Commons"
   vi) a process such as filling out a college or job application or an income tax return
   vii) a process such as taking and successfully completing a course
   viii) a structure such as the house or apartment building you live in

a) For each system you consider, what are the parts of the system? What is included in the system and what are the variables that must be used to describe the state of the system?
b) For each system you consider, identify the important inputs and outputs. Name each of the connecting flows and characterize or classify it as fully as possible.
c) For each system you consider, identify some parts of the surroundings that are definitely not in the system, but that participate in the input and output exchanges.
d) Do any of the systems you have considered have some variables that are regulated or controlled? If so, is this done by feedback?
e) Draw a schematic block diagram to describe each of the systems you have considered. Indicate how the internal parts are connected as subsystems.
f) Compare your responses to the above with those of some of your classmates. Are there differences that involve omission or error? Do you need to improve your knowledge of the system in question? Are there differences that arise from different perspectives on the function of the system? Has your own perspective been clarified by working through this exercise?

2. Draw a systems block diagram of a home heating system where the temperature is regulated by a thermostat. Label each of the pathways connecting the blocks in your diagram to clearly indicate what is present or flowing along each connecting link. Explain what negative feedback is and how it works in this system.
3. Form a systems model of the automobile similar to but different from that shown in the chapter. As a guide, take a perspective such as an interest in aesthetic design, concern for simplicity and ease of maintenance, a desire for interior space and comfort, low cost of manufacture, or some other outlook that intrigues you.
4. A piano is a system that contains a number of subsystems that are vibrators. Draw a labeled block diagram of a piano's major subsystems and how they are connected by inputs and outputs. Would the same diagram serve to describe a pipe organ? If not, what changes would be necessary? Could the same system diagram describe an electronic keyboard? If not, how is it that an electronic keyboard can sound like (perform the same function as) either a piano or an organ?
5. From your experience as a patient, try to describe how a doctor uses systems analysis to try to arrive at a diagnosis of a sick patient.
6. (a) Model a system in which photosynthesis is the process that is occurring. Indicate the nature of the inputs and outputs.
   (b) Repeat the above for a combustion process.
   (c) Couple the two processes together and indicate any feedback that is possible.

# Bibliography and Web Resources


Crane, H. Richard, "Some of the Physics of the Cardiac Pacemaker," *The Physics Teacher,* V. 24, Issue 4, April, 1986, pp. 248–250.

Feynman, Richard P. *What Do You Care What Other People Think?* W.W. Norton, Bantam, 1989.

Judson, H.J., *The Search for Solutions* (Chapter 5, "Feedback"), Holt, Rinehart, & Winston, 1980.

Paresgian, V.L., Metzler, A.S., Luchins, A.S., Kinerson, K.S. *Introduction to Natural Science, Part One: The Physical Sciences.* Academic Press, New York, 1968.

Biosphere II
**csf.colorado.edu/mail/ecol-econ/nov96/0117.html**

Medical Systems Biology in Health and Disease.
**www.euchromatin.org/systems1.htm**

Systems Analysis
**pespmc1.vub.ac.be/ASC/SYSTEM_ANALY.html**

Tacoma Narrows Bridge collapse.
**www.stkate.edu/physics/phys111/curric/tacomabr.html**

# Information and Coding

*"Man is the best computer we can put aboard a spacecraft—and the only one that can be mass produced with unskilled labor."*

**Werner von Braun**

In order to solve problems, we need information and information typically exists in some coded form. In fact, the chapter that you are reading is coded information. A code is a way of conveying or expressing information. We shall learn how information is coded in terms of vertical lines or bars, dots and dashes, binary numbers, electrical signals, musical symbols, language, letters, and numbers. We shall learn to recognize that the patterns formed by groups of these symbols carry information that we need to solve problems. A code we shall determine is a way in which information may be stored and transmitted.

Knowledge is factual material that is stored in our brains. Information can be thought of as knowledge in transit in the form of data, which are written in the form of some code.

## Goals

After studying this chapter, you should be able to:

- Define information.
- Distinguish between knowledge and information.
- Understand the U.S. Postal Service zip code.
- Understand the binary code.
- Have some familiarity with musical coding.
- Understand that language is a code.
- Be familiar with the Universal Product Code (UPC).
- Be aware of some of the concerns society has with automation.
- Identify examples of how scientists use codes and their applications in technology.

## Codes

In our attempt to understand scientific and technological issues, we have focused on problem solving. We have considered *process* over *content* in order to grasp the big picture and in order to be able to ask meaningful questions and judge the responses critically. (This statement does not minimize learning basic scientific concepts; the greater the knowledge base the better. After all, we never know what information we'll need in the future.) We have looked at some tools used in problem solving, including modeling and systems analysis. We now examine another: **coding.** In order to solve a problem, information is needed. Sometimes that information is not obtained in a form that is readily understood. Understanding codes and coding can help us to gain that information.

Often when we hear the word "code" we think of something that is associated with a secretive or covert activity. While there are in fact secret codes, the word "code" implies more than a clandestine endeavor. The only thing that makes a code secretive is our inability to understand it. Part of the process of science is the search for understanding the codes in nature.

Physicists for example have come to understand the coded information about the chemical composition of distant stars that is contained in their spectral emission patterns. Biologists have come to understand the functioning of the human nervous system by determining the code used in nerve cell signaling.

A code is simply a way of conveying or expressing information. The kind of code employed depends on the nature of the situation involved in the information transmission. It may be necessary to transmit the information secretly (secret code), or in an economical and efficient way (smoke signals, semaphore), or to another individual (language, mathematical equations) or to a machine or device (Morse code, bar code, binary code) or to protein producing centers in our bodies (genetic code). A code in the broadest sense is a way in which information may be stored and/or transmitted.

## Information

Information may be defined as knowledge in transit. Knowledge is what we have in our heads. We transmit information in the form of data. Data are encoded or written in terms of some set of symbols. A person, after receiving the information encoded in terms of data symbols, converts it into knowledge. When Native Americans communicated by smoke signal, the information was coded in puffs of smoke and the pattern, which contained the information, consisted of a unique series of puffs of smoke rising into the air over a period of time. Knowledge was obtained after the observer's brain converted the data carrying the information.

## Zip Code

The United States Postal Service (USPS) created the zip code as a means of speeding up the processing and delivery of mail. Automated machinery would be used to read and sort a five-digit code written on envelopes and labels whether they were typed or handwritten. The machine readers that were used at first were far from perfect. It was decided by the USPS to encode the information in a form that would be easily machine recognizable. The series of vertical lines, tall and short, that are printed along the bottom of envelopes illustrates this method (Fig. 7.1).

The lines contain the same information as the printed zip code for both the original five-digit zip code and the enhanced nine-digit version. Can you break the code (Fig. 7.2)?

l...llll...l..l.ll......llll...l
Iona College
715 North Avenue
New Rochelle, NY 10801

**Figure 7.1** Address label.

l...llll...l..l.ll......llll...l

**Figure 7.2** Five digit zip code.

**USPS Bar Code System**

| Digit | Bar Code | Digit | Bar Code |
|---|---|---|---|
| 1 | . . . l l | 6 | . l l . . |
| 2 | . . . . l | 7 | l . . . l |
| 3 | . . l l . | 8 | l . . l . |
| 4 | . l . . l | 9 | l . l . . |
| 5 | . l . l . | 0 | l l . . . |

**Figure 7.3** U.S. postal service bar code.

Note that each code sequence begins and ends with a tall bar. Note that five vertical bars represent one digit. We might guess that these are the start and stop markers that tell the machine where to begin and end the reading of a sequence. Another clue is to realize that the bars contain information for one more digit than there is in the zip code. This is a "check" digit, which tells a reader when an error has been made in reading off the information, and is located at the end of the sequence just before the stop marker. It is called the check sum digit. It is the number that when added to all the other digits, will make the sum a multiple of 10. For example, with 10804, $1 + 0 + 8 + 0 + 4 = 13$, so the check sum digit is 7. Thus a sequence for a 5-digit zip code contains bars for six digits (five zip digits plus one check digit) together with start and stop marker bars at the end of each sequence (Fig. 7.3).

**Example**

l...llll...l..l.ll......llll...l
1 0 8 0 1 (0 is the check digit)

**Example**

l...llll...l..l.ll....l..ll...ll
1 0 8 0 4 (7 is the check digit)

We are living in the Information Age and the media symbols and coding schemes in which information appears is varied and has to be optimized so as to provide the information appropriate to the users.

**Figure 7.4** Pony express. *http://www.aoc.gov/cc/art/cox_corr/w_exp/pony.htm*

## Pony Express

In an attempt to speed the U.S. mail across the United States the Pony Express (Fig. 7.4) was formed and service began in April 1860. Horseback riders traveled at top speed from one relay station to the next. The relay stations were spaced about 10 to 15 miles apart, where the rider jumped from his horse to a fresh horse, which was saddled and ready to travel. The mailbags were transferred and the rider was on his way in about two minutes. The trip of 1,966 miles took 10 days to cover but this was 12 to 14 days shorter than the time taken by stagecoaches of the Butterfield Overland Mail.

## Morse Code

In October 1861 the Pony Express ended because at that time a new technology made it obsolete and the financial backers of the Pony Express were ruined. The new technology that Samuel Morse patented in 1837 that replaced the Pony Express was the telegraph. In May of 1844 Samuel Morse, after stringing wire from the U.S. Supreme Court in the Capitol to Baltimore, Maryland, sent the coded message "What hath God wrought." He demonstrated that people could send messages using a simple code of dots, dashes, and spaces. His code became known as the Morse code (Fig. 7.5), which was used to transmit letters of the alphabet, punctuation, and numbers using electrical impulses traveling along a wire. However, there was still a need to have a telegraph operator at the sending end to code the messages and a telegraph operator at the receiving end to decode them. It was a major technological breakthrough, which allowed instantaneous communication between individuals over long distances.

The telegraph (Fig. 7.6) basically operates an electromagnet that is turned on and off creating a

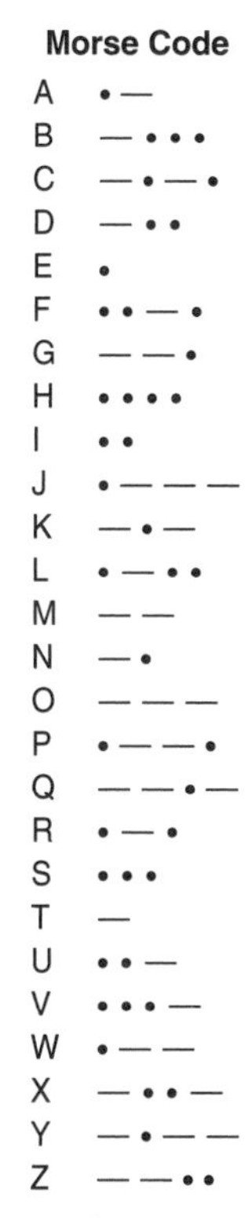

**Morse Code**

| | |
|---|---|
| A | •— |
| B | —••• |
| C | —•—• |
| D | —•• |
| E | • |
| F | ••—• |
| G | ——• |
| H | •••• |
| I | •• |
| J | •——— |
| K | —•— |
| L | •—•• |
| M | —— |
| N | —• |
| O | ——— |
| P | •——• |
| Q | ——•— |
| R | •—• |
| S | ••• |
| T | — |
| U | ••— |
| V | •••— |
| W | •—— |
| X | —••— |
| Y | —•—— |
| Z | ——•• |

**Figure 7.5** Morse code. *http://www.nsf.gov/od/lpa/nstw/teach/activity/mcode.htm*

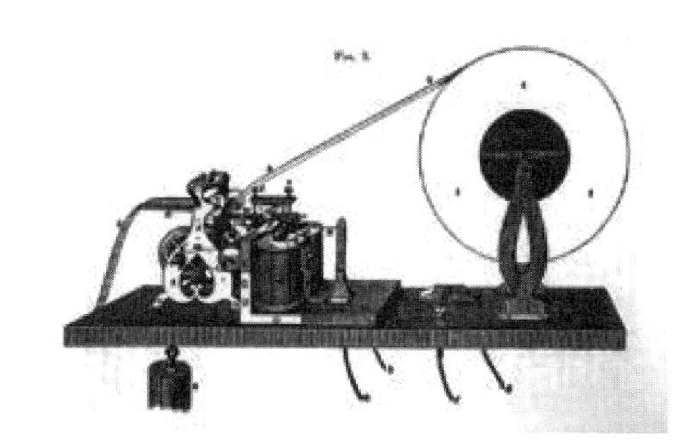

**Figure 7.6** Morse telegraph. *http://www.nlm.nih.gov/onceandfutureweb/database/seca/case2.html*

clicking sound. A short click is interpreted as a "dot" and a long click as a "dash." Different combinations of dots and dashes are used to represent the letters of the alphabet in code. For example, the universal distress signal SOS would be a sequence of 3 dots, 3 dashes, and 3 dots, ". . . - - - . . ." (Fig. 7.5).

## Typewriter

At about the same time, a new technology, which would revolutionize the way business was conducted, came to the marketplace. In 1874 E. Remington and Sons, best known as a manufacturer of guns, brought the typewriter to market. It allowed records and correspondence to be written on a typewriter, which produced printed characters on paper. The need for legible handwriting and the endless hours of training people to write clearly was minimized.

In fact, in the early 1900s there was a shortage of educated people who were needed to operate this new machine. Business then turned to a pool of well-educated college graduates (that were presently staying at home baking cookies, playing the piano, and caring for children) and college-educated women. Women were sought out for the business environment in order to use the new technology efficiently. Some believe that this technology, which resulted in bringing women into the business environment, led to the Women's Movement between 1910 and 1920, leading them to acquire the right to vote, among other rights.

It almost seemed obvious that the typewriter could be modified in some way to translate the dots and dashes of Morse's telegraph into printed messages. In the early 1900s this was accomplished with the invention of an electromechanical typewriter that could translate the electrical impulses into a printed message. The code that was developed for the teletype machines was the American Standard Code for Information Interchange, which was based on the binary number system.

# Binary Code

Today computers and synthesizers both use microchips for their operation. Information inside the computer is stored as bits or binary digits. Bits and combination of 8 bits, called bytes, are the building blocks of the code used by today's computers.

Machine language that computers use for their operation is coded in binary. Binary code uses two symbols, 0 and 1, to code meanings. Our decimal number system has ten symbols, 0, 1, 2, 3, 4, 5, 6, 7, 8, 9, that may be used to code meanings. We call the decimal system a base 10 system. The number used in the base 10 system starts at zero and ends at one less than the base. For example, in a base eight system the numbers would go from zero to seven but there would be eight symbols, numbers (available for coding).

All information in a computer is somehow represented using binary values. Each storage location in a computer has a low voltage (zero) or high voltage (one). A storage location cannot be empty but must contain a zero or one. In a present day computer, a combination of eight zeros and ones are used to represent common symbols like letters of the alphabet, decimal numbers, and other symbols like the "=" sign.

Suppose we tried to represent a letter of the alphabet using two bits. Could we devise a code of zeros and ones for all twenty-six letters of the alphabet?

If the length of a word were restricted to two characters in length, then the number of meanings we can code would be:

00
01
10
11

Only four different meanings can now be coded.

Three-character-length words allow us to code eight different meanings:

000 011
001 101
101 110
100 111

The basics of a **pattern** begin to emerge. Remember that **pattern recognition** is one of the key strategies used by scientists and engineers in their work. The pattern may be used to make a general statement regarding character length and the number of meanings that can be represented.

- For one-character length word, we associate two different meanings.
- For a two-character length word, we associate four meanings, or $2 \times 2 = 2^2$.
- For a three-character length word, we associate eight meanings, or $2 \times 2 \times 2$ or $2^3$.

A little thought would reveal that a four-character-length word would allow $2 \times 2 \times 2 \times 2 = 2^4$ or 16 meanings. Using exponents allows us to express the results compactly.

How many characters long would our word have to be if we wanted to code each of the letters of the alphabet in terms of ones and zeros? Now you will have to write a code that will be able to symbolize 26 meanings. A four-character-long word allows us to code $2^4$ meanings, or 16 meanings. For a five-character length word, the number of meanings would be $2^5$, or 32 meanings, more than enough to code the letters of the alphabet.

The first microcomputers used words that were eight characters long, or eight binary digits (eight bits) long. Eight bits altogether is called a byte. Now $2^8$, or 256, meanings can be coded, which means all the letters of the alphabet, the 10 decimal digits, the dollar sign, the plus and minus sign, etc., can be coded in terms of ones and zeros. Today the Unicode, which is able to code $2^{16}$, or more than 65,000, meanings is being adopted for international use so that every character in every language used in the world, including Chinese ideograms and even some of the symbols of dead languages, may be coded.

## Binary Numbering Systems

The binary numbering system works just like our everyday decimal system. The word decimal itself takes its root from the Latin number for ten, *decem*. It implies that there are 10 different symbols that are used in expressing numbers in our system. Their value is intimately connected with their position in the number. For example, 287 with a 7 in the units position, an 8 in the tens position, and a 2 in the hundreds position tells us we have two hundreds, eight tens, and two units for a total of two hundred and eighty-seven. Using powers of ten, we can write this as

$$287 = 2 \times 10^2 + 8 \times 10^1 + 7 \times 10^0$$

where $10^0 = 1$. Any number in our base 10 or decimal system can be decomposed in this way.

In the binary system, the base is two instead of 10. We have only two symbols to work with, 0 and 1. Under these circumstances a number like $101_2$ when decomposed becomes (note that the base is indicated by a subscript number, below the line):

$$1 \times 2^2 + 0 \times 2^1 + 1 \times 2^0 = 5_{10}$$

in the decimal system. An algorithm or recipe for converting any decimal number to binary involves dividing it by two and keeping track of the remainder after each division.

### Example

Convert $14_{10}$ to binary. The recipe is:

14 divided by 2 = 7 with remainder 0.
7 divided by 2 = 3 with remainder 1.
3 divided by 2 = 1 with remainder 1.
1 divided by 2 = 0 with remainder 1.

So, $14_{10} = 1110_2$, reading the remainders from the bottom up.

Notice that to convert $1110_2$ to base 10 we would write:

$$1 \times 2^3 + 1 \times 2^2 + 1 \times 2^1 + 0 \times 2^0$$

or

$$8 + 4 + 2 + 0 = 14_{10}.$$

### Example

Convert $12_{10}$ to binary.

12 divided by 2 = 6 with remainder 0.
6 divided by 2 = 3 with remainder 0.
3 divided by 2 = 1 with remainder 1.
1 divided by 2 = 0 with remainder 1.

Therefore, $12_{10} = 1100_2$.

Checking: $1 \times 2^3 + 1 \times 2^2 + 0 \times 2^1 + 0 \times 2^0 = 8 + 4 + 0 + 0 = 12_{10}$.

### Example

If you were asked to convert a number to base eight, the recipe would be the same. For example, convert $457_{10}$ to base 8.

Divide 456 by 8, or

457/8 = 57 with a remainder of 1.
57/8 = 7 with a remainder of 1.
7/8 = 0 with a remainder of 7.

So reading from the bottom up, $457_{10}$ is equal to $711_8$. Checking: $7 \times 8^2 + 1 \times 8^1 + 1 \times 8^0 = 457$.

### Odometer Example in Base 10

Consider the odometer, the mileage indicator, on a car. Suppose it has room for four digits. When the car is delivered new, it reads (or close to it):

0000.

After the wheels slowly turn in the last digit after one mile, it reads:

0001.

After two miles the wheels slowly turn and it reads:

0002.

After the car travels nine miles:

0009.

After the tenth mile the last digit becomes a zero and the second digit in the tens position reads one, or 0010.

### Odometer Example in Base 2

If we had only two symbols to work with as in the binary system, after the first mile it would read:

0001,

and after the second mile, the last digit would become a 0 and the next to the last a 1. So it becomes

0010.

After three miles,

0011,

and after four,

0100, and so on (See Table 7.1).

**TABLE 7.1 Decimal Numbers and Binary Equivalents**

| Decimal Numbers | Binary Numbers |
|---|---|
| 0 | 0000 |
| 1 | 0001 |
| 2 | 0010 |
| 3 | 0011 |
| 4 | 0100 |
| 5 | 0101 |
| 6 | 0110 |
| 7 | 0111 |
| 8 | 1000 |
| 9 | 1001 |
| 10 | 1010 |
| 11 | 1011 |
| 12 | 1100 |
| 13 | 1101 |
| 14 | 1110 |
| 15 | 1111 |

## Digital and Analog

You are probably aware that the codes used by digital computers are binary codes. Devices that use binary values to store and manipulate information are less expensive and more reliable than those that are based on other number systems. All computers and other electronic devices that manipulate and transfer digital information operate with codes using only two symbols. The two symbols could be high and low voltages, a magnetized and an unmagnetized state, a light reflecting and light scattering surface, an electrical conductor and an electrical insulator, or any other pair of physical states that are easily distinguishable.

In the binary code we used the one (1) and zero (0). Such a coding system that allows only a fixed, countable number of meanings to be assigned is said to be digital as opposed to analog coding, where essentially an infinite number of meanings or values can be assigned to the physical state of a variable.

# DNA

In the 1950s, one of the most important patterns of information was analyzed in determining biochemical reactions in the cell. The problem that faced researchers at the time was the question of how information carried by genes concerning heredity was implemented in the formation of the organism. Researchers had already determined that the cell contained many molecules in the form of fats, sugar, proteins, and genes made up of deoxyribonucleic acid (DNA). They already had determined that the genes carried the information needed to fabricate the proteins.

| | |
|---|---|
| AA | CA |
| AC | CC |
| AT | CT |
| AG | CG |
| TA | GA |
| TC | GC |
| TT | GT |
| TG | GG |

**Figure 7.7** Two-character-length words using four symbols.

In 1954, Watson and Crick discovered the structure of DNA. They found that the coding region of DNA was made up of four chemicals called bases (i.e., adenine, thymine, cytosine, guanine), and the bases were arranged in a double helix. The next question to be answered was how the DNA specifies the structure of a specific protein. We find that a protein strand, of the type found in living cells, is made up of a string of amino acid building blocks. Only 20 amino acids are necessary to produce the thousands of different proteins of life. What distinguishes one protein from another is the sequence in which these amino acids are placed.

How does DNA, with four bases to act as information symbols, analogous to having a message written in a alphabet of four letters, specify the language needed to code 20 different meanings for amino acids?

Using a coding system that has four symbols, what is the minimum length a word has to be to code 20 different meanings?

If the length of the word was one character, we could code four different meanings:

A G T C

If the length was two characters, the number of meanings we could code would be $4 \times 4 = 16$. Still, it is not enough to code the 20 amino acids needed to make up the genetic code (see Fig. 7.7).

If we use words three characters long, we have $4 \times 4 \times 4 = 64$ meanings, more than enough to code the 20 needed. Hence, the sequence for three-letter words made up of the bases A, C, T, and G along a strand of DNA can carry the information in a cell. Since there are 64 possible "words" for the 20 amino acids, we can think of three letters (standing for three bases) as identifying a particular amino acid—in a sense naming it.

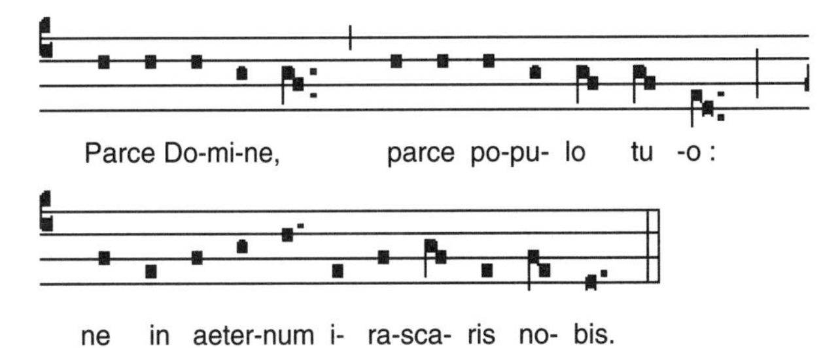


**Figure 7.8** Four–line staff with square notes.
*http://comp.uark.edu/~rlee/missal.html*

The meaningful words in genetic material are three-symbol arrangements of the nucleotide bases in DNA. They are called *codons*.

Molecular biologists have completely deciphered this code and produced a genetic dictionary. With the scientific understanding of which three-letter DNA code specifies each of the 20 amino acids, we are now able to modify DNA in the laboratory and produce "designer proteins" with desired functions. This is a clear example of the progression from scientific understanding to technological application.

## Musical Coding

The Western system of musical notation has evolved over many centuries. It seems that in the sixth century, Pope Gregory the Great devised the first good system of notation for coding Gregorian chant. A new monk had to learn the melody for the prayers sung at various times during the day. To help him acclimate quickly, a coding scheme for music was developed. He used a four-line staff with three spaces on which he could place specific symbols to represent notes or pitches (Fig. 7.8).

Because of the relatively small range of pitches used in the Gregorian chant, the staff proved adequate in coding melodies.

However, in time an expanded staff was needed and the staff evolved into the five-line staff we know today (Fig. 7.9). The symbols for notes were egg-shaped and depending whether they were hollow, filled, and had a stem, their duration changed. Each line and space on the staff was labeled with a letter of the alphabet, a, b, c, d, e, f, g. Placing the symbol for the note on the staff denoted its pitch and duration.

Using clef symbols, G-clef, 𝄞 to designate a staff for higher pitched notes or the melody, and a staff for the lower pitched notes or harmony, satisfied the need for a larger staff.

The treble and bass clefs can be joined into the great staff (Fig. 7.10). Each pitch, depending on its

**Figure 7.9** Five–line staff.

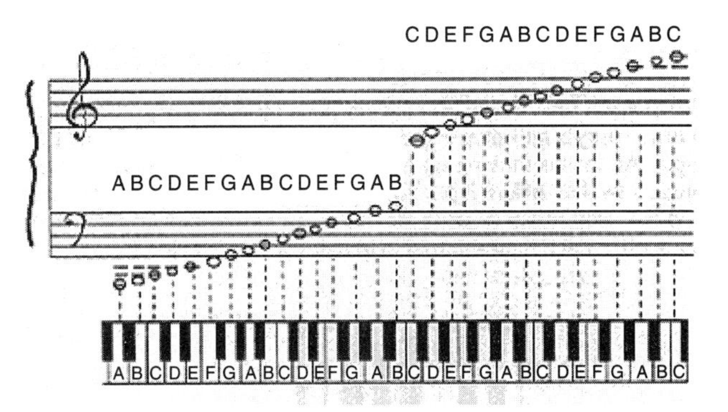


**Figure 7.10** Great staff and notes.

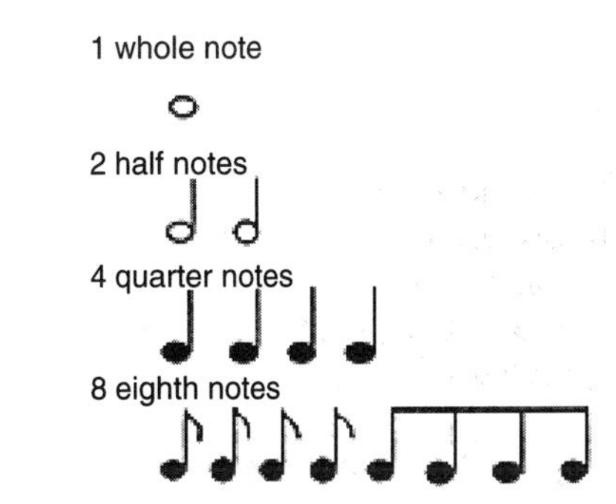


**Figure 7.11** Musical note names.

name and placement, corresponds to a black or white key on a keyboard.

The duration of a pitch is designated by an appropriate symbol. Time is measured relatively and expressed in terms of beats. For example, a whole note lasts four beats, and a half note lasts two beats, and so on. The duration (Fig. 7.11), when combined with different combination of pitches, allows the composer to create music consisting of melody and harmony.

## Language Coding

Consider the written statements:

La puerta es roja.

La porte est rouge.

Die tur ist rot.

Coding is about information and communication. Usually we don't think of language as a code but in fact it is. It is a way of representing information.

**TABLE 7.2 Russian alphabet**

| Russian letter | English analog | Russian letter | English analog |
|---|---|---|---|
| Аа | A | Рр | R |
| Бб | B | Сс | S |
| Вв | V | Тт | T |
| Гг | G | Уу | U |
| Дд | D | Фф | F |
| Ее | E | Хх | H |
| Жж | ZH | Цц | TS |
| Зз | Z | Чч | CH |
| Ии | I | Шш | SH |
| Йй | I-short | Щщ | SCH |
| Кk | K | Ъъ | - hard |
| Лл | L | Ыы | - (maybe y) |
| Мм | M | Ьь | - soft |
| Нн | N | Ээ | E |
| Оо | O | Юю | IU |
| Пп | P | Яя | IA |
| Andrei Latychev 1998-1999 | | http://travel.to/petersburg | |

(http://www.travelto.spb.ru/alphabet.html)

Porte e rossa.

Doren er rod.

A porta e vermelha.

Unless you know one or more languages (Spanish, French, German, Italian, Norwegian, Portuguese) you might have a problem solving what is being communicated. Here the statements all note the same thing: the door is red.

When the language is spoken, since we have memorized the sound for each word and what it represents, we can quickly relate the sound to the meaning and gain understanding.

In written form each word in our language is a code for an object, an activity, an action, etc., for which we have memorized all the symbols collectively, and can therefore decipher. If we are confronted with a new or unfamiliar word, we can sound out the word since letters are codes for sounds.

## Russian

For example, what do these words mean?

РЕСТОРАН

БаНк

ПеПСИ КОЛа

РОССИ

Actual translation of these words requires a bit more information. Perhaps if you are familiar with Greek or Russian or other languages that use the Cyrillic alphabet, the task at hand would be a bit easier (Table 7.2). Actually the words all sound like their English counterparts when then are pronounced using their corresponding English sound for the Cyrillic letters. They are:

Restaurant

Bank

Pepsi cola

Russia

Try to use this information to generate a listing of correspondence between the letters in English and those in the Cyrillic alphabet.

Did you come up with the following?

Russian Р = R

Russian С = S

Russian Н = N

Russian И = I

Russian П = P

Russian е = e

Russian A =A

| We have conquered our enemies | Nin hokeh bi-kheh a-na-ih-la |
| --- | --- |
| All over the world | Ta-al-tso-go na-he-seel-kai |
| On land and on sea | Nih-bi-kah-gi do tah kah-gi |
| Everywhere we fight | Ta-al-tso-go en-da-de-pah |
| True and loyal to our duty | Tsi-di-da-an-ne ne-tay-yah |
| We are known by that | Ay be nihe hozeen |
| United States Marines | Washindon be Akalh Bi-kosi-la |
| To be one is a great thing. | Ji-lengo ba-hozhon |

**Figure 7.12** Navajo language. *http://www.history.navy.mil/faqs/faq61-4.htm*

Translate this: CHAPΛ. (Hint) I heard the dog CHAPΛ.

The complete correspondence between the letters is noted in the table.

## NAVAJO

Finally, there is an example where a language was also used as a secret code. In fact this code is reputed to have significantly aided in the victory in the Pacific in World War II. The language is Navajo, and Navajo Soldiers who spoke and understood the language used it. Because of the extremely limited number of individuals knowledgeable in the language, the language could be used over walkie-talkies without fear of the enemy intercepting and translating the messages. The code was never broken. Figure 7.12 is an example of Navajo (see the end of the chapter for a reference).

## SMOKE SIGNALS

Transference of information at a distance requires additional technology than just the use of the human voice. A few hundred years ago, methods used for this purpose were limited. Voice amplification via echoes (yodeling) was one method. Smoke signals was another.

Smoke signals are a somewhat effective way of communicating over distance if atmospheric and lighting conditions are favorable. In fact, smoke signals are still used today, although under very special and rare conditions. They are used to communicate the results of ballots cast by sequestered Roman Catholic cardinals in papal elections.

## SEMAPHORES

A slightly "higher tech" code involves the use of semaphores. Semaphores involve the use of two flags (Fig. 7.13) held at various positions to indicate different letters in the alphabet. It is a sign language meant to be seen over distance while rapidly communicating and maintaining radio silence. The flags are usually square and divided diagonally into red and yellow triangles. As the flags are held with the arms extended, they represent letters of the alphabet. The pattern is like the face of a clock. For more details see the Web resource noted in the figure caption.

A.

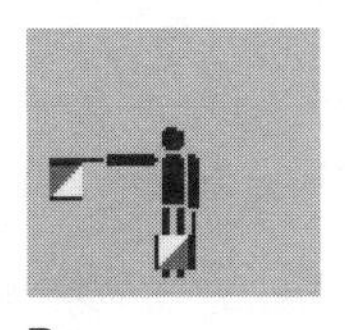
B.

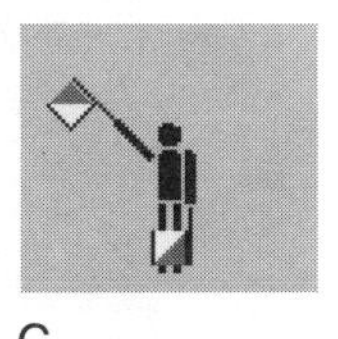
C.

**Figure 7.13** Semaphores. *http://www.anbg.gov.au/flags/semaphore.html*

## LICENSE PLATE CODES

There are a wide variety of codes that are used for personal identification and services. You are familiar with many of these, such as credit card numbers, automatic teller PINs, social security numbers, etc. These are strictly numerical codes based on our decimal system of digits. Some of the more interesting coded sequences make use of the full complement of alphanumeric symbols (numerical digits plus the letters of the alphabet). Two of these are auto license plates and telephone numbers. The lengths of the sequences used here are limited because of space restriction and the limited abilities of the human mind to manage easy recall of information.

Many states in the past (for example, New Jersey) limited the normal pattern on a license plate to six characters. How many different plates can be produced for a series that employs three letters followed by three numbers (Fig. 7.14)? The answer:

$$(26 \times 26 \times 26 \times 10 \times 10 \times 10) = 17{,}576{,}000$$

This must be reduced, practically speaking, by one thousand for every three-letter combination that is censored out for reasons of decorum and good taste.

**Figure 7.14** License plate I.

**Figure 7.15** License plate II.

There remains something on the order of 17 million combinations. While this is sufficient to last a small state for a good number of years, it is inadequate to handle the number of registrations in a large state such as California or New York, unless additional plates are to be issued every year with different sequences. The three-letter/three-numeral arrangements are also only marginally adequate for states with intermediate sized populations, such as New Jersey, Florida, or Massachusetts. Indeed, New Jersey has gone through two such series of licenses already since adopting a "permanent" plate system. The second series differed from the first by having three numerals followed by three letters (Fig. 7.15).

How could or should a new series of plates be encoded to provide a greater number of possibilities without going beyond the six-character length? Many options could be offered. Any other pattern of three letters and three numbers would add another set of about 17 million plates, which could coexist with the first two versions without duplication or too much confusion. Alternating letters with numbers would be one way of doing this.

Actually, New Jersey has chosen to select a new series that uses four letters and two numbers in the pattern—three letters, two numbers, and one letter. This provides a series of plates with 2.6 times as many combinations as either of the earlier versions. Because the four letters are separated, the censor is not quite so busy removing objectionable four-letter words from the set. This series of about 44 million different license designations will last for a decent interval even in a state with a rapidly growing number of auto registrations. Of course, new variations on the scheme can continue to add series to meet future needs.

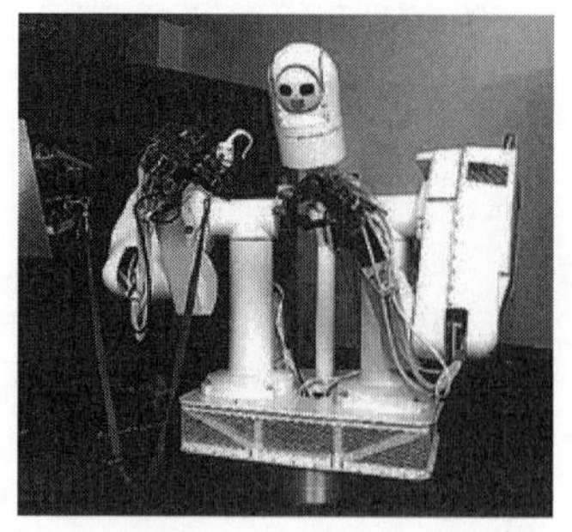

**Figure 7.16** Dexterous anthropomorphic robotic testbed (DART). *http://vesuvius.jsc.nasa.gov/er_er/html/dart/index.html*

Add to these possibilities the rising popularity of vanity plates, where the six alphanumeric symbols are arranged according to the car owner's wishes without any other constraints (except that they cannot duplicate another plate already issued), and we can see that, without going beyond six symbols per plate, the available combinations can accommodate most states' needs for an indefinite period of time.

## Automation Technology

Automation is the replacement of the human being by a machine (see Fig. 7.16) in decision-making tasks. Automation has brought great changes to our society and involved us in no small number of problems and disagreements, including those of workers' rights and job security.

Yet, automation has some commendable objectives, which include the advantages of lower costs, improved products or services, and the more humane use of people. However, it should be clear that many examples of automation exist that allow tasks to be performed in environments that are too uncomfortable, too unsafe, or simply impossible for human workers to do the job.

In automated systems there must be:

- stored information,
- recognition that a decision is to be made,
- comparison to some standard or other criterion with the current conditions,
- the ability to choose the "best" option from the available alternatives.

Thus for example, an automated coin-operated machine (Fig. 7.17) has stored within itself the ability to initiate action when a coin-shaped object rolls down the chute. It measures some size, weight, electrical conductivity, and/or magnetic

**Figure 7.17** Vending machine.

properties and compares them against stored data on nickels, dimes, and quarters. It then registers the proper amount of money deposited to the customer's credit, or it rejects the object as a slug or some inappropriate kind of coin, depending on the results of the comparison.

Most automated systems make at least some use of information in feedback paths. The control that results from passing information from output to input reduces errors by the system and allows complex decisions to be handled as a series of simpler choices, each contingent on the choice made earlier in the sequence.

Thus, our coin machine can decide whether to unlock the product selector and allow the customer to get his granola bar if it feeds back information about the total in the credit register to the unlocking mechanism. It also must feed back the information that the customer has exercised his option and received the product. This information returns to the input or the credit register so that the amount recorded can be returned to zero. This is the end of the process unless one of the information transfers fails, in which case the customer doesn't get his bar, and the machine receives immediate feedback from its external surroundings in the form of several well-placed kicks—but now we're no longer talking about automation.

## Retail Checkout Automation

Retail stores, such as supermarkets and discount stores, which work on a very small profit margin and count on sales volume to provide a reasonable profit, are obviously interested in any technology that can trim costs and improve the efficiency of their operation. When laser bar code readers (Fig. 7.18) were teamed with computer systems to automate the cash register operation, those stores were eager to put the systems in operation.

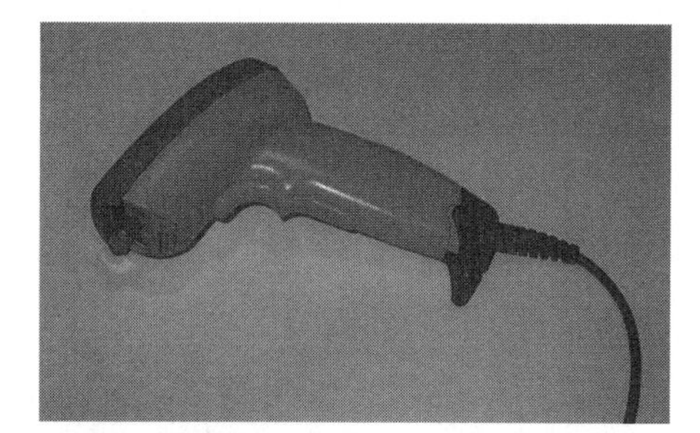

**Figure 7.18** Hand scanner.

These systems are automated because the reader is able to recognize the identity of the article being purchased from the imprinted code on the package. It then communicates with the central computer, obtains the correct sale price of the item, and passes the information along to the register, where both the item identification and the price can be recorded and printed out. If the item is taxable, eligible for a rebate, or on special sale, such information can also be entered and dealt with automatically within the system.

### Advantages

At almost the same time that the system is recording the sale, the information about the quantity and type of item being sold can be transferred to a portion of the central computer that is dedicated to keeping track of inventory. The sale is subtracted from the amount of the product in stock and a query is made regarding the need to reorder. If the amount in stock has fallen below what has been predetermined as a desirable level, the inventory program can initiate the reorder, or at least alert the store manager. The ability to manage inventory more efficiently produces the greatest economy for the store and is generally considered by management to be the biggest advantage of the automation.

### Other Advantages

Other advantages of the checkout systems, from the store's point of view, are faster, more efficient checkout procedures, reduced cashier errors (both unintentional and intentional), ease of updating prices and managing special sales, and the opportunity to assess the effects of marketing efforts such as special ads, sale prices, or rebate programs. The buyer can hope to share the advantages of reduced errors at the register, as well as some part

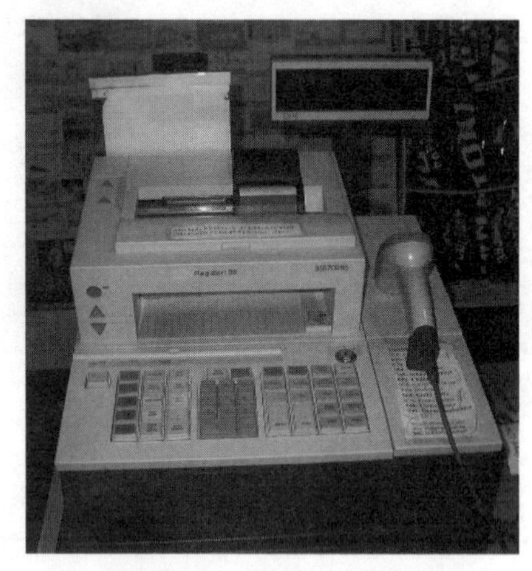

**Figure 7.19** Register scanning system.

of the overall cost savings realized by the store. Customers receive a printed receipt, which can be a much more detailed record of their purchases than was previously possible (Fig. 7.19).

Finally, any time saved in the checkout line is greatly appreciated and makes for a happier customer. From the cashier's point of view, the new systems make the job no less repetitious, but do aid accuracy and take away some of the stress involved with satisfying customers, especially at peak times.

## Customer Acceptance

Once the stores had studied the cost effectiveness of the new systems, many were ready to introduce them in the mid-1970s. However, customer acceptance did not come overnight, and stores in a competitive environment cannot afford to adopt an innovation that may drive away business. The public had to come to an understanding of the new systems and to think of them as advantages rather than merely strange. The central sticking point, which led to considerable consumer resistance, was the issue of item pricing. Since the laser readout system, Fig. 7.19, can obtain the price from the computer after reading the bar code, there is no longer any need from the store manager's point of view, for each item to be labeled with an individual price tag. Workers would no longer have to be paid for the relatively slow, labor-intensive job of attaching printed price stickers or tags to each package.

However, customers saw this as a reduction of the information that they were receiving from the seller. Most were reluctant to trust the automated pricing scheme unless they had a way to check product price against register receipt while they were standing at the checkout. Enough errors were made in entering store prices into computers in the early days of checkout automation that consumer confidence was not high.

Now state or local regulations are in place to govern abuses, and various schemes to post prices have been established. Consumers seem to have adjusted and accepted the systems, perhaps more in a spirit of resignation than with enthusiasm. The perception may be that, on balance, the advantages of the new technology are more on the side of the retailer than on the side of the consumer.

## Technology Development

Once automated cash register systems began to be used in retail stores in a big way, the commercial opportunity involved stimulated the continued development of this kind of technology. The original type of laser readers, which are fixed into the counter so that products can be passed above their field of view, have been improved to be more reliable and sensitive. More recent laser readers have smaller active regions (heads) and may be hand held, or even portable with a wireless transmitter type of coupling to the computer or computerized cash register.

New computer applications software allows the storekeeper many more options in analyzing and profiting from the data records of sales. Improvements in computer hardware and increased numbers of sales of systems have reduced costs and made this automation available even to smaller businesses. Next-generation systems have been designed where the manual work now performed by the cashier in passing the product code across the head of the laser reader will be entirely replaced by an automated arrangement completely integrated into the system.

One version of this type of system is similar to a mammoth automatic vending machine. The sales region of the store may be simply a set of compartments in which examples of each product may be viewed. The customer, identified by a code on a credit or debit card, enters a request for each product through input terminals located periodically through the store or perhaps models that can be hand-carried.

The goods are gathered automatically from coded locations in the warehouse and transported to the checkout site so that when the customer enters a final transaction command into the terminal, the order will be packaged and ready for pick-up almost immediately. The order may be paid for in cash, or by immediate electronic transfer of funds from the customer's bank, using a preapproved debit card.

Certainly, such a system will change the way we shop for produce and will place new demands on the packaging of fragile things. It is a retailing technology, which may or may not be desirable or economical, but it is a technology that is now feasible.

## Societal Impact

A technological innovation in business, like checkout automation, always has implications for the labor force. Since the initiative for adoption of the technology usually comes from management, which authorizes capital expenditures, employees normally find themselves reacting to the imminent changes rather than being part of the problem-solving design team that developed the technology. In the case of retail store automation, it is clear that adoption results in a reduced need for price labelers and checkers. On the other hand, there will be an increased need for technical maintenance and computer systems people.

As usual, the skills demanded for the new jobs are not the same as those possessed by the workers who will be displaced. The introduction of automation technology has often, in such situations, led to bitter confrontations between management and labor and the question of economic justice and the equitable control of the shaping of new technology is one that our society has yet to resolve.

## Technology Components

The automated checkout system needs three major technology components: the digital computer to handle the flow of information and process it, the reader to translate the information about the product into the digital electronic form that a computer can deal with, and a coding system that can present information compactly and reliably to the computer. The digital computer was certainly well established when automated checkout systems began to move off the drawing boards.

The novel elements that allowed the automation to be implemented were the development of lower-cost lasers for optical detection systems and the adoption of the Universal Product Code (UPC) as a standard way of encoding manufacturer and product identification data. With these innovations it became possible to take such information directly from a package and insert it into the input of a computerized system without any mechanical contact being necessary between the object and the reader.

# Universal Product Code (UPC)

The UPC is arguably the most widely used bar coding scheme in existence. Like the Postal Service Code, the information is presented in vertical light and dark lines or bars. Thus it is basically a binary code. The scanning, focused laser light passes over the pattern and is either reflected and returned to a detector in the reading head (when it strikes a light colored bar) or absorbed and not returned to the detector (when it strikes a dark bar).

**Figure 7.20** UPC bar code.

A typical UPC symbol consists of two parts—the array of light and dark bars, which appear to have various widths, and a series of numerical digits, which are printed, for the most part, just below the bar pattern. The numbers are present so that the cashier can enter the information manually in cases where the automatic reader does not accept the input. (Frost or moisture on the surface of a package often gives these systems problems.) The bars are part of a code that has been cleverly designed to allow rapid and accurate reading of the information by means of optical readers.

Consider the UPC symbol shown in Figure 7.20. This represents a set of twelve numbers. The first number describes the type of product; in this case it is a grocery item identified by 0. The next five digits appear in order under the left half of the symbol. These correspond to the manufacturer, Tropicana, which has been assigned the designation 48500.

The five digits under the right half of the symbol, 00100, represent one of this manufacturer's products, the one-quart size of fresh orange juice. Finally, the twelfth and last digit, 4, appears. This is a check digit to help the automatic system avoid reading errors.

The check digit is not always written out underneath, but it will always be present so that the reader can compare the earlier eleven digits with the twelfth according to the defined error check formula. If the reader encounters a twelfth digit that differs from what it expects according to formula, it rejects the data and gives the operator an error signal.

## READING THE CODE

To read the code, imagine the symbol to be divided into adjacent vertical spaces, each as wide as the narrowest dark bar seen in the pattern. If dark bars are interpreted as binary 1, and light bars as binary 0, the pattern begins with a three-bar starting signal consisting of 101, or dark-light-dark. The next seven bars code for the numeral 0, which is the type or character number for this symbol; its code is 0001101. This appears as a broad light bar (triple), followed by a narrower dark bar (double), followed in turn by a single light and a single dark. Note that each numeral requires seven bars.

On the left half of the symbol, all numerals begin with a binary 0 and end with a binary 1; that is to say, begins with a light bar and ends with a dark. On the right half of the symbol, the opposite to this rule is true. These conditions mean that the scanning reader will always know whether it is traveling across the UPC symbol from left to right or from right to left.

The guard bars and the first numeral in the symbol we are decoding are represented by bars that are taller vertically than those that appear to their right—at least until we get to the center of the pattern where another set of tall bars stand as center markers. The five manufacturer's digits begin with a shorter light bar, which is the first binary digit in the code for the numeral 4. The code equivalents for the left-hand half of a UPC symbol is shown in Figure 7.21. Use these to decode the pattern shown in Figure 7.20.

In the center of the symbol, between 48500 and 00100, stands a center bar pattern of the form 01010. To the right of this the numbers are coded by the exact complement (0s replaced by 1s and 1s by zeros) of the rules listed in Figure 7.21; that is, number 1 is represented by 1100110, 2 by 1101100, etc. Another requirement on the left half of the symbol is that a number always is represented by a code with an odd number of 1s. It is said that these representations have odd parity. On the right half, the codes have even parity.

Finally, each arrangement that is allowed by the coding scheme has exactly two light and two dark regions. This type of restriction again eases the recognition task of the optical reader and reduces the probability of error in a reading.

The use of seven binary bars to code for a number makes available 128 combinations. The restriction that the ten digits begin with 0 and end with 1 on the left half of the symbol reduces the possibilities to 32. Requiring odd parity further limits the allowed arrangements to 16. Of these, only 10 are combinations with two light and two dark regions. Thus, from an abundance of possibilities, the code designers have cleverly limited those that are actually used by adding criteria that help to increase the reliability of the decoding system.

| Number | Binary Bar Code |
|---|---|
| 1 | 0011001 |
| 2 | 0010011 |
| 3 | 0111101 |
| 4 | 0100011 |
| 5 | 0110001 |
| 6 | 0111011 |
| 7 | 0111011 |
| 8 | 0110111 |
| 9 | 0001011 |
| 0 | 0001101 |

**Figure 7.21** Binary UPC bar code.

**Figure 7.22** Decoding the bar code.

# Reflections

By considering retail store automation and UPCs, we have made concrete some of the issues that touch on the involvement of information in modern technology. As we look for technology to undertake tasks for us, which involve more and more complex decision making, the information component of the technological system plays a greater and greater role.

Hence, we see more introductions of "smart" instruments, factories, and manufacturing lines using robotics or "distributed intelligence," "expert systems", and other examples of an information technology, which recognize information as a con-

**Figure 7.23** U.S. post office.

cept and a quantity with behaviors that we must understand more fully.

We began this chapter by considering the USPS Code. Using this bar code and automated readers, the postal service (Fig. 7.23) is attempting to improve the accuracy and efficiency of its service. However, unlike the retail checkout example, the postal service attempt is severely beset by labor problems, which arise because of the workers who will be displaced by the new technology. The problem is staggering in size. Since 1980, the number of pieces of mail handled annually by the USPS has risen by more than 55%, to about 160 billion. In the same period, the number of postal service employees has risen to about 800,000—approximately a 12% gain. The costs of labor are now about 83% of the total costs of the operation. Failure to do something dramatic to improve service could, in the light of a proposed rate increase, lead to a considerable loss of the most profitable segment of the business to alternative carriers.

The postal service's automation scheme requires fewer skills for some sorting personnel than have previously been required for manual sorting. Management wishes to contract both the skilled maintenance and the sorting requirements out to private contractors, who will hire low wage, nonunionized labor to handle the sorting that cannot be managed by the technology. Current postal service sorters, who are unionized, would be redundant in the new arrangement and many would lose their jobs. And so the full implementation of the technology that USPS has already begun to bring on line is in danger of crashing into the rocks of a labor management impasse. Both partners in the debate carry considerable influence over what will eventually happen. Thus, not only must the technology be found to be appropriate and cost saving, but also an equitable solution, which protects job security without carrying nonproductive employees, must be found. The USPS has encountered an old truth: technology can make certain options available; whether or not they are truly workable involves addressing many other human and social concerns that arise from the proposed actions.

# Cryptology

We have spent most of our attention on aspects of the coding of information. Although we are not going to dwell on the topic of secret codes, it would seem incomplete not to mention some of the amazing efforts that have gone into the making and breaking of codes when the purpose of the activity is either to restrict access to certain information or to gain access to information we are not authorized to have.

Approaches to the problem of keeping information secure involve creating secret access codes, which are required to unlock the flow of information (e.g., computer passwords or data file "lock" codes), or coding schemes, which encrypt the information so that it can be translated only if the special scheme (cipher) is known.

## Ancient Codes

Secret codes go back to ancient times. The "Caesar Cipher" shifts each letter of the alphabet by three:

$$A \rightarrow D$$
$$B \rightarrow E$$
$$X \rightarrow A$$
$$Y \rightarrow B$$
$$Z \rightarrow C$$

It is secure only as long as the enemy does not know, or guess, the idea.

A brief article in *Smithsonian* magazine, in June 1987, provides a revealing insight into the careers and accomplishments of William and Elizabeth Friedman as well as an introduction to ciphers and code breaking. The Friedmans were an American couple who brought the ancient skill of cryptology into the scientific age. Reading about code-makers and code-breakers in competition will provide you with a good example of an area of human endeavor where having two opposition viewpoints on a matter create a competition involving technology vs. technology. The routine procedures used in cryptology as well as the actual machines and devices which may be employed also indicate clearly the presence of both software and hardware aspects of technology in the field.

# Information as a Resource

Although it is technically correct to think of information as a quantity, the amount of information present in our universe is not restricted or limited by any known law of nature. Thus we are free to discover or create new information continually. Since information is also of great practical use to people, this means that in this quantity we have an unbounded resource. We can make use of this resource to help us anticipate and resolve problems, to clarify our differing desires and objectives, and to communicate what we value or hold sacred.

The well of stored information can help us to appreciate the aesthetic dimensions of life, as happens when we remember great dramatic or musical performances. Having an unlimited resource in a world where most of what we recognize as resources is not only limited, but also very finite, should come as a happy revelation. Now we have to use this resource well.

Our information technologies must be developed to extract the greatest human benefits, the greatest advantages for our species. This can emerge from continued progress in the knowledge base called information science and the wise and clever development of information technologies. The potential of information as a resource and the meshing of technology with the special nature of the human mind are dealt with for a perspective, which focuses on the closing years of this century, by Nobel laureate Arno Penzias in his book, *Ideas and Information.*

# Review Questions

1. What is information? How does it differ from knowledge?
2. What is data?
3. What is a zip code?
4. Why did the Pony Express fail after only one year?
5. What is the Morse code?
6. Covert 35 to binary code.
7. Convert 11001101 from binary to decimal.
8. What is a semaphore?
9. What are the advantages of retail automation?
10. What impacts does automation have on society?
11. Decode:
12. Decode:
13. Decode:

14. Decode:

# Multiple Choice Questions

1. Information is _____.
   a. data
   b. knowledge
   c. code
   d. knowledge in transit
2. Data are _____.
   a. knowledge
   b. information
   c. code
   d. knowledge in transit
3. The original US zip code had _____ digits.
   a. five
   b. four
   c. three
   d. six
4. Morse code is written in terms of _____.
   a. vertical bars
   b. dots and dashes
   c. ones and zeros
   d. Navajo words
5. The binary number system has _____ digits.
   a. one
   b. two
   c. three
   d. four
6. The decimal number system has _____ digits.
   a. ten
   b. nine
   c. eight
   d. seven
7. If one had five different symbols to work with in forming words three characters in length, how many different meanings could be encoded?
   a. 15
   b. 25
   c. 125
   d. 250

8. If New Jersey used only six numbers to make up its license plates then the number of different plates would be _____.
   a. 600
   b. 1,000,000
   c. 6,000
   d. None of the above.
9. The most widely used bar coding system in existence is probably the _____.
   a. ABC code
   b. zip code
   c. UPS code
   d. UPC code
10. Cryptology is _____.
    a. the study of tombs
    b. the study of secret codes
    c. the study of automation
    d. technology used in automation
11. The number 10112 in decimal is _____.
    a. 5
    b. 10
    c. 11
    d. 13
12. The decimal equivalent of the binary number 1000 is _____.
    a. $1 \times 10^3$
    b. 0.9
    c. 9
    d. 8
13. The characteristic that enables DNA to be used to carry a code is the fact that _____.
    a. the sequence of phosphates can be unique
    b. the sequence of ribose can be unique
    c. the sequence of nitrogen bases can be unique
    d. None of the above.
14. "–.–. .– –" translates as _____.
    a. cat
    b. dog
    c. just a bunch of dots
    d. None of the above.
15. ПеПСИ КОЛа translates as _____.
    a. soft drink
    b. Russian bear
    c. vodka brand
    d. Caspian sea caviar
16. How many possible arrangements exist if a state issues license plates with two letters?
    a. 26
    b. 52
    c. 1,000
    d. 676
17. How many possible arrangements exist if a state issues license plates with two numbers?
    a. 26
    b. 52
    c. 1,000
    d. 676
18. How many possible arrangements exist if a state issues license plates with one letter and one number (assume zero and the letter o are different?)
    a. 260
    b. 520
    c. 1,000
    d. 676
19. WKLV LV DQ HALSOH RJ is _____.
    a. Cyrillic
    b. Navajo
    c. gibberish
    d. Caeser cipher
20. UPC stands for __________.
    a. universal production corporation
    b. universal product symbol
    c. uninhibited Peruvian cattle
    d. universal product code
21. *Washindon be Akaih Bi-kosi-la* is code for _____.
    a. U.S. Marine
    b. a Navajo home
    c. the Atlanta Braves.
    d. None of the above.
22. If one has three symbols, A, B, C, how long would each word in the code have to be in order to code 16 meanings?
    a. 2
    b. 3
    c. 4
    d. 5

# Exercises

1. The Postal Service bar code employs five vertical lines in a binary code to represent the 10 digits of our decimal number system. Obviously, such a code provides 32 patterns from which the meaningful ones could be chosen.

   Inspect the list of assignments for the USPS code and try to determine what additional conditions have been specified by the code designers to reduce the 32 possibilities to the 10, that are actually used. How could these added requirements be used to help a reading machine avoid errors or check when an error has occurred?

2. One frequently used binary code for representing the alphanumeric characters with which we are familiar is the American Standards Code for Information Interchange, or ASCII. ASCII is commonly used with computer input devices such as keyboards. It uses seven bits with an eighth often included to serve as a check digit (which may be selected according to either an even or odd parity convention).

   Find a source that will provide you with the assignments used in ASCII. How many *meanings* can be handled by the code? How many are ordinarily used, beyond the alphabet and the decimal numbers?
3. In studying the examples of coding schemes presented in this chapter, you have been calling on your skills in thinking. Use these skills you have been practicing to answer the following questions.
   a) A digital electronic display uses components that accept a binary signal and interpret it as a decimal digit. How many wires must enter a display unit with the capability of displaying a single digit, if each wire can carry only a 1 or a 0, and the simultaneous pattern of 1s and 0s is to represent the digit (from 0 to 9) that should be displayed?
   b) Assume you have invented an "organic" computer that uses a four-symbol molecular scheme (A, T, C, and G) to form *meaningful* patterns. How many bits would be needed per byte in your computer if it were to be as powerful as an 8-bit binary machine? If it were to replace a 16-bit binary computer, how many bits would be needed?
   c) Suppose a form of life were discovered for which the genetic information is coded by 5 nucleotide bases—adenine, thymine, guanine, cytosine, and the newly discovered ionamine. How long would the codon sequences be if they still had to code for only 20 amino acids?
   d) How could you use DNA to write messages? (Hint: use the 64 different codon combinations to express the alphabet and punctuation, etc.)
4. a) Consider chemical molecules, which are arrangements of atoms, to be patterns having recognizable and significant differences. If there were only four different types of atoms (e.g., the four elements of the Greek thinkers) and the largest molecules that could be formed contained four atoms, how many different substances could exist? Assume any kind of atom could link up with any other kind of atom.
   b) Realize that there are 92 different naturally occurring elements and determine the number of substances that could exist if the other assumptions of part a) still held.
   c) Extend the thinking you have just done by discussing, without calculating, the implications of the following additional facts: i) Molecular sizes are not limited to four atoms. Some molecules contain literally thousands of atoms. ii) Chemistry has revealed that not all atoms link up with one another. There are both preferred and restricted combinations that comprise the "laws" of chemical bonding. Thus the number of molecules that actually form is quite different than the number of arrangements of atoms we can generate by combinational reasoning. iii) Some combinations of atoms that have not yet been observed to exist are not forbidden by the known laws of chemistry. These represent new substances that might be brought into existence through the technologies of chemical synthesis.
5. One automated system that has gained great popularity in a small time is the ATM, or automatic teller machine. Reflect on what happens as you use the machine to make a cash withdrawal. How does it know you are there to be served? How does it identify you, decide what service you want, check for sufficient funds in your account, count out the proper bills, debit your account, create a record of the transaction, and sign off?

   Try to sketch a block diagram indicating what the main parts of the system and the information flows must be. Are there any feedback paths in your description of the process? What would be different, if anything, about the information exchange if you completed the same type of transaction through a human teller? What are the benefits of this automation? What are the associated problems and trade-offs?

# Bibliography and Web Resources

Truxal, John G., *Supermarket Automation,* Research Foundation, SUNY at Stony Brook, 1978.

Truxal, John G., *Feedback - Automation,* A Sloan Foundation New Liberal Arts Program Monograph, Research Foundation, SUNY at Stony Brook, 1989.

Weaver. The Newsletter of the Council on the Understanding of Technology in Human Affairs, *Information Technology* issue, Fall, 1989.

Jimmy King, a Navajo instructor, translated the Marine Hymn.

**http://www.history.navy.mil/faqs/faq61-4.htm**

Bar codes.

**http://www.ean.jedco.gov.jo/products.htm**

Semaphores.

**http://www.anbg.gov.au/flags/semaphore.html**

The following Web site translates words into Morse code sounds.

**www.soton.ac.uk/~scp93ch/morse/trans.html**

## Readings

Hardin, Garrett, "Coding the Mechanism," in *Writing about Science,* Bowen, Mary Elizabeth, and Joseph A. Mazzeo, Eds., Oxford University Press, 1979.

Morowitz, Harold, "The Six Million Dollar Man," in *The Wine of Life,* St. Martin's Press, 1979.

Penzias, Arno, *Ideas and Information,* Simon and Schuster, 1989. (Read Chapters 1 and 2 at least.)

# Information and Coding in Living Systems

*"Biology has at least 50 more interesting years."*

**James D. Watson**

We shall learn about the nervous system in this chapter and how it allows animals to have an awareness of the environment around them. The nervous system also allows an animal to respond to environmental signals and take any necessary action.

One of the greatest, if not the greatest, discovery of the 20th century was deciphering the structure of the DNA molecule by Watson and Crick. It led to a whole new scientific area of inquiry known as molecular genetics and a new area of technology application known as genetic engineering. What sort of promises for the future does genetic engineering give us? What are some of the dangers? The answers to these questions will become clear as we study the content of this chapter.

## Goals

After studying this chapter, you should be able to:

- Understand the structure and function of the living cell.
- Explain how information is coded and transmitted by the nervous system.
- Understand how neurons transmit information.
- Understand how feelings of hunger develop.
- Know what DNA is.
- Know how information is coded in DNA.
- Explain what is meant by genetic engineering.
- Know some of the implications of genetic engineering.

## The Living Cell

Living organisms, be they plants, mushrooms (a type of fungus), birds, fish, and humans are complex systems comprised, at the most basic level of life, of cells. Just as matter is comprised of atoms, the smallest units that retain the properties of an element, living organisms are comprised of cells, the smallest unit that retains the properties of life. Some life forms consist of single cells and nothing more. These are referred to as unicellular organisms and include the prokaryotic (no nucleus) bacteria and blue-green algae, or the eukaryotic (possessing a cellular nucleus) amoeba and paramecia. Complex organisms, such as humans, are made up of many trillions of cells that function together in a coordinated way. These are referred to as multicellular organisms.

A cell is a living system capable of sensing, responding to, and interacting with its environment. The materials that make up the cell are ordered into a highly complex arrangement. In order for the cell to accomplish this high degree of organization it must utilize energy obtained from its environment. Using energy and raw materials from its environment, a cell can build new living matter where none previously existed, a process known as synthesis. It is through this process, stimulated by exercise, that we build new muscle proteins, produce new bone matrix, and synthesize new aerobic enzymes.

A cell is capable of reproducing itself. Through a process known as mitosis a cell is able to produce an exact copy of itself. In a slightly different process, meiosis, sex cells produce copies that are not all exactly alike.

Since the inside of this living system is so different from its surroundings (ordered versus disordered) some physical barrier is necessary to keep the two apart. The barrier must not be absolute, however, since the selective exchange of materials between the two compartments is necessary. For example, oxygen in the environment must be able to enter the cell, and carbon dioxide produced within must be able to leave. The complex regulatory barrier between the intracellular (within the cell) and extracellular (outside of the cell) environments is the **plasma membrane** (Fig. 8.1).

The plasma membrane surrounds all cells and controls the flow of substances into and out of the cell. It is comprised of dual layers of phospholipid molecules, or a phospholipid bilayer, that is impermeable to water, and to atoms, ions, and small molecules that are dissolved in water (these are called hydrophilic molecules; *hydro* = water, *philic* = loving). Hydrophobic molecules, those that are not soluble in water (*hydro* = water, *phobic* = fearing), can pass freely through the cells' plasma membrane. (What implications do you think this might have for those developing new drug compounds that must enter cells easily?) Atoms and molecules that cannot enter the cell because of their polarity (the primary determinant of their hydrophilic or hydrophobic property), or because of their size (large molecules cannot easily diffuse across the cell membrane), can be carried across by active transport pumps, membrane proteins that use cellular energy to carry molecules across.

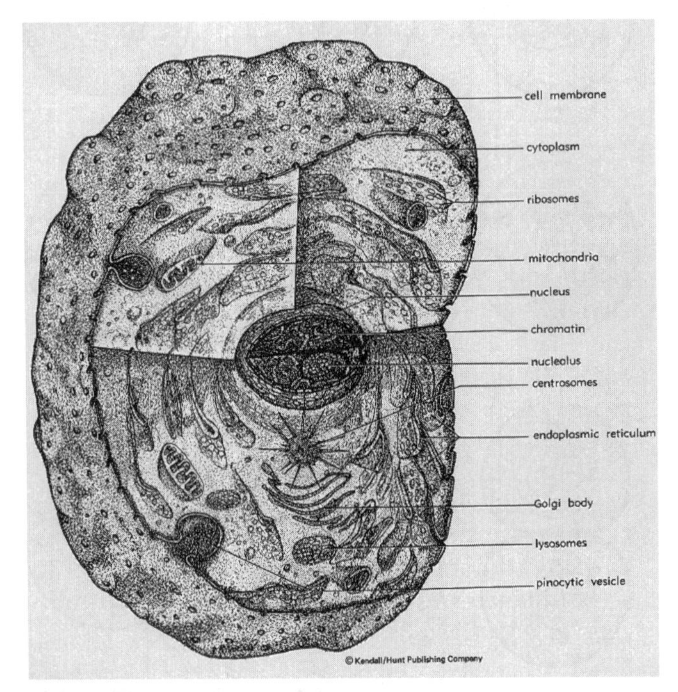


**Figure 8.1** Animal cell.
(Biology plate series, Part I Zoology, Kendall/Hunt, figure 11–1 1974).

The plasma membrane also possesses receptors that detect molecules in the environment and allow the cell to "be aware" of what is going on around it. It is through these receptors that our cells can sense and respond to hormones, neurotransmitters, drugs, and other molecules. Plants, and some bacterial cells, have an additional cell wall surrounding the plasma membrane. The cell wall provides structural support and acts as an additional barrier between the cell and its environment.

Contained within the plasma membrane is a viscous, highly ordered material called **cytoplasm.** The cytoplasm contains many enzymes and substrate molecules necessary for the chemical reactions that sustain life. Suspended within this cytoplasm are numerous organelles or cellular substructures with specific functions. Among these are the **mitochondria,** which produce the energy carrying ATP (adenosine triphosphate), **lysosomes,** which contain digestive enzymes important to cellular feeding and nutrition, **endoplasmic reticulum,** which is the machinery on which new protein and fat molecules are synthesized, and the **nucleus,** the repository of the cells' genetic information (Fig. 8.1).

## Information Coding in the Nervous System

The ability of animals to have an awareness of the environment around them, to react based upon this information, and to take action in response to environmental signals are all functions accomplished by the nervous system. For example, the ability to detect odors and to determine whether those odors represent food, potential mates (via airborne chemicals known as pheromones), or danger (the smell of smoke from a fire, spoiled food, or toxic chemicals) is critical for the survival of the animal and the species.

Information within the nervous system is coded for and transmitted as electrical signals, which are generated by nerve cells when triggered by a stimulus.

The nervous system is composed of cells called neurons, which have the ability to detect and respond to stimuli and to transmit that information to other neurons in the nervous system. Consider this simple example that involves the sense of smell. The sense of smell depends upon small airborne molecules that are given off by objects in the environment. Our ability to smell bread baking in a neighborhood bakery depends, first, upon small molecules of baking bread escaping from the oven on hot air currents (Fig. 8.2) and traveling throughout the neighborhood to reach the olfactory (smell) receptors in our nose (Fig. 8.3). Had the wind been blowing in the opposite direction, we would not be aware of the smell of the baking bread. Within our nasal lining are approximately 350 different odorant receptors, each one capable of detecting one specific odorant molecule. Because most odors consist of complex combinations of molecules (in our example, wheat, yeast, etc.), these 350 receptors are capable of detecting and discriminating among millions of combinations of odor molecules that characterize different smells.

**Figure 8.2** Freshly baked croissants.

Every neuron, whether it is an olfactory receptor, a conducting neuron, or a neuron triggering a muscle response, transmits its information in the same way; as a sequence of electrical signals flowing along the neuron fibers (Fig. 8.4). A single electrical signal pulse, known as an action potential, is generated when sodium ions flow into the cell from the surrounding body fluid causing the cell's interior to momentarily change from its resting state of having a slight negative charge to its excited state of having a slight positive charge. Within milliseconds, another positive ion, potassium, flows out of the nerve cell's interior, reestablishing the internal negativity. After a brief period in which the original sodium and potassium levels are reestablished, the cell is able to conduct another impulse and will do so if the stimulus is still present.

The signal-generating event depicted in Figure 8.4, in which a stimulus generates a localized change in the electrical charge of a neuron, is referred to as an **action potential.** The action potential, once generated, then travels along the neuron's length as a wave of positive charge, reaching the neuron's terminal end as a single bit of information. Coding within the nervous system is, in part, accomplished by modulating the frequency and pattern of these action potentials, or bits of information. For example, a weak stimulus may generate an intermittent stream of only 5,000 action potentials per second, where a

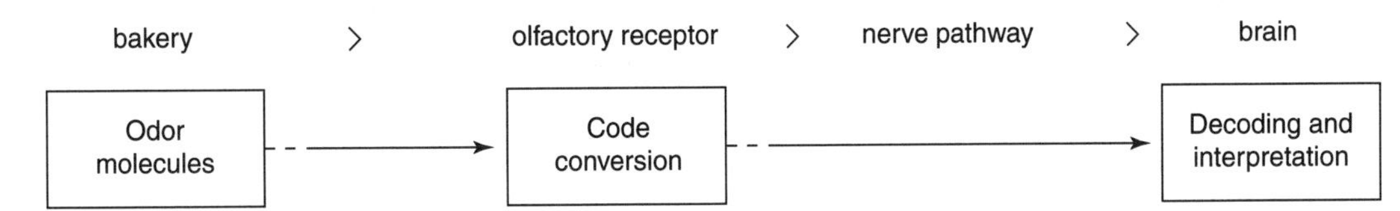


**Figure 8.3** Flow chart for odor propagation and reception.

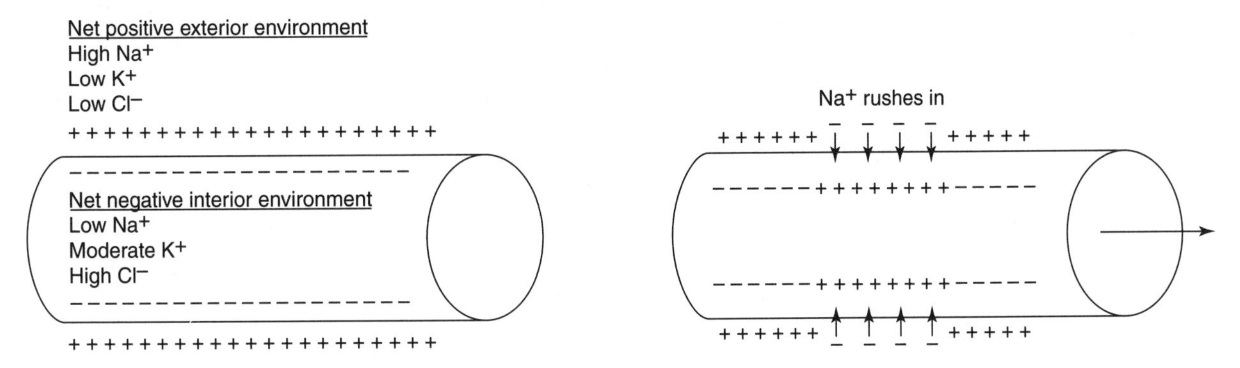


**Figure 8.4** Stimulated neuron.

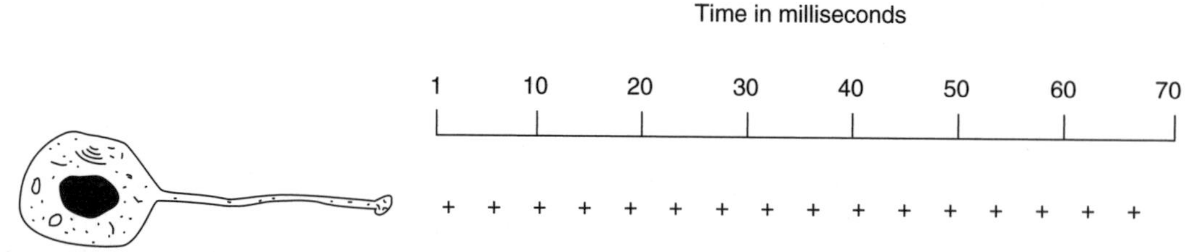


**Figure 8.5** Signal generated by a weak stimulus.

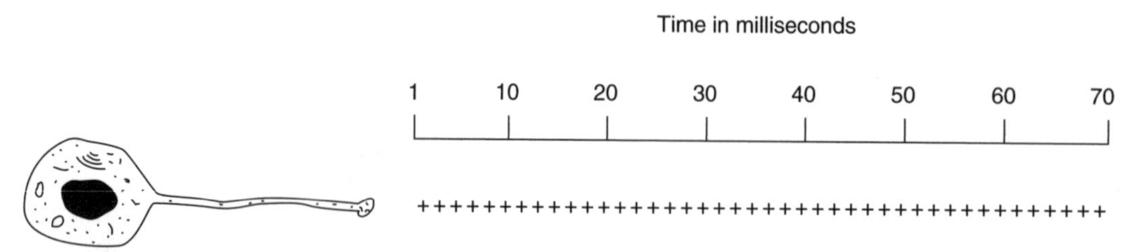


**Figure 8.6** Signal generated by a strong stimulus.

strong stimulus may be coded for by a steady stream of 20,000 action potentials per second. This method of coding, relying on the timing of action potentials, is referred to as **temporal** coding. Illustrations of temporal coding are shown in Figure 8.5 and Figure 8.6.

When the odor of the baking bread, actually a unique combination of drifting molecules, reaches our nasal lining, the molecules bind to complementary-shaped receptors on some of the 350 different odorant receptors. A different smell, say that of frying bacon, would have bound to a different yet characteristic set of receptors. When a molecule of odorant binds to a receptor, it opens channels in the neuron's cell membrane, allowing an inward rush of sodium and the generation of an action potential. Each action potential flows along the nerve fibers ultimately reaching the brain. It is there, in the olfactory area of the brain, that each unique pattern of signals is interpreted as a corresponding unique smell. It is also within the brain that an association is made with the smell of baking bread and the stored memory of the pleasant taste of warm, fresh, baked bread.

The olfactory pathway just described is an example of an excitatory pathway (Fig. 8.7), that is, one in which a neuron triggers an information-carrying action potential in an adjacent neuron (known as an EPSP, or Excitatory Post-Synaptic Potential). In some neurons, the activation of the sensory receptor causes a potassium channel to open first, not a sodium channel. In this case, positive potassium ions rush out of the neuron's interior and move it further away from the generation of an action potential. A nerve pathway that carries

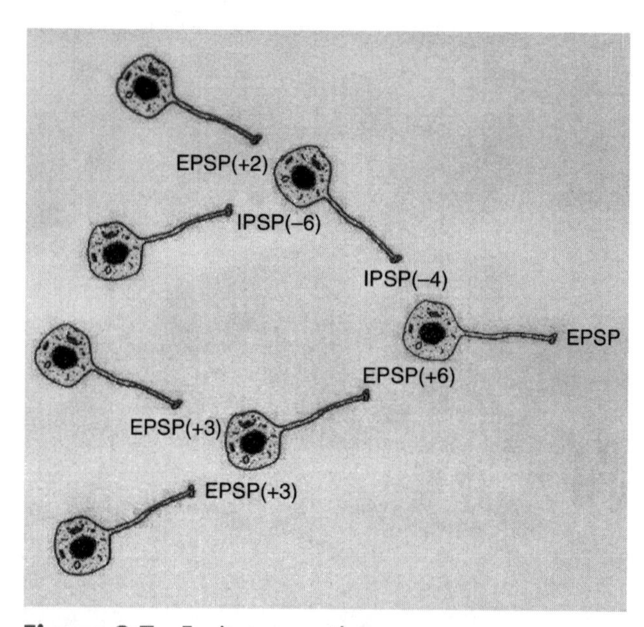


**Figure 8.7** Excitatory pathway.

this type of signal is called an inhibitory pathway (an IPSP, or Inhibitory Post-Synaptic Potential).

At a simplistic level, such a decision-making process may happen as follows: As several hours have passed since the last meal, and blood glucose has been drawn out of the blood by working cells within the body, sensory receptors detect a drop in the blood glucose level from its normal concentration of 90 mg/deciliter to 65 mg/deciliter. This initiates a signal to the hypothalamic region of the brain to generate the perception of hunger and drive a person to seek food. Although this might generate a positive stimulus in the brain center responsible for making the decision to eat, it is not sufficient to generate a signal above the minimal threshold level, so the decision to eat is not yet made.

Let's imagine an overly simplified situation in which the decision to eat requires a threshold signal of at least +10. Assume that an individual in our example is standing in front of a bakery and can both see and smell freshly baked pastries. Olfactory sensors in nasal lining detect the odor molecules and generate excitatory action potentials that help push the decision-making neurons toward threshold, perhaps with a stimulation level of +4, and visual receptors viewing the tasty selection further contribute to this process by adding an additional stimulation of +4.

The summation of signals from these two neurons brings the stimulatory signal to a level of +8, below our threshold of +10. Although we may see and smell food, our brain has not yet reached the decision to eat. If, in our example, our subject has not eaten in several hours and has a low blood glucose level, the glucose receptors may add an additional +4 stimulation, bringing our total summative stimulus to +12 and triggering the decision to eat.

## Information Coding in DNA

The physical characteristics, or traits, of all living things, be they humans, fish, insects, plants, or bacteria, are determined by an instruction set contained within each (Fig. 8.8) that exists in the form of DNA.

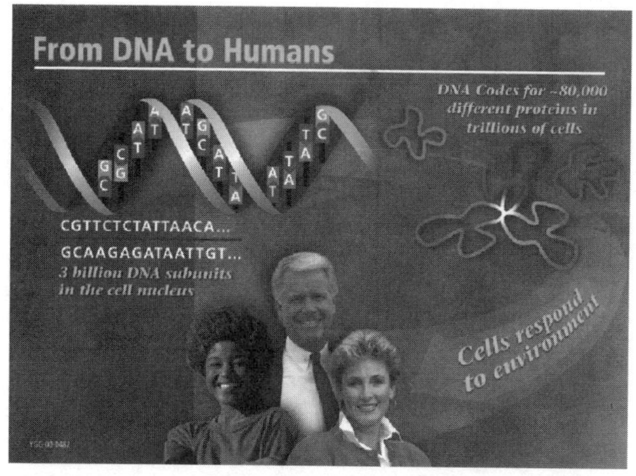


**Figure 8.8** Human Genome Program. *U.S. Department of Energy Human Genome Program, http://www.ornl.gov/hgmis*

If you were to look around your classroom you would notice that, while all of your classmates possessed the common characteristics of human organisms—upright posture, forward-facing eyes, ears positioned on the side of head, four limbs, etc.—there were many noticeable physical differences between and among individuals.

Why is it that all humans (assuming they are "genetically" healthy) are born with four limbs, while insects are born with six and spiders with eight? Why is it that, although all humans have two forward-facing eyes, they vary widely among individuals with respect to color and shape and function (note that some of your classmates wear eyeglasses and others do not)? The answer is both simple and complex. Simply, we look and function in accordance with the way that we are genetically programmed.

The complexity sets in as we try to understand the nature of how this information is encoded and just what the instructions say. This is an area in which science has made significant progress.

Every human cell has coiled within its nucleus 23 pairs of chromosomes. Each chromosome consists of a helical strand of the molecule deoxyribonucleic acid (DNA) wrapped around a protein scaffolding. Contained on the chromosomes within each cell is the complete instruction set to make a human being.

The DNA macromolecule consists of two opposing chains, each containing a sequence of coding molecules known as nucleotide bases. These nucleotide bases, of which there are four varieties, are adenine (A), thymine (T), cytosine (C), and guanine (G). Because of their complementary chemical characteristics, an adenine (A) base appearing on one strand will always align with a thymine (T) on the adjacent strand, and a cytosine (C) base will always pair with an opposing guanine (G) on the adjacent strand.

If this double helix molecule were to be straightened out and arranged end-to-end, the fine, filamentous DNA molecules within a single human cell would measure almost one meter in length. If we could picture this straightened double helix as being like a railroad track, and increased in scale to railroad size, its proportional length would be about 1,000,000 miles.

Arranged along the length of each DNA molecule is an instruction set consisting of approximately 3 billion letters of a four-letter language, or code, comprised of the A, T, C, and G molecules. Of the approximately 3 billion base pairs that make up the human genome, something less than 2% actually contain useful information. The remaining 98% of our DNA consists of noncoding regions commonly referred to as "junk" or "nonsense" DNA. The precise function of this random DNA is unknown. There is evidence that the nucleotide sequence may not be random at all. For example, there are numerous regions of the regularly repeating nucleotide sequences.

If we continue with our ladder analogy (Fig. 8.9), we would note that on each of the railroad ties was

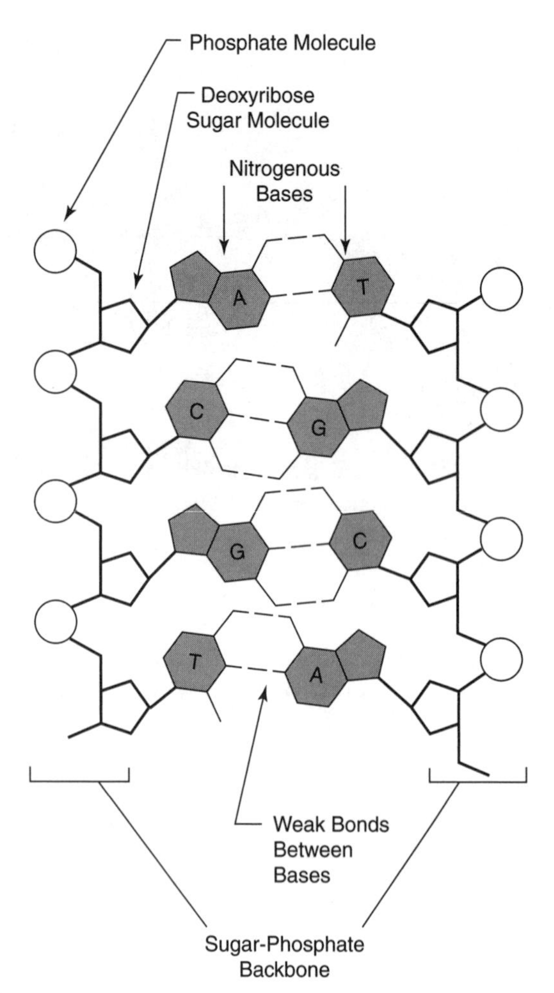


**Figure 8.9** DNA structure. *http://www.ornl.gov/sci/techresources/Human_Genome/graphics/slides/images1.shtml*

stamped a pair of letters, one on the right side of the railroad tie and one on the left.

As we travel along the track, we would be able to read the sequence in which the letters have been arranged and, if we understood the language, would be able to read the information contained by the sequence of letters. Doing so should not be foreign to us: you are doing that right now as you read this text. For example, if we encountered a code comprised of the 11 characters, or letters:

a, d, e, h, i, n, o, r, s, t, w,

and if these characters were arranged exclusively in groupings of three or four, we might find the following arrangements to be possible, each conveying specific information:

-read this now -

- his and her own snow hats -

- she rode that horse -

Although this is a very simple example, we can see how instructions can be contained within such a code, as long as we understand the language. Similarly, the four characters in the genetic code,

A, T, C, G

when arranged in groupings of exclusively three characters, can form the following arrangements, or words, if you will:

| | |
|---|---|
| ATC | GAT |
| ATG | CGA |
| TCG | CAA |
| GCA | etc. |
| TAG | |

In fact, there are 64 possible combinations formed by these four characters when structured into three character groupings. Rather than having to form the words of a complex language, such as English, that contains thousands of different words, the genetic code has to be only complex enough to represent the 20 different amino acids from which our cells build proteins. It is these proteins, both the structural proteins from which we build our physical structure, and the catalytic proteins, or enzymes, which direct our internal chemical processes, that result in our unique genetic traits.

The genetic instructions on the DNA molecules are contained within the cells' nuclei and are physically isolated from the protein producing ribosomes in the cell cytoplasm. We can, perhaps, liken this to the operation of a shoe factory, in which the blueprints, or instructions, for making shoes are stored within a file cabinet in the plant's office (the cell nucleus). With the machinery for actually producing the shoes located out on the shop floor (ribosomes out in the cell cytoplasm), we have to rely on a messenger to carry the instructions from the office to the shop floor. In the cell, this messenger is the RNA molecule, and what it carries is a copy of the instructions, not the instructions themselves. The genetic code is known and the dictionary is presented in Table 8.1.

The code in Table 8.1 is one that can be used to directly translate the sequence of RNA bases, and, therefore, indirectly reflects the DNA code. It describes the specific base sequences that code for each of the 20 amino acids from which proteins are built. For example, the three-base sequence C-A-G would code for the amino acid arginine (ARG), while the sequence C-A-C would code for

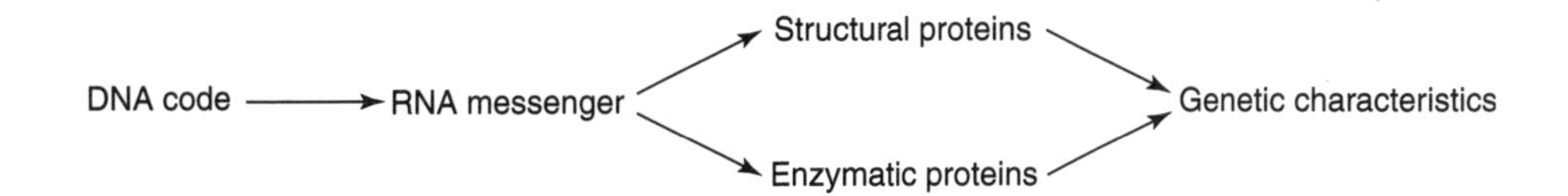


the amino acid Tyrosine (TYR). Therefore, if we encountered the following RNA sequence,

A-U-G-C-U-A-G-A-G-C-A-U-G-C-A-U-A-A

the resulting protein would be:

Methionine(MET)—Leucine(LEU)—
Glutamate(GLU)—Histidine(HIS)—Alanine(ALA)

The final coding sequence, U-A-A, would not be translated into an amino acid added to the protein chain but, rather, would code for protein chain termination (STOP).

Information coded for on the DNA molecule exists in the form of sequential segments called genes. The genes occur sequentially along a continuous strand of which there are 22 complementary pairs. Each DNA strand is linked to other molecules of protein and RNA to form physical structures known as chromosomes. Therefore, each chromosome in a cell contains a single DNA strand on which are located numerous genes, or information coding regions.

Each gene can be thought of as a separate chapter in the instruction book to build a human and contains approximately 3,000 bases. The DNA in each human cell contains about 30,000 genes responsible for regulating every human function and controlling or influencing every human characteristic. Collectively, these 30,000 genes comprise the genome, or the complete instruction set for an organism.

**TABLE 8.1**

Ser
Tyr
tRNA
U C A anticedon
A G U codon
A U G
U A C
V
mRNA y

2nd base in codon

| 1st base in codon | U | C | A | G | 3rd base in codon |
|---|---|---|---|---|---|
| U | Phe | Ser | Tyr | Cys | U |
| | Phe | Ser | Tyr | Cys | C |
| | Leu | Ser | STOP | STOP | A |
| | Leu | Ser | STOP | Trp | G |
| C | Leu | Pro | His | Arg | U |
| | Leu | Pro | His | Arg | C |
| | Leu | Pro | Gln | Arg | A |
| | Leu | Pro | Gln | Arg | G |
| A | Ile | Thr | Asn | Ser | U |
| | Ile | Thr | Asn | Ser | C |
| | Ile | Thr | Lys | Arg | A |
| | Met | Thr | Lys | Arg | G |
| G | Val | Ala | Asp | Gly | U |
| | Val | Ala | Asp | Gly | C |
| | Val | Ala | Glu | Gly | A |
| | Val | Ala | Glu | Gly | G |

The Genetic Code

*http://nai.arc.nasa.gov/library/images/news_stories/before_genetic.gif*

## Human Genome Project

Recognizing the importance of the DNA instruction set to a range of human diseases, the United States government began, in 1990, a 15-year multi-billion dollar project to interpret the entire three-billion base pair code of the human genome. The U.S. National Institutes of Health and the Department of Energy were the lead agencies in the project. The project, known as the Human Genome Project (Fig. 8.10), was intended to determine the precise location and coding sequence of each of the 30,000 human genes, develop and store this information in easily accessible computer databases,

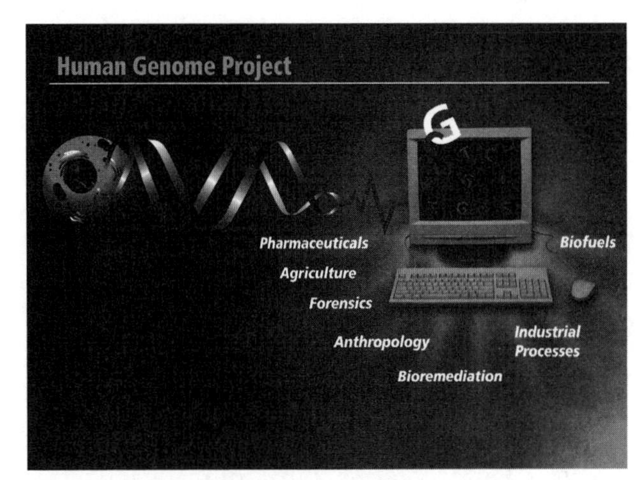


**Figure 8.10** Human Genome Project. *U.S. Department of Energy, http://www.ornl.gov/TechResources/Human_Genome/graphics/slides/images4.html*

**TABLE 8.2 Genes and Associated Diseases**

| Disease | Gene | Chromosome |
|---|---|---|
| Alzheimer's Disease | AD3 | 14 |
| Alzheimer's Disease | AD4 | 1 |
| Breast Cancer | BRCA1 | 17 |
| Breast Cancer | BRCA2 | 13 |
| Colon Cancer | MSH2 | 2 |
| Colon Cancer | MLH1 | 3 |
| Cystic Fibrosis | CFTR | 7 |
| Malignant Melanoma | CDKN2 | 9 |
| Obesity | OBS | 7 |

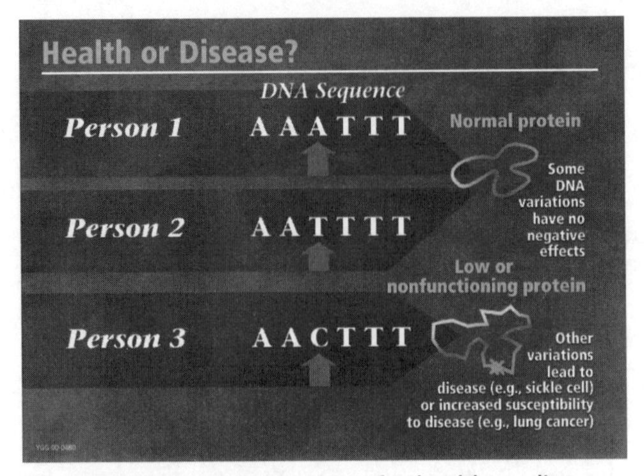


**Figure 8.11** DNA sequence for health or disease. *U.S. Department of Energy Human Genome Program, http://www.ornl.gov/hgmis*

transfer this knowledge and technology to the private sector, such as pharmaceutical companies, so that they may lead to advances in disease prevention, diagnosis and cure, and to address the complex ethical issues that arise from this technology. In November of 1999, the one-billionth base pair was sequenced and on December 1 of that year, almost 10 years into the project, sequencing of the first full human chromosome, chromosome # 22, was completed. The completion of four more chromosomes followed the next year (# 5, 16, 19, 21) and in 2003 the project was finished.

With the complete nucleotide base sequence known for every gene, and utilizing the genetic code dictionary described earlier, scientists can now determine the specific protein coded for by each gene and can determine the biological function of each protein. For those proteins associated with disease processes, a further step will be the development of ways to regulate the activity of the gene or the action of the protein. The full benefits from this project are still decades away.

The genes for a number of diseases have now been identified and sequenced, including those identified in Table 8.2.

## Single Nucleotide Polymorphisms (SNPs) and the HapMap Project

With the knowledge gained through the Human Genome Project, we now have a complete map of the human genome that includes the specific nucleotide sequences for every gene. What has been discovered in the process is that for many individuals, there exist minor variations in these nucleotide base sequences, differences that often amount to only one nucleotide substitution per several hundred bases, the most common being the substitution of a single cytosine (C) for a thymine (T). When such a variation occurs, it is referred to as a **single nucleotide polymorphism,** the term *polymorphism* derived from the Latin for "other forms."

While these SNPs are not substantial enough to have major effects on the gene's action, they do confer minor changes that influence both disease susceptibility and the individual's response to specific drugs. Because of this it may be possible through genetic screening technologies to determine an individual's likelihood of developing certain diseases as well as designing specific therapies that would work best for their specific haplotype. A haplotype is a set of closely linked genes that tend to be inherited together as a unit.

## Commercial Applications

There are about 1.4 million positions along the human genome where such SNPs have been found, accounting for much of the individual variations among people.

With the knowledge that we have gained regarding the genetic code there have been numerous technological innovations, some of which have deep social consequences. On the simple side, we now have the ability to insert into microscopic bacterial cells the genetic coding sequences that direct the production of important human proteins such as the hormone insulin. By inserting the insulin-making instructions into the bacterial cell, it will now go about the business of synthesizing and releasing human insulin. If we cultivate these bacteria in the proper nutrient environment, they will multiply rapidly with all of the offspring carrying the genetic instructions to produce human insulin.

Given the ease and the rate of bacterial growth under the proper conditions, pharmaceutical producers can easily raise hundreds of billions of bacterial cells in large vats, which will produce human insulin in fairly large quantities. Whereas in years past people suffering from insulin deficiency disease, diabetes, were treated with extracted insulin from animals, today they can be treated with the more pure, and more effective, human insulin produced by genetically altered bacteria.

We also have the ability to genetically manipulate larger animals. In 1980s, the journal *Science* reported the development of a strain of mouse that had implanted into it the growth hormone gene from a rat.

As expected, the mouse grew to the size of rat. Subsequently, food producers attempted to insert the growth hormone gene of a cow into a pig, hoping to develop a line of pork-producing pigs that grew as large as cows. Interestingly, they were successful. However, the pigs' bodies grew to such large and heavy sizes that the pigs could not support the weight on their legs, which eventually collapsed under the weight of their bodies.

A patent application has been filed in Europe for the production of chickens that contain cow growth hormone genes. Are such modifications justified? Are they ethical? These are the types of questions that new technologies often create.

## Review Questions

1. What is a neuron?
2. What is a pheromone?
3. What is an action potential?
4. How does coding take place in the nervous system?
5. What is temporal coding?
6. What are inhibitory pathways?
7. What is DNA?
8. Name the nucleotide bases.
9. What are ribosomes?
10. What is the Human Genome Project?
11. What are genomes?
12. What are genetic characteristics?
13. What is genetic engineering?

## Multiple Choice Questions

1. The nervous system is composed of cells called _____, which have the ability to detect and respond to stimuli and transmit information.
   a. photons
   b. klingons
   c. neurons
   d. phonons
2. A single electrical impulse generated when sodium ions flow into the cell from surrounding body fluid is known as a(n) _____.
   a. neuron
   b. olfactory receptor
   c. action potential
   d. shock
3. The traits and characteristics of all living things are determined by an instruction set in each called _____.
   a. RNA
   b. DNA
   c. FDA
   d. enzymes
4. Adenine, thymine, cytosine, and guanine are called _____.
   a. neurons
   b. olfactory receptors
   c. pheromones
   d. nucleotide bases
5. The Human Genome Project was intended to determine the precise location and _____ sequence of each of the 30,000 human genes.
   a. information
   b. feedback
   c. output
   d. coding
6. The Human Genome Project allowed scientists to determine the specific _____ coded for each gene.
   a. protein
   b. cell
   c. chromosome
   d. nucleus
7. Genetic engineering is performed by modifying an organism's own _____, introducing new _____ to perform desired functions.
   a. cells
   b. RNA
   c. DNA
   d. neurons
8. The sequence of three nitrogen bases in RNA is called a(n) _____.
   a. codon
   b. anticodon
   c. polymer backbone
   d. genome

9. The sequence of the three nitogen bases "uracil cytosine uracil" in RNA codes for the protein _____.
   a. alanine
   b. leucine
   c. histidine
   d. serine

10. If a codon contained a sequence of four nitrogen bases rather than three, one would be able to code for how many possibilities?
    a. 4
    b. 256
    c. 64
    d. 128

11. There are how many amino acids essential for living?
    a. 10
    b. 15
    c. 20
    d. 64

12. Proteins are composed of _____.
    a. DNA
    b. RNA
    c. deoxyribose
    d. amino acids

13. Altering the sequence in which amino acids are attached _____.
    a. will have no effect
    b. will result in the formation of a different substance
    c. cannot occur
    d. is typical in one cell plants

14. A gene is _____.
    a. a chromosome
    b. a section of DNA that codes for a protein
    c. RNA
    d. None of the above.

15. The Human Genome Project is _____.
    a. make work activity.
    b. central to the study of dwarfs
    c. of no practical value.
    d. a project to interpret the entire base pair code of humans

16. The approximate number of genes responsible for regulation of every human function and characteristic is _____.
    a. 3,000
    b. 10,000
    c. 30,000
    d. 1,000,000

17. Genetic diseases are caused by _____.
    a. errors in placement (sequence) of nitrogen bases in DNA
    b. exposure to bacteria
    c. consumption of synthetic food additives
    d. a diet low in fiber

18. The basic coding in cells is given by _____.
    a. a molecular sequence of four different acids arranged in a double helix
    b. a sequence of six different acids arranged in a double helix
    c. a single helix of four different molecules
    d. a chain of six different molecules

## Bibliography and Web Resources

The Nervous System. Subdivided into the sympathetic and parasympathetic systems. Description: An overview of nerve systems, impulse transmission, neurotransmitters, the brain, etc.
**www.emc.maricopa.edu/faculty/farabee/BIOBK/BioBookNERV.html**

GreenpeaceUSA—Genetic Engineering: "Go Organic on GE Food."
**www.greenpeaceusa.org/index.fpl?article=76&object_id=6986**

Genetic Engineering: "Defining Our Children's Traits in Genetic Engineering" (1985 and 1989 versions). Description: Considerations and arguments on several scientific procedures that could improve human life.
**www.jpreason.com/science/gene.htm**

Genetic Engineering (FDA Consumer Reprint). Genetic Engineering: "Fast Forwarding To Future Foods" by John Henkel.
**www.fda.gov/bbs/topics/consumer/geneng.html**

# Energy

*"$E = mc^2$."*

**Albert Einstein**

Energy is involved in all the processes that maintain life and civilization. A major thrust of technology has been to provide more energy and different kinds of energy. With more people alive on the planet than at any other time in the earth's history, the need for food energy and energy to carry on agricultural efforts is enormous. In the past two hundred years there has been an explosive growth in world trade, travel, and communications that has been energy intensive.

In this chapter we shall explore the *energy code,* what some deem to be the premier physical resource in the universe. Recent imbalances between the supply and demand for this valuable resource have given rise to the expression "energy crisis."

## Goals

After studying this chapter, you should be able to:

- Use the appropriate units when making measurements.
- Know the difference between scalars and vectors and how they are used.
- Know what force, work, and energy are and be able to distinguish among them.
- Know what kinetic and potential energy are.
- Describe the beginning of the universe.
- Understand the first law of thermodynamics and its role.
- Understand energy conversions and efficiencies.
- Understand the second law of thermodynamics and limitations it imposes on energy conversions.

## Standards and Units

Biology, physics, and chemistry are experimental sciences and are based on measurement. Scientists make observations on nature around them and look for patterns and characteristics that relate them. In order to do this, they make measurements and describe their results using numbers. How much do you weigh? How tall is the largest building in the world? How far away is the sun? What is the circumference of the earth?

In order to define certain quantities we define them by describing how to measure them (Fig. 9.1). Such a definition is called an operational definition. Measurement involves comparing the physical attribute of an object against some standard. To say someone is 2 meters tall, we are saying that that person is 2.0 times the length of a meter stick. The meter stick is the standard. To say someone is 2.0 (no unit) tall is inaccurate and communicates no meaningful information. Scientists and engineers

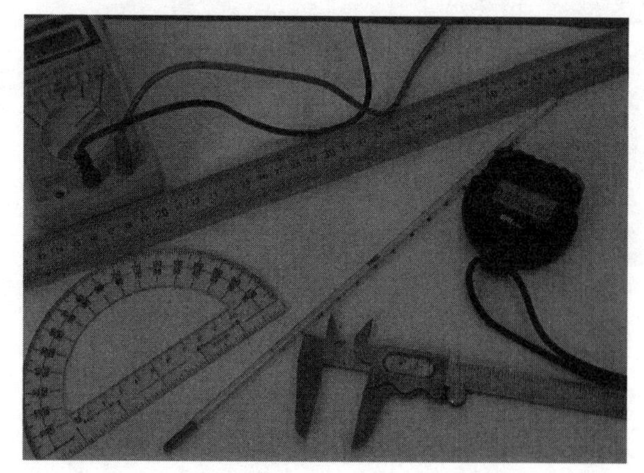

**Figure 9.1** Measuring tools.

around the world have agreed upon a system of units that they all use for measurement and it is called the International System (SI).

In the SI system of units, four of the fundamental units are defined operationally by telling how to measure them: length, mass, time, and current. It is useful for us at this time to know the fundamental units of length, mass, and time in the SI system and also in the British system of units. Length is defined as the meter (m), which is the distance light travels in a vacuum in 1/299,792,458 second. Time is defined in terms of the second (s), which is the time required for 9,192,631,770 cycles of radiation from a cesium atom. The standard unit of mass is the kilogram (kg), which is defined to be the mass a cylinder of platinum-iridium alloy that is locked in a vault in France.

The British system of units, interestingly enough, is no longer used by the British but only by the United States and a few other small countries in the world. The British units are officially defined in terms of SI units today. The unit of length is the inch (abbreviated as in), which is defined to be 2.54 centimeters (cm) and the pound, which is 4.448 Newtons (N) (which is the SI unit of force). The second or unit of time is defined the same way as in the SI system.

Other units that scientists use are combinations of the fundamental units and are usually referred to as derived units. A car moves at 4 meters/second. The speed is therefore measured in units of m/s. The metric unit of force the Newton (N) is the name given to the combination of units kg m/s$^2$. Prefixes are used to multiply the fundamental units by powers of 10.

Length

1 centimeter = 1 cm = $10^{-2}$ m

1 millimeter = 1 mm = $10^{-3}$ m

1 kilometer = 1 km = $10^{3}$ m

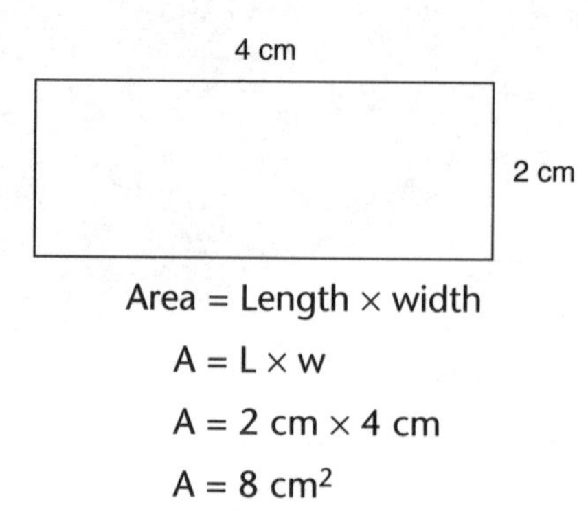


**Figure 9.2** Rectangular area.

Mass

1 milligram = 1 mg = $10^{-3}$ g = $10^{-6}$ kg

1 gram = 1 g = $10^{-3}$ kg

Time

1 millisecond = 1 ms = $10^{-3}$ s

1 microsecond = 1 μs = $10^{-6}$ s

It is important to note that the data recorded by scientists usually contains a number and a unit, and when performing arithmetic or algebraic operations using these quantities, the units must be consistent. For example, you cannot add a height and a weight. The units are different. When calculating with numbers and units the calculations should carry through the algebra of the units at each stage of the calculation. To find the area of a rectangular sheet 2 cm by 4 cm, one would use the formula shown in Figure 9.2.

# Force

Physical quantities in classical mechanics are divided into two subsets: scalars and vectors. Scalars are physical quantities completely specified by a magnitude (number) and, of course, an associated unit. The mass of an object given as 5 kg would be a scalar. The speed of a car at 88 km/h would be a scalar, as would the time interval to travel from New York to Boston.

Vector quantities are specified by a magnitude and a direction. Stating that a force is acting requires one to know in what direction. Up, down? It makes a difference. A plane approaches an airport at 400 mi/h. The pilot must also radio the direction to receive proper landing instructions. When drawing a diagram of vectors that are present, they are usually represented by directed line segments (arrows). When writing vectors, **bold face** is used on a letter of the alphabet to represent a vector such as **a, b C, D.** On the black-

**Figure 9.3** Sir Isaac Newton. *http://antwrp.gsfc.nasa.gov/apod/ap960707.html*

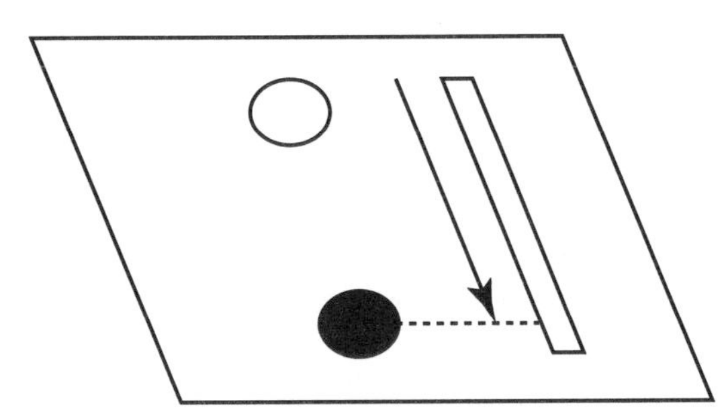

**Figure 9.4** Projection of a hockey puck.

board an arrow is placed over the letter designating a vector **r, s, t** ($\vec{r}$, $\vec{s}$, $\vec{t}$).

To have a deeper understanding of the definition of energy, which is what this chapter focuses on, we have to know how force is defined. Yes, most of us have an intuitive idea of a force as being the nature of a push or pull that can be exerted on an object. Sir Isaac Newton (Fig. 9.3) in the 17th century made a major contribution to science by summarizing the work of scientists before him in terms of three laws that became the foundation of classical mechanics.

Newton's first law may be stated simply in that a body left to itself maintains its velocity unchanged. Velocity is a vector quantity and therefore has magnitude and direction. It is easy to imagine a body at rest. Newton says it will remain at rest for all eternity, if left alone. Also one can imagine a hockey puck projected along a frozen but glazed lake. Yes, friction will eventually slow it down but suppose there was no friction. Then it would continue in a straight line if it is left alone and nothing acts upon it. It is moving in a straight line with a constant vector velocity (Fig. 9.4). Newton's first law is sometimes called the law of inertia.

Newton's second law tells us what happens if a force acts upon a stationary body or a body

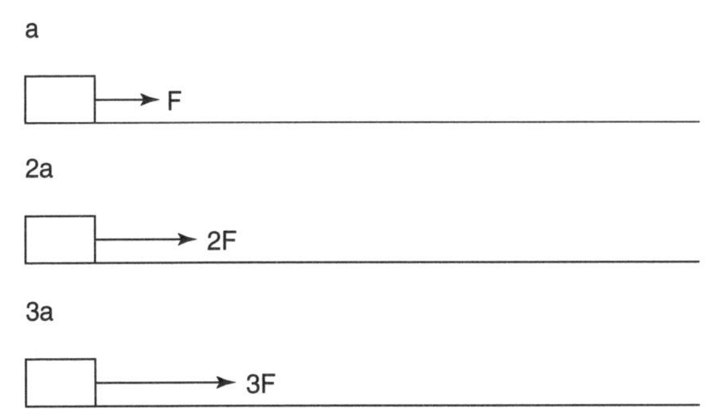


**Figure 9.5** Acceleration proportional to force.

moving with constant vector velocity. We either change its direction or speed or both. Doing any of these acts produces a change in velocity in a given time interval. This is exactly what happens when you change speed or direction while driving your car and you step on the accelerator pedal. You accelerate the body or produce a change in velocity. In order to produce this change in velocity, we would have had to exert a force on the body. Newton's law tells us that the direction of the acceleration will be in the direction of the applied forces exerted on the body (Fig. 9.5).

**F** is proportional to **a.**

In mathematics we proceed from a proportional relationship to equality by inserting a proportionality constant. In this case we shall use m, which represents the mass of the body. It is the number associated with the inertia of the body, which is a measure of its resistance when we try to change its state of motion. It is associated with the inherent amount of matter in the body. Thus,

**F** = m**a**, or Newton's second law of motion.

Note that the units for force are a combination of the units for mass and acceleration: mass = kg, **a** = $m/s^2$, or **F** = kg $m/s^2$, which is defined as the Newton, a derived unit. So in the SI system we measure forces in Newtons.

Newton's third law, or as it is sometimes called the law of action and reaction, tells us for every action there is an equal and opposite reaction. Picture a book sitting on a table. Gravity acts on the book. If that were the only force, then according to Newton's second law, the book would accelerate downwards in the direction of the gravitational force. However it doesn't. There must be a counterbalancing force. The table exerts this counterbalancing or reaction force on the book upwards (Fig. 9.6).

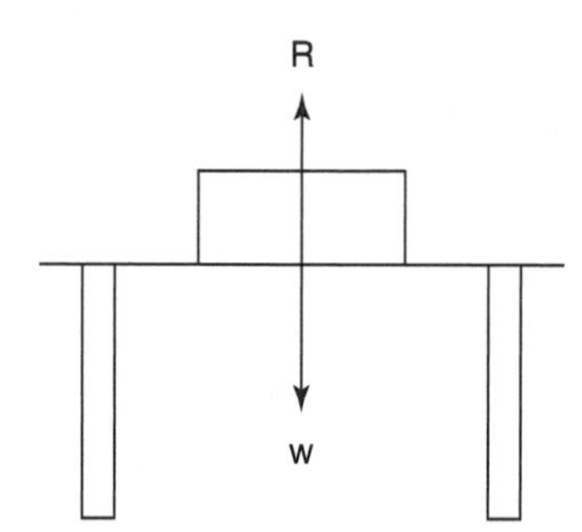


**Figure 9.6** Book on table.

**Figure 9.7** Steam-engine-powered locomotive. *http://www.blm.gov/education/00_resources/articles/steel_rails_and_iron_horses/classroom.html*

# Energy

Before we can talk about energy we have to define a number of terms, such as *work* and *power,* to fully understand what energy is.

## Work

We are all familiar with the concept of work and associate it with time-consuming efforts to accomplish different tasks. Most of us have to go to work every day—wake up, eat breakfast, and commute to a place where we will work all day with others. In the early 1800s most people at the time after awakening, feeding and caring for the farm animals, and eating breakfast could then sit on the front porch and watch the grass grow. Then, in the late 1700s James Watt designed a workable steam engine (Fig. 9.7). This new technological device powered a revolution, the Industrial Revolution, which resulted in the fact that most of us now had to travel to work.

The steam engine was large, required a large amount of capital to build, and had to be housed in its own building, the factory. It was realized that heat from the burning of wood or coal could be applied as an input to a system that was capable of producing useful work as an output. Since then, most people over the last 150 years have had to commute to work every day. To make commuting easier, steam engines were mounted on wheels and used to pull trains to take us to work.

**Figure 9.8** Inputs and outputs of a heat engine. *http://www.phm.gov.au/exhibits/exib_perm/boult.htm*

It is only most recently that a new technology, the microcomputer, has allowed many to stay home once more and not to have to go to work. It is a device that has led to another revolution in the work place, the Information Technology Revolution. Now one may write a book in the woods of New Hampshire and send it to the publisher on Madison Avenue in New York City electronically without having to travel to work.

The meteorologist can receive all the information needed for the weather forecast at home and broadcast the forecast on local radio using telephone lines. The microcomputer once more has allowed numerous people to avoid commuting to work.

The recognition of energy as an important concept took place during the Industrial Revolution. Useful work could be obtained as output from a steam engine. The steam engine was a type of heat engine taking in heat from the combustion of fuel and putting out work in the form of a rotation of a shaft or the strokes of a rocker arm attached to a piston (Fig. 9.8). Heat engines yielded work for the industries of the 19th century from heat input obtained by burning fuel. Not all of the heat energy is converted to work by an engine. Some fraction of the input appears as output in the form of heat leaks or exhaust and is wasted. The most important insight into the nature of heat engines and thermodynamics revealed that no amount of energy appeared to vanish between the input of the engine and the output. Whatever was not delivered as useful work was discharged as some amount of waste

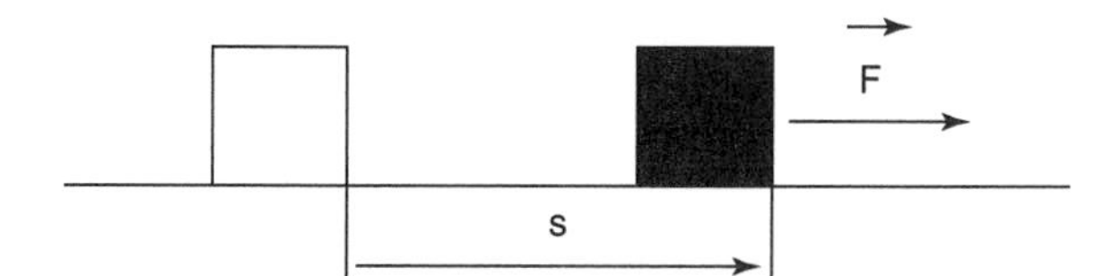


**Figure 9.9** Work done by a constant force.

**Figure 9.10** James Prescott Joule (1818–1889).
*http://www.eia.doe.gov/kids/pioneers.html#Joule*

heat. A heat engine does not create or annihilate energy, it merely converts some of the energy in an available form to work, a form useful to propel the machines of industry. Heat engines such as steam engines are energy conversion systems.

Although we use the word *work* everyday, in science it has a special meaning. It has a narrow definition when the force is constant. Work then requires a constant force to act and move an object through a displacement parallel to its line of action, as in Fig. 9.9. Pushing on a wall all day will certainly make an individual feel tired but would result in no work being done by the pushing force as long as the wall didn't move.

$$\text{Work} = \text{Force} \times \text{Displacement}$$

This formula will give us the work done by a force acting in the direction of the displacement (Fig. 9.9). If the force acts at an angle with the displacement, then we can still have work done by using the part or component of the force that acts in the direction of the displacement and multiplying it times the magnitude of the displacement (i.e., Work = $F\cos\theta \times s$). When the force is variable such as that needed to stretch a spring, calculus must be used to calculate the work done (i.e., Work = $\int F\cos\theta\, ds$ from $s_1$ to $s_2$).

As you know, science is based on measurement and we associate units with the quantities we measure with our scales, rulers, and watches. In the metric system the unit of force is called the Newton (N) and the unit associated with displacement is the meter (m). So the unit of work is the Newton meter (Nm). This combination of units or derived unit is given another name, the joule (J).

$$1 \text{ Joule} = 1 \text{ Newton meter}$$

International groups on weights and measures that assign names to the units we use have been using memorialize scientists like Newton and Joule (Fig. 9.10) by naming the derived units after them so that their names will not be lost in history.

In the English or American system, force is measured in pounds (lb) and displacement in feet (ft). So work would be measured in ft lb. It does not have a derived name of its own. Slowly but surely the English or American system of units is being phased out around the world.

## Power

If one were to enter an elevator and press the button for the 20th floor, the elevator motor would do work in that a force would be generated and the elevator would be moved from the first to the 20th floor. One elevator might take a minute to do it while another could take three minutes. Which one does more work? According to our definition, they both do the same amount of work but yet one does it faster than the other. There is something different. To measure this difference, the rate at which work is done is defined as power. Power is defined as the work per unit time.

$$\text{Power} = \text{Work/Time}$$

The unit for power in the metric system would be the Joule/second (J/s), which is given the derived name Watt (W). So, 1 J/s = 1 W. In the British or American system the unit for power would be the ft lb/s. Historically, when steam engines were first used their power output was compared to the work done by horses and the unit horsepower (hp) was used for power measurement. One horsepower turns out to be equivalent to 550 ft lb/s (Fig. 9.11).

Consider the power required to move an elevator weighing 100 N through 20 meters in 5 seconds.

$$P = W/t$$
$$P = (100\text{N} \times 20\text{m})/5\text{s}$$
$$P = 400\text{J/s or } 400\text{W}$$

Now we are ready to define what we mean by energy. Energy by definition is the capacity or ability to perform work. We say something has energy if it can do work. If a system has no energy, it cannot do any work.

A speeding bullet has the capacity to do work. When it hits something, it generates a force that is

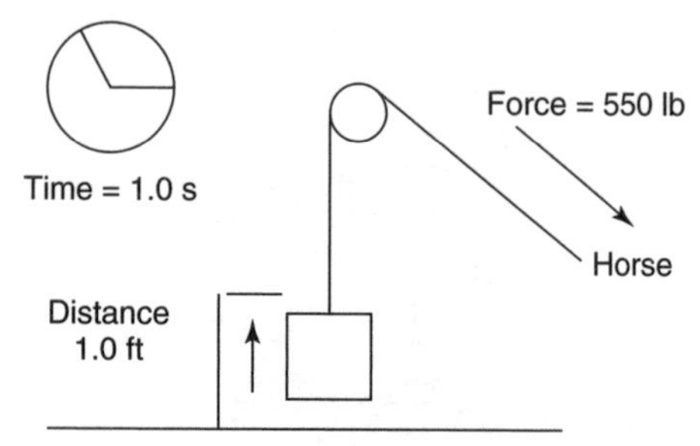


**Figure 9.11** Definition of horse power.

capable of moving something through a displacement. So we see it possesses energy in a motional form. A compressed spring in a pellet gun when released will exert a force on a pellet moving it down the length of the barrel. So it too can do work. In this case the energy is stored in an elastic form in the compressed spring.

Energy is found in all systems in nature and normally divided into two broad categories: kinetic energy and potential energy. Kinetic energy is defined as energy a body or system possesses by virtue of its motion. Heat energy and energy carried by wave motions is included in this group.

Potential energy is defined as energy a body possesses by virtue of its position or configuration and includes energy stored in elastic, chemical, electrical, magnetic, and gravitational forms.

## Kinetic Energy

The work done by the expanding gas or spring in the barrel of a gun on the pellet goes into giving the pellet a velocity. The energy of the moving pellet, which is moving with velocity **v** after leaving the gun barrel, has resulted from the work done on it, as shown in Figure 9.12. It is measured in terms of a formula:

$$KE = 1/2\ mv^2$$

Notice that the units of mass are kg and units for velocity squared are $m^2/s^2$. We can rearrange them into (kg $m/s^2$)m or N m, which are Joules or energy units.

## Potential Energy

Potential energy of an object stored in gravitational form can be measured by asking about the amount of work the object *could* do because of its position. For example, how much work could be obtained from a 30 N book sitting on a table 2 m above the floor? Well, the amount of work done in raising the book from the floor to the table is

$$W = Fd$$
$$W = 30\ N \times 2\ m = 60\ J$$

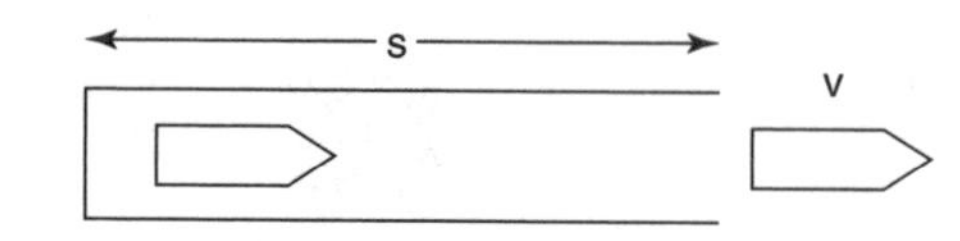


**Figure 9.12** Moving bullet.

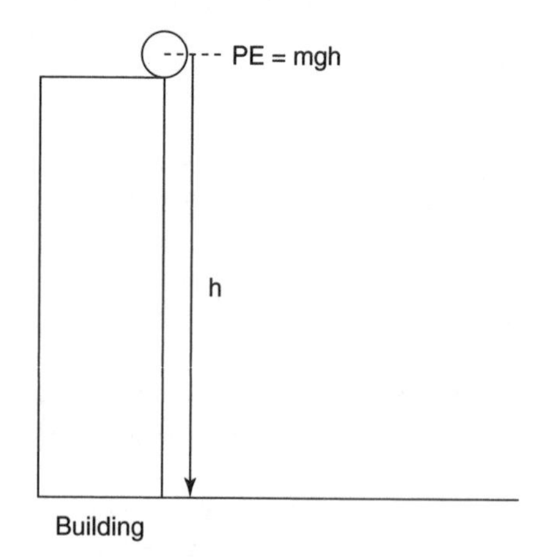


**Figure 9.13** Potential energy of a ball.

This is the potential energy of the book as it sits on the table. This is the amount of work that *would* be done on the book by gravity if it falls off the table. It transforms itself into kinetic energy, which can now do work on an object below.

# Energy Forms

In physics there are four known force fields that give rise to different forms of energy. They are: gravitational, electromagnetic, strong nuclear, and weak nuclear force fields. The different forms of energy that result from these forces would be mechanical, elastic waves, thermal, chemical, electrical, electromagnetic, and nuclear.

**Mechanical energy** is possessed by a piece of matter because of its motion (kinetic energy). Examples would be speeding cars or falling rocks. Mechanical energy is also possessed by a piece of matter because of its position or configuration (potential energy) (Fig. 9.13).

**Elastic waves** are a form of mechanical energy where the kinetic and potential energy is traveling as a disturbance through matter without carrying the material along with it. Examples would be water waves and earth tremors. Waves are sufficiently different from other mechanical forms of energy to have their own listing.

**Thermal energy** is a form of energy that is internal to matter. The kinetic and potential components are contained as internal energy at the level of individual atoms and molecules that make up the object. The essential characteristics of this form of energy are that it is submicroscopically distributed and the distribution is random and disorderly. Examples of thermal energy are the internal energy of steam and of hot pizza.

**Chemical energy** is stored in the bonds that join atoms in combinations in a substance. Energy, which has been added to the reactants in a chemical reaction, can be stored in the internal structure of the molecules, which are formed as products. Fossil fuels such as coal or oil have this type of energy. A battery stores energy in chemical form ready to make it available as electricity when called upon.

**Electrical energy** occurs when charged particles are caused to move. They carry with them energy associated with their electrical nature as well as ordinary kinetic energy. This is why we can transfer considerable amounts of energy by conducting electrons around the wires of an electrical circuit even though the mass and therefore the kinetic energy of the particles is miniscule. The energy, which lights your lights, toasts your toast, and powers your TV, has been transferred by and transformed from the electrical form.

**Electromagnetic energy** is a traveling waveform of energy consisting of both electrical and magnetic quantities. Such disturbances can travel through many materials and also the emptiness of vacuum in outer space. There is a whole spectrum of electromagnetic waves. They can be distinguished by properties known as frequency or wavelength and range from relatively low-frequency radio waves through microwaves, infrared, visible light, and ultraviolet, to high-frequency X-rays and gamma rays.

**Nuclear energy** is derived from the tiny but massive and ultradense nucleus of an atom. The nucleus is composed of protons and neutrons and rearrangements in its structure can store or release energy in a manner similar to the way in which chemical energy is stored or released when atomic or molecular bonds change. The energy is much greater in the case of a nucleus. Measurable amounts of mass appear or vanish whenever nuclear energy is stored or released and is governed by Einstein's famous equation $E = mc^2$.

## Origins of Energy

In the beginning there was energy. The picture of the origin of the universe that currently has the widest acceptance is that at time zero, the universe consisted of a hot, dense, chunk of matter. In this chunk was concentrated all the matter and energy of the universe. Such a concentration of mass and energy was highly unstable and existed for only a few moments before erupting in a cosmic "big bang."

In the beginning energy was concentrated in this dense chunk in its two major forms: kinetic energy and potential energy. Kinetic energy is associated with motion. The kinetic energy was concentrated in the radiant energy emitted and absorbed by particles of matter and in the energy of motion of the particles. Potential energy or stored energy was localized in the mass of the particles in the chunk. Because of the high density of the initial chunk of matter, it is believed that the constituent particles were neutrons.

Normally matter is made up of atoms. An approximate way of thinking of an atom is to visualize it as a miniature solar system. Instead of a sun at the center, there is a positively charged nucleus circled by orbiting negatively charged electrons. The two main constituents of the positively charged nucleus are positively charged protons and neutral particles of about the same size called neutrons. The electrons in orbit are about 1,800 times smaller than the protons or neutrons.

A few moments after the explosion, matter was flying in all directions. As the temperature dropped and the space between the neutrons increased, many neutrons were changed into protons and electrons, forming hydrogen, the first atom. A hydrogen atom consists essentially of a single proton circled by an electron.

## First Law of Thermodynamics

It is well known by scientists that in such a closed system energy can neither be created nor destroyed, only changed from one form to another. The law is known as the conservation of energy and also as the first law of thermodynamics. Einstein assures us that mass is equivalent to energy in his famous equation, $E = mc^2$, where E represents energy, m represents mass, and c is the velocity of light. So the mass present in this primordial chunk represented stored energy. As far as can be known, a dense chunk of matter and energy that existed for a few moments more than 10 billion years ago was such a closed system. It is from this primeval fireball and cosmic "big bang" that all our energy flows.

An interesting consequence of the law of conservation of energy is that humans will never "use up" all available energy. However, in converting it from one form into another, not all the energy present can be utilized. Some is degraded in the process of conversion and changed into a form where it is unavailable for use. It is also a well-known fact that all mass particles in the universe attract one another. In fact, this attractive force would tend to slow down particles expanding from the explosion and perhaps ultimately cause the universe to contract according to one theory. The consequence would be a contraction into a single lump of mass, another cosmic "big bang," and another beginning.

## A Star Is Born

Early on, following the "big bang" tens of millions of years later, the hydrogen gas cloud was still expanding. Fluctuations in the density of the cloud gave rise to local condensations. As the density increased within one of the local condensations, it attracted together most of the gas atoms in the immediate neighborhood. Slowly shrinking in size, it became more spherical under the forces of gravity. It also began to rotate, leading to a flattened disk-like shape with a dense center, trailing great spiral arms forming one of the first galaxies (Fig. 9.14). New eddies within the pinwheel-like cloud formed.

The kinetic energy of the falling hydrogen within an eddy resulted in a large number of collisions, thus increasing the random motion or heat energy of the atoms. Soon the temperature of the gas reached about 20,000,000 degrees Fahrenheit and the hydrogen began to "burn" in a thermonuclear reaction (which is explained later). A star was born.

The thermonuclear reaction had its origin in the gravitational potential energy that was converted to heat energy. The atoms of gas lost their

**Figure 9.14** Spiral galaxy.
(Debra Meloy Elmegreen [Vassar College] et al., and the Hubble Heritage Team (AURA/ STScI/ NASA).

electrons. Protons came close enough together to react and fuse into larger nuclei. A newly formed nucleus had a mass slightly less than the combined masses of the original nuclei. The missing mass according to Einstein's formula, $E = mc^2$, appeared as energy in the star. The new source of energy, fusion, resulted in an increase in internal pressure, thus balancing out the gravitational contracting forces. Our sun is such a star.

Much of the energy released in the "big bang" was converted to gravitational potential energy. Some was left in the form of kinetic energy of motion of the dispersing atoms. Some of the gravitational potential energy was in turn converted to kinetic energy in forming the stars and this in turn was eventually given up to the random motion of heat energy.

When the thermonuclear, or fusion, reactions began, a conversion of nuclear potential energy into radiant and heat energy took place. Every second, the sun converts more than four million tons of matter into energy and should continue to do so for many billions of years. The earth, 93 million miles away, receives about 20,000 calories of energy per minute on each square meter of area. This energy is then stored or radiated.

# Energy Sources

Of the total energy reaching the top of the atmosphere from space, much is converted to heat energy by absorption in the atmosphere and by the earth's surface. A small fraction is converted into chemical potential energy in the process of photosynthesis. In photosynthesis plants manufacture carbohydrates from water and carbon dioxide using sunlight. Under certain conditions the remains of these plants are converted into fossil fuels (coal, petroleum, natural gas), storing a small amount of energy for future use.

In addition to the solar radiation reaching the earth, the earth itself has its own sources of energy originally derived from the "big bang." Besides the stored energy reserves and the chemical potential energy found in fossil fuels, it has energy stored in the nuclei of certain atoms, in geothermal energy seeping up from the hot interior of the earth, and also in gravitational potential energy (tidal energy) stored in the earth-moon system.

To be used by humans, energy that is stored as potential energy must be converted to kinetic energy. Known pathways for the conversion of energy from primary form to intermediate form to final form are shown in Figure 9.15.

## Energy Conversions and Efficiencies

Converting forms of one type of energy to another was stimulated by James Watt's steam engine. How could a better engine be designed to produce more labor-saving output using the same amount of heat generated from the wood and coal burning? How could the efficiency of such an engine be improved? Efficiency is defined as the work output divided by the work input:

$$E = W_{out}/W_{in}$$

Were there any limits to how good or efficient a machine could be made? The branch of engineering science known as thermodynamics was born to answer these questions. What followed were the first law of thermodynamics (conservation of energy) and the second law of thermodynamics (the law of entropy).

The steam or heat engine does not create or destroy energy but converts some of the heat available from combustion to work, which then propels the machines of industry.

The intermediate forms are all kinetic forms and the end uses may be kinetic or potential. For example, chemical potential energy of gasoline is converted to thermal energy by burning and then to mechanical energy. Finally, it is put to some end use.

There are, however, limitations on the conversion of energy from one form to another. In each conversion there is some inherent heat loss whether it be in terms of friction, radiation, or exhaust. For example, in heating a boiler to produce steam to heat a house or drive a turbine, most of the heat energy goes up the chimney and is dissipated into the atmosphere radiating out into space. Opportunity to perform work with the exhaust heat has been lost.

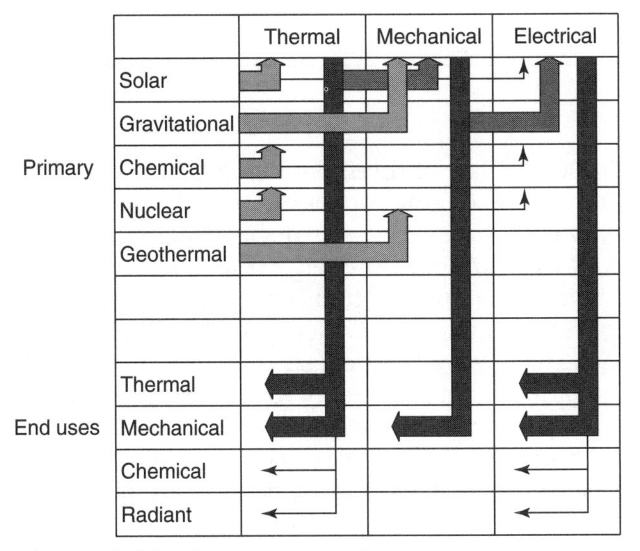


**Figure 9.15** Energy conversions.
(*Energy and the Environment,* John M. Fowler, McGraw-Hill, 1975).

One of the main concerns in utilizing any energy source is the efficiency of the conversion process (Fig. 9.16). To compute the efficiency of any conversion process, the ratio of the energy output to the energy input is used. If a gallon of gasoline containing 100,000 BTU (a BTU, or British thermal unit, is essentially the amount of heat required to raise a pound of water one degree Fahrenheit) of chemical energy is burned in an engine and the mechanical energy output is 25,000 BTU, the efficiency would be E = work output/work input, or 25,000 divided by 100,000 BTU, or 0.25. This is equal to 25% efficiency.

One class of conversions involves the conversion of energy to potential form or storage energy. For example, water is pumped by an electric utility during the night into a reservoir at high elevation. The water then flows back down to the turbines when it is needed to generate electricity. In doing this type of work the net lifting force has to be greater than the gravitational force on the water or it couldn't be lifted into the reservoir. The starting and stopping work done by the lifting force is not stored. So the amount of energy input in getting the water up into the reservoir will be greater than the energy stored. The starting and stopping work is wasted as heat in the air.

Converting to and from electrical energy is another type of energy transformation in which some energy is lost. There is always friction in a generator, which converts mechanical energy into electrical energy, and in a motor, which converts electrical energy into mechanical energy. Since friction can never be eliminated completely, there will always be heat losses in conversions involving mechanical energy. However, it is possible to attain close to 100% conversion efficiency in heating water with an electric wire. Table 9.1 shows some typical conversion efficiencies.

Efficiencies can be improved but there is a limit of 100%, which is absolute. To evaluate efficiencies, the system efficiency of a multistep conversion process should be considered. To power a car, gasoline must first be pumped as oil from a well, transported to a refinery, refined, transported to a gasoline station, pumped into an automobile, converted into heat in a combustion process, and the heat finally converted to mechanical energy. In each step some energy is lost and cumulative efficiency is less than that of any step.

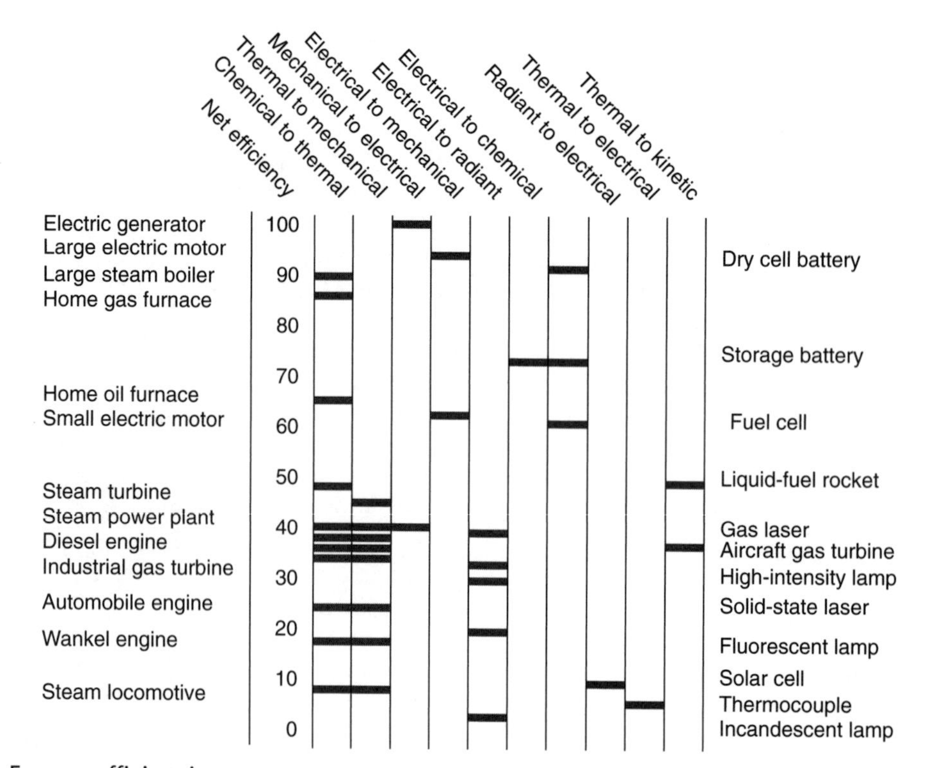


**Figure 9.16** Energy efficiencies.
(*"The Conversion of Energy,"* Claude M. Summers, *Scientific American,* September 1971).

**TABLE 9.1 System Efficiency of Electric Lighting (from coal-fired generation)**

| Step | Efficiency of step (percent) | Cumulative efficiency (percent) |
|---|---|---|
| Production of coal | 96 | 96 |
| Transporation of coal | 97 | 93 |
| Generation of electricity | 33 | 31 |
| Transmission of electricity | 85 | 26 |
| Lighting, incandescent (fluorescent) | 5 (20) | 1.3 (5.2) |

(*Energy and the Environment,* John M. Fowler, McGraw-Hill 1975.)

Table 9.1 and Table 9.2 show system efficiencies for electric lighting from coal-fired generation and for water heating. Although the step of converting electricity into heat can approach 100% efficiency, the system efficiency involving electrical heating may be low, as seen in the data.

# Second Law of Thermodynamics

Ninety-six percent of the energy used by humans goes through a conversion to thermal energy. Some of it, when used to produce mechanical work, is converted to thermal form immediately, and eventually it all ends up in thermal form. It is the second law of thermodynamics that assures us that every engine operating on a cycle must exhaust heat energy. No device operating on a cycle can extract heat energy from a reservoir and convert it completely to mechanical work. This, of course, reduces the efficiency of the engine and the exhaust or waste heat represents a loss to us of the potential of the energy for productive conversion. It is a loss of opportunity.

Every process that takes place in nature, whether it be mechanical, electrical, chemical, or biological, proceeds in conformity with the first and second laws of thermodynamics. The second law of thermodynamics is also referred to as the law of entropy.

Entropy is a property of a system that measures the degree of disorder associated with the energy of

**TABLE 9.2 System Efficiency for Water Heating**

| Step | Efficiency of step (percent) | Cumulative efficiency (percent) |
|---|---|---|
| *ELECTRIC (coal-fired)* | | |
| Production of coal | 96 | 96 |
| Transportation of coal | 97 | 93 |
| Generation of electricty | 33 | 31 |
| Transmission of electricity | 85 | 26 |
| Heating efficiency | 92 | 24 |
| *GAS* | | |
| Production of natural gas | 96 | 96 |
| Transportation of natural gas | 97 | 93 |
| Heating efficiency | 64 | 60 |

(*Energy and the Environment,* John M. Fowler, McGraw-Hill 1975).

**TABLE 9.3 Energy Ranking Based on Entropy**

| Form of energy | Entropy per unit energy (ev) |
|---|---|
| Gravitation | 0 |
| Energy of rotation | 0 |
| Energy of orbital motion | 0 |
| Nuclear reactions | $10^{-6}$ |
| Internal heat of stars | $10^{-3}$ |
| Sunlight | 1 |
| Chemical reactions | 1–10 |
| Terrestrial waste heat | 10–100 |
| Cosmic microwave radiation | $10^{4}$ |

("Energy in the Universe," Freeman J. Dyson, *Scientific American,* September 1971.)

a system. Energy flows in such a direction that the entropy of the universe increases. In fact, energy sources can be arranged into an order showing their associated entropy per unit of energy, as in Table 9.3. An increase in entropy is a measure of the loss of opportunity to perform work.

The degree of disorder of a form of energy varies approximately as the temperature associated with that form. Gravitation has no associated temperature and therefore its entropy is zero. Since energy flows from higher to lower levels in the table, which is in the direction entropy increases, cosmic microwave background radiation appears to be the ultimate heat reservoir. No way is known in which to further degrade or convert this energy into any other form.

The concept of entropy allows us to view the limitations on heat engines in a broader context. It is the concept that governs the degradation of energy from a form in which it is available to do work into a form in which it cannot be transferred to work. Change in entropy of the universe is an irreversible process and all spontaneous natural processes are such.

In any conversion or transfer of energy within a closed system, the entropy of the system increases. It is possible to produce a decrease in entropy in part of the system. An ice cube is a more ordered state for water molecules than in liquid form. The entropy of the water sample has decreased. If the total system involved in producing the decrease in kinetic energy of the water is considered, there is a net increase in the disorder of the universe.

The total system would include the work done in supplying fuel for generating steam and the work done by the steam in generating electricity to power the refrigerator that freezes the water forming the ice cube.

## Loss of Opportunity

The first law of thermodynamics assures us that the energy of the universe is constant and the second law tells us that the entropy of the universe is constantly increasing. Although energy can't be lost, its mere possession has no value. It is valuable only if it can be converted into work. However, in converting it into work some of its ability to do work is lost. Some loss of opportunity to do work is lost whenever energy is used to produce work.

The law of entropy should be used as a guide in selecting conversion processes so as to maximize the amount of work that can be extracted from a given amount of energy. Energy sources should be matched to the tasks so as to avoid wasting high quality (low entropy) energy on low quality tasks. For example, electricity may be used efficiently to operate the motor of an electric refrigerator. It is not a sound use of electricity to use the heat generated by the refrigerator coils for heating a room. Heat for a room can be obtained more efficiently by burning fuel directly in a central heating system.

The use of a source of energy should be based on the task to be performed so that the two can be matched for maximum efficiency. The entire system involved in the conversion process, from beginning to end, must be considered in order to obtain a true picture of the efficiency of the net conversion.

For the most part, the energy shortage is a shortage of sources of low entropy energy. In utilizing the potential energy sources available, a conscious effort should be made to utilize them in a fashion so as to minimize the increase in entropy.

Considering the universe to be a closed system, its entropy must increase as ongoing processes bring about change and evolution. As a consequence of these processes the entire universe may attain a state of absolute uniformity of temperature at some time in the future. There would still be no change in the energy of the universe, but all physical, chemical, and biological processes would cease. This state of affairs has been described as the heat death of the universe.

# Boundary Conditions

In the beginning there was energy. It evolved in time into different forms but always the total remained constant. Efforts at conserving energy should not be based only on the first law of thermodynamics but also on the second law of thermodynamics, the law of entropy. The conversion efficiency of the entire system must be examined to determine the best use of an energy source. The laws of thermodynamics place general limitations on the conversion of energy from one form to another and are of central importance to all of them. It is therefore necessary for us to keep these boundary conditions in mind when evaluating energy proposals for energy conversions and conservation.

## Barry Commoner's View

It is interesting to examine the role that energy plays in our world through the interplay of several systems. Barry Commoner, a scientist and an environmental and political activist, holds the view that all human activity is on the surface of the earth and is governed by complex interactions among three basic systems: the ecosystem, the production system, and the economic system.

Summarizing Commoner's arguments he states that all the resources that support human life and activity are provided by the ecological cycles on the surface of the earth and the minerals within it. The production system, consisting of agricultural and industrial processes, converts these resources into goods and services—food, manufactured goods, transportation, and communications. The economic system, which receives the real wealth created by the production system, transforms that wealth into earnings, profit, credit, savings, etc.

These three systems are intimately connected, and, according to Commoner, the controlling influence should flow from the ecosystem, through the production system to the economic system. In actuality the environmental crisis tells us that the ecosystem has been adversely affected by the modern production system, which grew up with little or no concern for the protection of the environment or the efficient use of energy (which was historically seen as being cheap and a readily available commodity once it was recognized as being useful).

The origins of the faulty design of the production system lie in the operation of an economic system that views factories and offices as installations, which are designed solely to produce profits rather than to be seriously concerned with the environment or the best use of resources.

The relationships among the three seem to Commoner to be the reverse of what they should be. The economic system seems to be driving the production system, which in turn acts on the ecological system.

Commoner presents his criticisms of the political and economic realities in his book, *The Poverty of Power*[1]. Regardless whether or not we can all agree on Barry Commoner's prescriptions for curing the problems, planet earth is certainly sending us clear signals that continuing past practices of resource consumption, waste production, manipulating supply and demand is not a promising route to the resolution of the problems of the ecosystem and point toward more troubles for the future.

# Review Questions

1. Why are units in science important?
2. What is the difference between a vector and a scalar?

[1]Commoner, Barry, *The Poverty of Power,* New York, Knopf, 1976.

3. What is energy? How is it related to work?
4. What is the difference between kinetic energy and potential energy?
5. What is the first law of thermodynamics?
6. In light of the first law of thermodynamics, how can there be an energy shortage?
7. What is the second law of thermodynamics?
8. What is meant by entropy?
9. What was the "big bang"?
10. Discuss Barry Commoner's views in terms of interacting systems.

# Multiple Choice Questions

1. Newton's law of inertia states that a particle left to itself maintains its _____.
   a. acceleration
   b. mass
   c. velocity unchanged
   d. force
2. Work done by a force in science is computed by multiplying the force in the direction of a displacement by the _____.
   a. displacement
   b. acceleration
   c. mass
   d. velocity
3. The unit used for work is the _____.
   a. newton
   b. meter
   c. joule
   d. watt
4. The main difference between work and power is that power is the _____ at which work is done.
   a. rate
   b. acceleration
   c. force
   d. energy
5. Kinetic energy is energy a particle has by virtue of its _____.
   a. position
   b. configuration
   c. motion
   d. potential
6. Potential energy is the energy a particle possesses by virtue of its _____.
   a. position
   b. configuration
   c. motion
   d. a. and b.
7. Einstein, in his famous equation $E = mc^2$, tells us that energy and _____ are different forms of the same thing.
   a. power
   b. force
   c. mass
   d. speed of light
8. The entropy of a system measures the amount of _____ in the system.
   a. mass
   b. energy
   c. disorder
   d. waste
9. Energy has no intrinsic value. It is valuable only if it can be converted to _____.
   a. work
   b. power
   c. force
   d. electricity
10. Barry Commoner believes that all human activity is governed by the interaction between _____.
    a. work and energy
    b. the ecosystem and man
    c. the ecosystem, production system, and economic system
    d. living and nonliving systems
11. Which of the following represents a source of potential energy?
    a. flowing water
    b. a stone on top of a cliff
    c. a tuba
    d. the periodic table
12. The potential energy of the universe _____.
    a. is increasing
    b. is decreasing
    c. is constant
    d. varies with temperature
13. Which of the following is not an energy unit?
    a. joule
    b. newton meter
    c. kilowatt-hour
    d. degree Celsius
14. All chemical reactions _____.
    a. give off energy
    b. absorb energy
    c. involve energy transformation
    d. result in an increase in entropy

15. An electric hair dryer rated at 1,000 watts is run for 200 seconds. The amount of energy consumed is _____.
   a. 200 joules
   b 200 watt-seconds
   c. 200 kilowatt-seconds
   d. 2000 kilojoules

16. Two objects of the same mass, shape, and density are each exactly 100 feet above the ground. Both are allowed to fall to the earth under the influence of gravity. Object A strikes the earth at the same time as object B strikes a very large inflated pillow. Which of the two possesses the greater potential energy before the objects were dropped?
   a. A
   b. B
   c. Can't tell.
   d. They are the same.

17. Which of the two in question 16 strikes the surface with a greater kinetic energy?
   a. A
   b. B
   c. Can't tell.
   d. They are the same.

18. Which of the two will strike the surface with more power?
   a. A
   b. B
   c. Can't tell
   d. They are the same.

19. Which of the two will possess more kinetic energy a fraction of a second after collision?
   a. A
   b. B
   c. Can't tell.
   d. They are the same.

20. Which of the two will possess more potential energy a fraction of a second after collision?
   a. A
   b. B
   c. Can't tell.
   d. They are the same.

21. A spotlight is rated at 1,000 watts. It was used for 0.1 hour. How much energy was used?
   a. 1000 watt hours
   b. 0.1 kilowatt hour
   c. 1.00 watts
   d. 10 kilowatt hours

22. What would be a correct expression for the second law of thermodynamics?
   a. In any open system, entropy always decreases.
   b. In a closed system, there is a tendency to go from order to disorder.
   c. In a closed system, there is a tendency to go from disorder to order.
   d. In an open system, order can never be restored.

## Bibliography and Web Resources

Abelson, Philip H. (ed.). *Energy Use Conservation and Supply,* AAAS, Washington DC, 1974.

Fowler, John M., *Energy and the Environment,* New York, McGraw-Hill, 1975.

Odum, Howard T., *Environment, Power, and Society,* New York, Wiley-Interscience, 1971.

*Scientific American,* "Cosmology+1," San Francisco, Freeman, 1977.

*Scientific American,* "Energy and Power," San Francisco, Freeman, 1971.

Introduction to Newton's laws of motion, in particular the concepts of force and Inertia.
**www-istp.gsfc.nasa.gov/stargaze/Snewton.htm**

Newton's Laws of Motion.
**www.ic.arizona.edu/~nats101/newton.html**

Laws of Thermodynamics. The laws of thermodynamics are stated simply.
**www.state.sd.us/deca/DDN4Learning/ThemeUnits/thermo/laws.htm**

2nd Law of Thermodynamics.
**theory.uwinnipeg.ca/mod_tech/node78.html**

Its fundamental principles are called the "laws of thermodynamics." The First Law of Thermodynamics (conservation of energy). The Second Law of Thermodynamics.
**www.deutsches-museum.de/ausstell/dauer/physik/e_thermo.htm**

Thermodynamics. There are three principal laws of thermodynamics, which are described on separate slides.
**wright.nasa.gov/airplane/thermo.html**

"Forgotten Fundamentals of the Energy Crisis," Albert A. Bartlett, published in the *American Journal of Physics,* September of 1978.
**faculty1.coloradocollege.edu/~shall/EV121General/bartlett.doc**

# A Primer

## Energy Sources: Their Nature and Use

*"You can't win.*
*You can't break even.*
*You can't even quit the game."*

**Allen Ginsburg**

The laws of thermodynamics are vital considerations in developing a philosophy and attitude toward energy usage. They can be summed up in the statement: "you can't win; you can't break even; you can't quit." In this chapter we shall study the energy resources that drive our society and sustain all living creatures: nuclear energy, solar energy and its variant forms, geothermal energy, and tidal energy. We shall examine their origins, their limitations, and their applications. In exploring solar energy, we shall discuss biomass energy, fossil fuels, wind energy, and hydropower energy sources.

The magnitude of available resources will be examined as well as the costs, environmental impacts, health, and safety issues and both their suitability and ease of use.

## Goals

After studying this chapter, you should be able to:

- Address energy issues in terms of the laws of thermodynamics.
- Understand the differences between radiation and radioactivity.
- Understand the origin of nuclear energy in terms of fission and fusion.
- Understand how a nuclear reaction produces energy.
- Understand the sun as an ultimate source of energy.
- Understand the role the sun plays as a source of energy in direct radiation, biomass, photovoltaics, wind, and hydropower.

## Thermodynamics

The laws of thermodynamics, as noted in the previous chapter, describe conditions and limits associated with the transformation of energy from one form to another.

Ginsburg's Theorem, as stated in the chapter introduction, succinctly affirms the fact that there is a finite amount of energy in the universe and the transformation of energy from one form to another will never exceed 100% (First Law—you can't win). In reality, the transformation of energy will usually result in an efficiency of less than 100%.

For example, the conversion of the potential energy of gasoline to kinetic energy to propel an automobile also results in waste heat being generated so the efficiency is less than 100% and has been observed to approach 100% only at best.

In reality, the second law—you can't break even—tells us more of what happens in energy transformations. The waste heat, which is generated in the potential to kinetic energy conversion, is dissipated to the atmosphere. It is unrecoverable and results in an increase in the random motion of molecules in the air as they are heated and hence leads to a decrease in the orderly arrangement of these air molecules. Another way of saying this is that there is an increase disorder, or randomness (or entropy) of the universe. We may try to be more efficient, but some of the waste heat energy will be gone forever as a useful resource. Hence our energy resources, which are limited, diminish with use since they cannot be totally recycled. As time progresses, more and more potential energy will be transformed to "unavailable or background" energy. Wise use of our energy resources extends the lifetime of its availability and makes good sense from a number of other perspectives as well.

We will not confront the third law (which many scientists treat as part of and a form of the second law) except to note that we are living, breathing, participants involved in the "energy game." As difficult as it may be to accept, after death, as we decompose through bacterial and other actions, the energy contained in the bonds of our molecules will be released for use by other organisms. Our material building blocks will also be recycled. That is to say, the third law states that only pure crystalline substances at absolute zero are perfectly ordered; anything else possesses entropy or "unavailable energy." Hence, like it or not, *we* are part of the energy equation; quitting is *not* an option.

The implications of the thermodynamic laws are vital considerations in developing a philosophy and attitude toward energy usage. However, while energy resources cannot last forever, there are also other more immediate considerations to be addressed in the examination of energy resource options. Besides the magnitude of the various available resources, cost, environmental impact, health and safety issues, and ease and suitability of use are also important.

In order to address these issues and develop a perspective and understanding concerning energy we will examine the energy resources in its various forms.

The potential and kinetic energy, which ultimately drives our society and sustains all living creatures, can be described as belonging to one of the following examples of energy resources

- Nuclear Energy
- Solar Energy
- Tidal Energy
- Geothermal Energy
- Fossil Fuels

Other forms of energy, such as hydropower, wind energy, or alternate fuels, are derived or encompassed in one of the resources just listed.

# Nuclear Energy

Nuclear energy is energy derived from processes occurring at the atomic and subatomic levels. One form of nuclear energy occurs when an atom of a heavier element is made to split into smaller segments. This process is called fission. It does not occur with all elements. It turns out that only certain of the heavier elements are fissionable. For example, the isotope of uranium atom mass 235 will undergo fission when a neutron, which has been slowed down, is made to collide with the uranium, as noted:

$$^{1}_{0}n + ^{235}_{92}U \rightarrow ^{236}_{92}U \rightarrow ^{90}_{38}Sr + ^{143}_{54}Xe + 3\ ^{1}_{0}n + \text{energy}$$

(Note that the number of neutrons and protons are conserved in the process.)

In the 1930s, Enrico Fermi first discovered that neutrons that were slowed down would be accepted or incorporated in the nucleus of an atom. He demonstrated that almost any element could undergo nuclear transformation via neutron bombardment and he could make new radioactive elements.

Atoms are radioactive if they emit certain kinds of radiation similar to X-rays, or high-energy particle beams. Radioactivity is indicative of an unstable atomic nucleus. The emission of particles or energy from the atomic nucleus, radioactivity, is the process whereby the nucleus tries to achieve stability.

Fermi's work involving neutron bombardment of the uranium atom, however, yielded results that were puzzling. Interpretation of these puzzling results led to the discovery of nuclear energy.

Duplication and analysis of Fermi's experiments involving neutron bombardment of uranium by Meitner, Hahn, Strassman, and others ultimately revealed that some of the uranium atoms that had gained a neutron, had split, totally unexpectedly, into other, lighter elements. As seen in the previous equation, $^{236}_{92}U$ breaks into two lighter parts, $^{90}_{38}Sr$ and $^{143}_{54}Xe$ (plus about three neutrons).

Lise Meitner deduced that the uranium atom had split into two large fragments. Nuclear fission had occurred.

Meitner recognized that the fission had resulted in a product, the aggregate mass of which was lighter than the reactants. There was a loss in mass. This is a big deal. All modern science up to this time had as one of its cornerstones the law of conservation of mass.

The law states that in ordinary physical and chemical transformations such as the melting of ice, or the reaction of different substances to form other products there is *never* a loss in mass. Mass is always conserved. When something is burned, the mass of the products and reactants is the same. Energy, which is generated from combustion, for example, comes from the potential energy associated with chemical bonds, which are broken, and new bonds formed. Fission, however, is not an ordinary chemical transformation. It is an example of the conversion of mass to energy. Fermi's experiments were an application of a prediction Albert Einstein had made in the early part of the 20th century.

Einstein had mathematically stated an equivalence between mass and energy, that mass and matter were potentially convertible. The famous equation

$$E = mc^2$$

allows calculation of the amount of energy one could obtain from matter that disappears in the transformation into energy.

Meitner recognized that this fission had resulted in the conversion of mass into energy.

How much energy can be produced by fission? Well, if one pound of a substance were completely converted to electrical energy, there would be sufficient energy to cover the electricity requirements for about three-fourths of a million homes for one day.

Nuclear fission of uranium, as noted in the equation, also yields more neutrons than is consumed. This fact leads to the possibility that a huge amount of energy can be generated very quickly via an uncontrolled chain reaction. That is, as noted in the equation, three neutrons are produced as the uranium atom is split. If the quantity of uranium is sufficient and properly arranged (a supercritical mass) the neutrons that are ejected can strike additional uranium atoms so that first one atom, then three atoms, then nine, etc., atoms undergo fission. (See Figure 10.1, illustrating a chain reaction.)

An uncontrolled nuclear fission of a supercritical mass yields a huge amount of energy in an extremely short time period. It is perfect for a bomb. If, on the other hand, the fission is made to proceed more slowly in a controlled fashion (so the reaction is still self-sustaining but the number of nuclei being split does not multiply), nuclear energy can be used as a resource. Controlled self-sustaining nuclear reactions require that for each neutron captured by a uranium atom, only one of the three produced be subsequently captured to cause additional fission. The other two neutrons must be prevented from participating further in the process. This is what occurs in nuclear reactors.

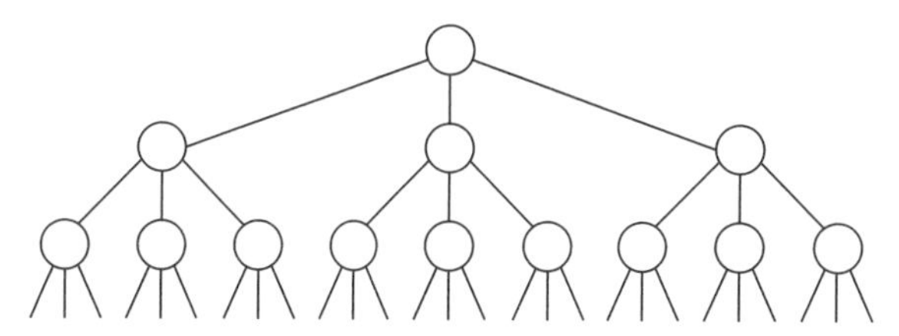

**Figure 10.1** Supercritical mass and a chain reaction.

The discoveries outlined above took place in Italy and Germany in the thirties as the German nation under the influence of the Nazis had become militarized and had expansionist designs on their neighbors' territory. The Germans embarked on a project to attempt to harness the power of the atom to augment their war effort: a German atomic bomb project. Scientists in the United States prevailed on Albert Einstein, perhaps the most famous scientist of the day, to contact the President of the United States in order to alert him and the government of potential dangers associated with an atomic bomb. A copy of the letter written by Dr. Einstein is shown in Figure 10.2.

Thankfully the German atom bomb project was unsuccessful, for had it been successful the course of history might have been altered. The American atomic bomb project resulted in the production of two atomic bombs; one used uranium 235 as the energy source, the other used plutonium 239, and both were used against Japan. The detonation of the atomic bombs heralded the dawn of the atomic, or the nuclear, age.

In the latter part of the 1940s great expectations were associated with nuclear energy. It was believed that the cost of nuclear energy would be so inexpensive that individual use would not be monitored; rather only a flat user fee would be paid. As it developed, there is a dark side to the use of nuclear energy.

How does nuclear energy work?

## Fission

A nuclear reactor, like an oil burner or a coal furnace or a fireplace, is basically a place where heat is generated. The latter more "conventional" devices use fossil fuel (oil, coal, gas, wood, etc.)

**Einstein's Letter to President Roosevelt - 1939**

Albert Einstein
Old Grove Road
Peconic, Long Island
August 2nd, 1939

F. D. Roosevelt
President of the United States
White House
Washington, D.C.

Sir:

Some recent work by E. Fermi and L. Szilard, which has been communicated to me in manuscript, leads me to expect that the element uranium may be turned into a new and important source of energy in the immediate future. Certain aspects of the situation which has arisen seem to call for watchfulness and if necessary, quick action on the part of the Administration. I believe therefore that it is my duty to bring to your attention the following facts and recommendations.

In the course of the last four months it has been made probable through the work of Joliot in France as well as Fermi and Szilard in America—that it may be possible to set up a nuclear chain reaction in a large mass of uranium, by which vast amounts of power and large quantities of new radium-like elements would be generated. Now it appears almost certain that this could be achieved in the immediate future.

This new phenomenon would also lead to the construction of bombs, and it is conceivable—though much less certain—that extremely powerful bombs of this type may thus be constructed. A single bomb of this type, carried by boat and exploded in a port, might very well destroy the whole port together with some of the surrounding territory. However, such bombs might very well prove too heavy for transportion by air.

The United States has only very poor ores of uranium in moderate quantities. There is some good ore in Canada and former Czechoslovakia, while the most important sources of uranium is Belgian Congo.

In view of this situation you may think it desirable to have some permanent contact maintained between the Administration and the groups of physicists working on chain reactions in America. One possible way of achieving this might be for you to entrust with this task a person who has your confidence and who could perhaps serve in an inofficial capacity. His task might comprise the following:

a) to approach Government Departments, keep them informed of the further development, and put forward recommendations for Government action, giving particular attention to the problem of securing a supply of uranium ore for the United States;

b) to speed up the experimental work, which is at present being carried on within the limits of the budgets of University laboratories, by providing funds, if such funds be required, through his contacts with private persons who are willing to make contributions for this cause, and perhaps also by obtaining the co-operation of industrial laboratories which have the necessary equipment.

I understand that Germany has actually stopped the sale of uranium from the Czechoslovakian mines which she has taken over. That she should have taken such early action might perhaps be understood on the ground that the son of the German Under-Secretary of State, von Weisäcker, is attached to the Kaiser-Wilhelm-Institut in Berlin where some of the American work on uranium is now being repeated.

Yours very truly,

*A. Einstein*

(Albert Einstein)

**Figure 10.2** Einstein's letter to President Roosevelt. *http://www.dpi.anl.gov/dpi2/hist_docs/einst_letter.htm*

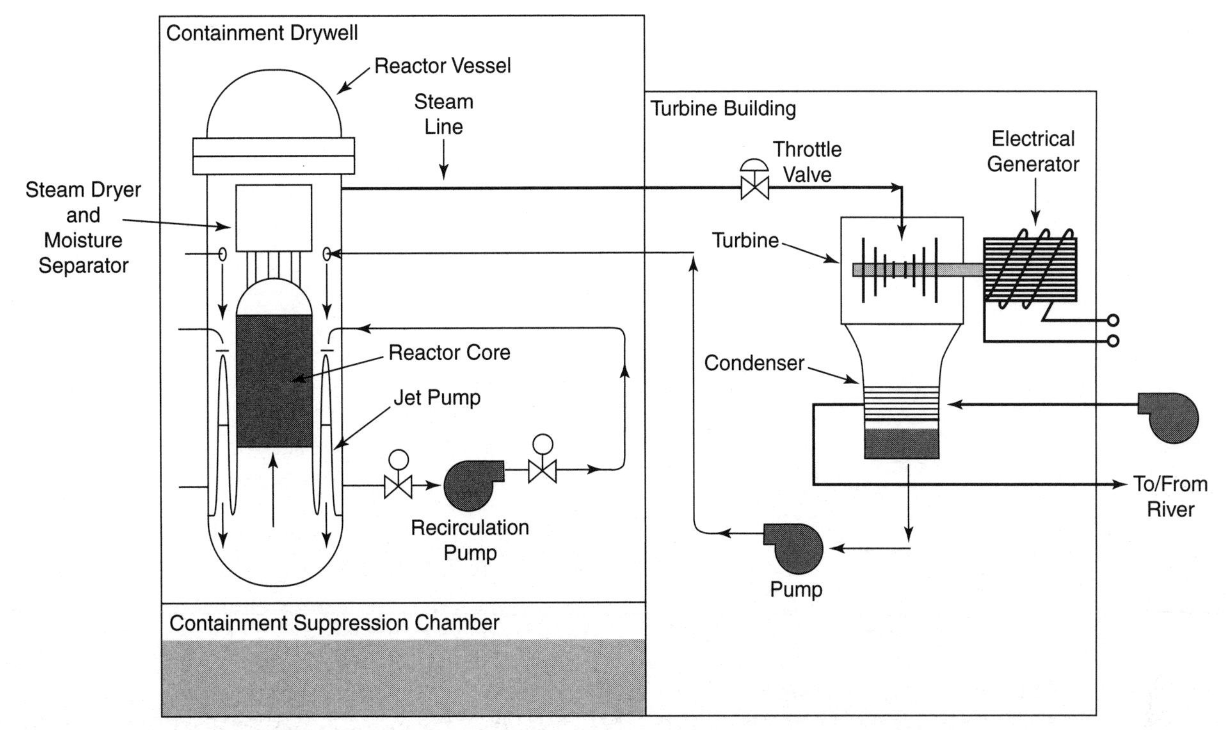


**Figure 10.3** Nuclear reactor. *http://eia.doe.gov/cneaf/nuclear/page/nuc_reactors/bwr.html*

and involve combustion to produce heat according to:

Oxygen + Fossil fuel → Carbon dioxide + Water + Heat/Energy (mass conserved).

Nuclear fission proceeds as noted:

Neutron + Nuclear fuel → Large fragments + Neutrons + Heat/Energy (loss of mass).

Just as the heat energy from conventional fossil fuel devices is captured to make steam, the heat from nuclear reactors can be made to produce steam. Once high-pressure steam is produced, regardless of the initial source of energy, the steam can be employed to drive turbines, which will enable electrical generators to rotate to produce electricity (see Fig. 10.3). In fact, electrical energy can be produced by any source or process that will cause turbines to rotate. (See the Web resources at the end of chapter.)

In the end then a nuclear reactor is just another form of a boiler. It is a device to make steam, but a whole new set of conditions and consequences are associated with its use.

For nuclear fission to work, a nuclear fuel is needed; isotopes such as uranium (U) 233 or 235 or plutonium (Pu) 239 can serve as nuclear fuels "capable of being split." In addition a source of neutrons, which have been "moderated," is also required. In order for neutrons to be absorbed by the nuclei and initiate the fission the neutrons must be slowed down. Moderators capable of accomplishing this task include ordinary water $H_2O$, heavy water $D_2O$ (where the isotope deuterium, containing one proton and one neutron in its nucleus, has replaced the hydrogen atom), and graphite. There are inherent problems associated with graphite reactors in that nuclear fission can occur in the absence of coolant, water, which is also a moderator. The Chernobyl reactor, which exploded, used graphite moderators.

Heavy water reactors do not require enriched uranium nuclear fuel. (The element uranium exists mainly as the isotope 238, which is nonfissionable with smaller amounts of [U] 235, which is fissionable). Light water reactors require that the nuclear fuel contain a greater percentage of the lighter isotope that is found in naturally occurring uranium. That is, the nuclear fuel must be enriched.

In order that the energy of nuclear chain reaction be limited and not reach explosive levels, control rods are installed in nuclear reactors. The purpose of these rods (generally made of cadmium or boron) is to absorb and limit the number of neutrons that are generated in the fission process so that a self-sustaining reaction is kept at desired levels. A coolant is also required to maintain a safe temperature of operation. The primary coolant is kept in a closed system owing to the need to minimize radiation leakage. See Figure 10.3 for a picture of a nuclear reactor.

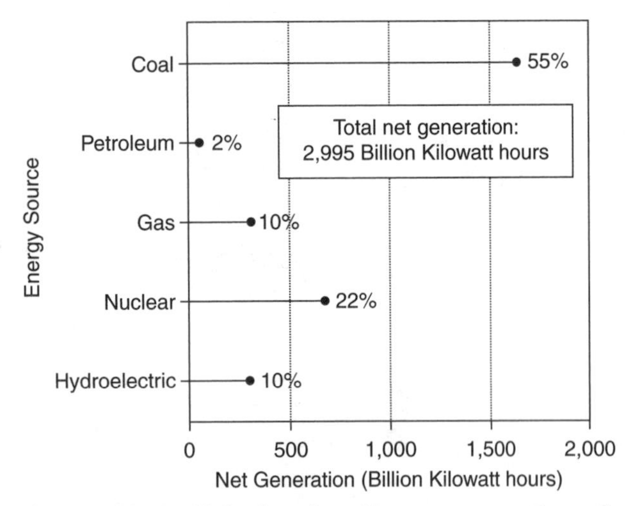


**Figure 10.4** U.S. electric utility net generation of electricity. *http://www.eia.doe.gov/cneaf/pubs_html/epa_1995/volume1/figures/fig2.gif*

The use of nuclear energy for electrical production is widespread. See Fig. 10.4. Propulsion of nuclear or atomic submarines, which previously could run underwater for only short periods using battery power, and large U.S. ships (aircraft carriers) that require huge amounts of electrical power, run efficiently on atomic energy. Worldwide different nations rely on nuclear power to different degrees. For example, 75% of the electrical energy needs in France and 20% of the world's electrical energy needs are obtained via nuclear fission. In the United States also about 20% of our electrical energy is obtained via this source and it is declining. Why such a difference? The cost of nuclear fuel is not the issue, but rather it is a matter of concern for safety in the public's perception.

There are several major concerns about international use of nuclear energy. The first is that weapons grade nuclear material can be produced by certain reactors in the process of "making energy." This means that "rogue" nations have the potential to make nuclear bombs and hold the world hostage. Terrorist organizations that lack the sophistication of making nuclear bombs can use reactor waste to make "dirty bombs," conventional explosives that will spread radioactive waste.

Another concern is associated with international terrorism. While modern nuclear reactors (of the nongraphite kind) are designed with fail-safe redundancies to contain radiation in case of an "event," there are no built-in defenses against a direct attack from the outside, as occurred on 9-11 on New York's World Trade Center, the Twin Towers, and the Pentagon in Washington, D.C. Nuclear reactors, to some, now represent new sites for potential targets.

Generally, the construction and configuration of nuclear power plants using water moderators is such that no nuclear explosion can occur. The nuclear fuel concentration is too low and the fuel is not confined. However, if there is an uncontrolled fission, significant overheating and meltdown is possible.

Nuclear reactors are housed in containment buildings to prevent spillage in the event of a nuclear incident.

Perhaps the biggest concern is associated with the disposition of the nuclear waste. At the end of the nuclear fuel cycle there are products that must be removed and disposed or stored in isolation. Many of these products are highly radioactive and have long half-lives.

Plutonium, one of the most toxic substances known, has a half-life of 24,400 years. (The half-life of a radioactive element is the time it takes for it to lose one half of its radioactivity and decay into another substance.) There are concerns by the public about the safety involved in reprocessing, transporting and, especially storing of nuclear waste. For storage of radioactive substances the question is, will the geologic stability of the disposal site be sufficient to maintain the isolation of long-lived radioactive substances? In this country, at least for the moment, it appears that the future of nuclear generated electricity is not in the ascent.

Radiation from radioactive sources may lead to significant health problems. Such radiation can alter the molecular structure of cells giving rise to genetic damage. As a final comment to this section, it might be noted in passing that radioactive isotopes have some salutary characteristics. They can be used medically in vivo procedures, imaging processes, and therapeutically. In addition, radioactive dating of objects as well as certain kinds of food preservation are dependent on radioactive decay processes.

## FUSION

There is a second possible route to nuclear generated energy. The process is called nuclear fusion and involves the union of nuclei of certain lighter atoms, which are made to coalesce into a heavier nucleus. The reaction is illustrated by the following equations:

$$^{1}_{1}H + ^{1}_{1}H \rightarrow ^{2}_{1}H + ^{0}_{1}e$$

$$^{1}_{1}H + ^{2}_{1}H \rightarrow ^{3}_{2}He$$

$$^{3}_{2}He + ^{3}_{2}He \rightarrow ^{4}_{2}He + 2^{1}_{1}H$$

$$^{3}_{2}He + ^{1}_{1}H \rightarrow ^{4}_{2}He + ^{0}_{1}e$$

Hydrogen isotopes (protinium $^{1}H$) first fuse to form deuterium ($^{2}H$) and ultimately helium ($^{3}He$ and $^{4}He$). In the process, as in fission, there is a loss in

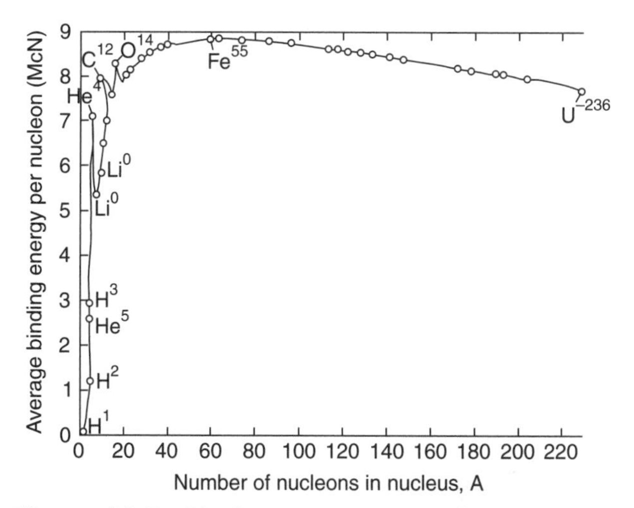


**Figure 10.5** Binding energy per nucleon plotted vs. atomic mass number. *http://imagine.gsfc.nasa.gov/Images/teachers/posters/elements/booklet/energy_big.jpg*

mass, which is converted to energy. As in nuclear fission, an enormous amount of energy is evolved.

Perhaps at this point the reader may be perplexed. How can fission, where atoms are split apart, and fusion, where nuclei are forced together, both result in a loss in mass with the attendant conversion of mass to energy? The two processes seem to be opposites. The basic answer is that different substances undergo fission and fusion. The lighter elements have the potential to undergo fission while the heavier elements have the potential for fusion. In reality very few elements have in fact been observed to undergo either of the two transformations.

The diagram in Figure 10.5 may help in understanding fission and fusion. The elements at the top of the graph, which are midrange in terms of their mass (Fe for example), possess nuclei of the greatest stability. Elements at either end of the graph can achieve greater nuclear stability if they are transformed to the "midrange mass." The lighter elements can achieve this condition via fusion, while fission is the route to nuclear stability for the heavier elements.

From a nuclear energy point of view iron (Fe) is an energy dead-end. In addition, the nature of the curve is such as to indicate that more energy can be generated in a fusion process compared to fission. It is no wonder therefore that for the last half century nuclear fusion has been anticipated as the ultimate energy source. In addition to the huge amounts of energy generated in fusion reactions, the power source, hydrogen, is cheap and plentiful and the products are nonpolluting (not even $CO_2$ is produced) and essentially nonradioactive.

So why aren't we there yet? There are enormous technical problems which must be solved before controlled fusion becomes a practical reality. For fusion to occur, to get the highly, densely charged positive nuclei to fuse, very high temperatures are required. A variety of approaches have been initiated without real success. Not only are extremely high temperatures required, on the order of 10–20 million degrees C, but also there are challenges in terms of practical containment of such temperatures. This is not to say that fusion is nonexistent. Fusion is and has been the driving force behind the sun's power.

In fact the sun, which powers the earth, providing us continually with about 1,000 watts/square meter of power, is a nuclear fusion reactor, converting hydrogen atoms present there to helium and energy. The massive gravity of the sun is a driving force in overcoming the repulsive force between the positively charged nuclei in getting the nuclei to fuse.

It is interesting to note that the element helium was first discovered on the sun, rather than on earth. It was new unidentifiable spectral lines in the solar emission spectrum that led to the postulation that an element existed on the sun that had no earthly counterpart. The element was called helium, Helios, Greek for sun. Helium was subsequently discovered on earth in mines.

Since very high temperatures are required to initiate fusion such processes are sometimes called thermonuclear reactions. Such reactions or processes have been brought about in an uncontrolled fashion in what has become known as the H-bomb. The device uses a nuclear fission reaction, and the associated very high temperature generated to provide the energy required to initiate fusion. It is clear this is not a suitable route for practical use of fusion energy to generate electrical power.

## Solar Energy

Most of the energy used to "run the earth" either directly or indirectly arises from the sun as the ultimate source. The sun, a nuclear fusion reactor, with a remaining lifetime estimated to be 10 billion years, is located the right distance away from earth, given our heat-sensitive biological nature. The "just right" intensity of radiation generated makes our planet habitable and enables life to be sustained.

The routes by which the sun provides us with energy occur in a variety of formats. These sources of energy derived from the sun include:

- Direct radiation
- Hydropower
- Wind
- Biomass
- Photovoltaics

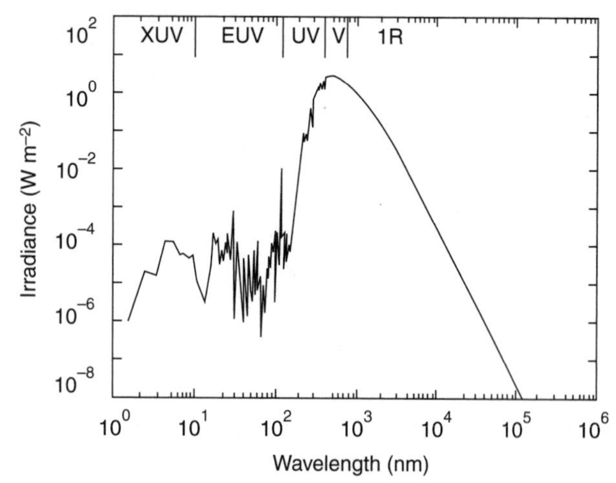


**Figure 10.6** Emission spectrum of the sun. *http://www.sec.noaa.gov/spacewx/images/Solar_Spectrum.jpg*

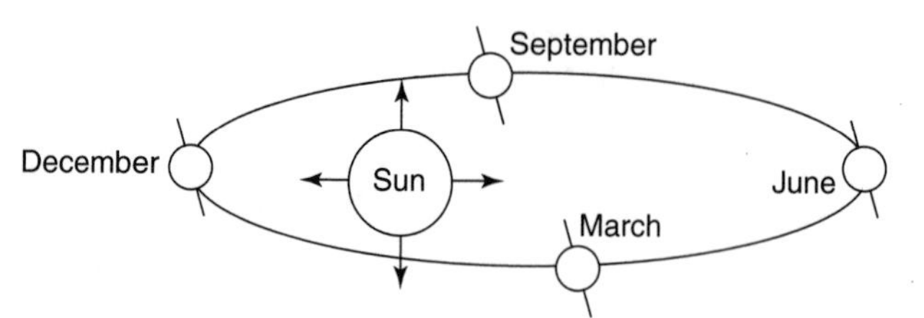


**Figure 10.7** Earth's seasons.

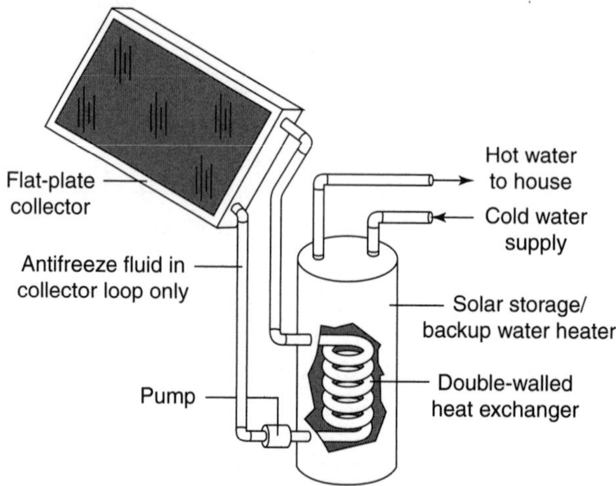


**Figure 10.8** Solar water heater (http://www.eere.energy.gov/solar/images/illust_fig.gif).

In addition, solar energy is directly responsible for production of fossil fuel, oil, coal, natural gas. While fossil fuel is a nonrenewable energy resource since it can be used only once, certain solar energy forms are called renewable energy because they can be regenerated. Solar energy forms tend to be more environmentally friendly and are sometimes called "green energy" (biomass). The connection between fossil fuel and solar energy will be considered later.

## Direct Radiation

The sun is about 90 million miles away from the earth. Yet a fraction of its energy, all generated as a result of nuclear fusion and radiated outwardly, strikes our planet. This energy at this fortuitous distance emanates as solar radiation and is translated to heat, plant growth, electron excitation, wind, and evaporation of water, the elements of solar energy.

The emission spectrum of the sun is noted in Figure 10.6.

Direct radiation is simply the sun's radiation striking the earth. When this occurs, a portion of the radiant energy is absorbed by the earth's surface (making the earth warmer) and a segment reemitted as heat and light. The more direct the radiation, the greater the amount of heat energy delivered and absorbed in a given area on earth. The fact that the northern hemisphere of the earth is hotter in the summer, when the earth is farther from the sun, than in the winter can be explained in terms of direct radiation and the tilt of the earth's rotation axis, as noted in Figure 10.7.

Direct radiation, which is transformed as heat, is of course, responsible for the earth's seasonal temperature variation and as such causes us to use variable amounts of energy. Direct radiation however may be employed to heat dwellings. Such solar homes require special orientation, construction, and insulation so that heat collection, storage, and distribution is adequate. (Depending on the location an auxiliary source of heat may be required.) Solar radiation is also used directly to make domestic hot water utilizing flat plate collectors. Such collectors can operate in both hot and cold climates.

Use of direct solar radiation for space or hot water heating enjoyed a boom in the late 1970s and early 1980s when tax incentives and the memory of recent energy shortages provided an impetus. While solar energy is "free," the collection apparatus must be maintained and there is a substantial outlay of initial costs before "payback" will begin.

Examples of solar heated home and solar hot water systems are noted in Figure 10.8 and Figure 10.9.

## Hydropower

As a consequence of the warming of the earth, the evaporation of water occurs from lakes, rivers, streams, and oceans. This ultimately leads to cloud formation and rain. The sun powers the water cycle,

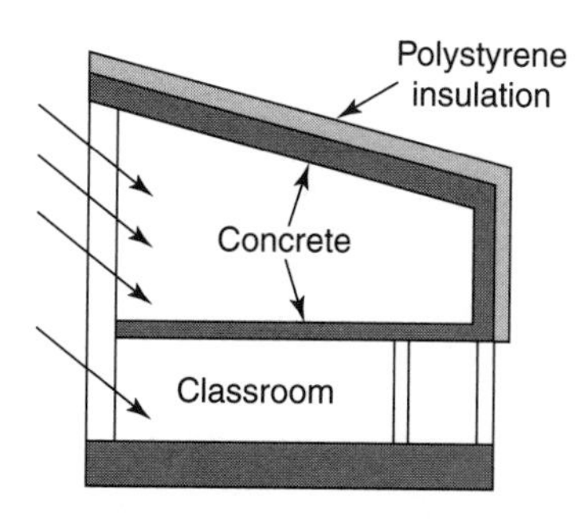


Coss-sectional view of the Wallasey annex to St. George's school. Sunlight enters through the large double-glazed south wall. Stable temperatures are maintained through the use of massive floors and walls and by the use of polystyrene insulation to help reduce building heat losses.

**Figure 10.9** Active and passive solar houses. (David K. McDaniels, *The Sun: Our Future Energy Source,* John Wiley and Sons, 1979 p. 201.)

which has just been outlined here. Any water at high altitudes represents a source of potential energy. It will seek to move to sea level if a path is provided. The flow of such water from a higher to a lower elevation is hydropower, and represents a very cheap and environmentally clean way to produce energy.

In former days grist mills and wood mills were set up in locations where the motion of the falling water could be used to turn the mill wheel and do the intended grinding or cutting task. On some rivers, such mills were so common that the rivers' name carries a reminder of this history, for example, Westchester County's "Saw Mill River," and the various "Mill" Rivers that can be found in Stamford, Connecticut, Mill River, Massachusetts, and Prince Edward Island, Canada. Now the power of water is harnessed to turn a turbine and produce less expensive electricity. In the United States hydropower produces enough electricity to supply almost 30 million residences. Up-to-date hydro turbines can be as high as 90% efficient in the energy conversion process. The cost per kilowatt hour is about one-third the cost of fossil or nuclear fuels and one-sixth that of natural gas.

## Wind Energy

The heating of the earth also results in certain geographical areas becoming hotter than others. The temperature differentials that occur result in a difference in the density of air masses, and so convection of the air masses, occur. This motion, in part, causes wind currents to develop. So in the end it is the sun's energy that is responsible for the wind's energy. In areas where there is a prevailing wind of some magnitude, the constant flow of the wind can be made to turn a windmill, and ultimately a turbine, and thus generate electricity. A photo of such

**Figure 10.10** Wind generator. *http://www.blm.gov/nstc/blmannual/annual97/serve.html*

windmills is pictured in Figure 10.10. Windmills can obstruct views and can be noisy.

A large number of windmills in a windmill farm can be used to supply the electrical needs of a community. On an individual basis windmills can be used as an energy resource and the excess power generated either sold back to the utility or stored in batteries.

## Biomass Energy

Biomass represents solar energy, which has been converted via photosynthesis into plant matter. The plant matter can be used as fuel directly to provide energy. Charcoal is prepared by heating wood at high temperature in a diminished oxygen environment. The product loses the volatile components of wood, is richer in carbon, and burns at a higher temperature. In addition, plant matter provides energy in the form of food either directly or indirectly to all inhabitants of earth.

Photosynthesis may be represented as a system:

**Input** ⟶ **Process** ⟶ **Output**

Sunlight

$CO_2 + H_2O \rightarrow$ Photosynthesis $\rightarrow O_2 +$ Plant Matter

Plant Nutrients (P,N,K) (ex. Glucose)

The energy from the sun drives the reaction whereby the chemical bonds that unite $CO_2$ and $H_2O$ are broken and new bonds are formed as the carbon, hydrogen, and oxygen are recombined to form oxygen and plant matter. Glucose ($C_6H_{12}O_6$)

is taken as a representative example of plant matter. The reaction is:

```
                                     H
                                     |
                           H H H O H H
                           | | | | | |
6 O=C=O + 6 H–O–H → 6 O=O + H–O–C–C–C–C–C–C=O
                           | | | | |
                           H O O H O–H
                             | |
                             H H
```

The representative photosynthetic reaction given here requires energy. (It is the reverse of a combustion or metabolism of glucose where energy would be generated.) As noted, the driving force for the process is the sun, which provides energy in the form of radiant light, which activates chlorophyll. (The asterisked terms represent energized molecules as the absorbed sunlight flows through the system.) That is:

Light + Chlorophyll → (absorption of light energy)
→ Activated chlorophyll*

The absorption spectrum of chlorophyll is given in Figure 10.11.

Note that chlorophyll absorbs light in the blue (400–500 nm) and red (600–700 nm) regions of the spectrum and so solar energy, which is useful in photosynthesis, must correspond to wavelengths of light, which the chlorophyll absorbs. The observation that chlorophyll itself is green or that plants containing chlorophyll are green is due to the fact that light in the region 500–600 nm is not absorbed by chlorophyll but either passes through the sample or is reflected by it. If someone were shining white light (all wavelengths) through a test tube containing chlorophyll, only the "green" wavelengths would pass through.

The photosynthetic reaction continues:

Activated chlorophyll* + $H_2O$ → $H_2O^*$
(Activated water) + Chlorophyll

2 $H_2O^*$ (Activated water) → 4 $H^+$ + $O_2$
+ 4 electrons

So oxygen, one of the products of photosynthesis, is generated.

The electrons, hydrogen ions, and carbon dioxide undergo a series of reactions, which result in the production of biomass. As noted earlier, biomass is ultimately responsible for sustaining all life on earth for living creatures. In addition, it has fuel value as in wood and charcoal. Biomass, which is organic matter, can be converted to more convenient fuels through various chemical processes. For example sugar, a plant product, can be fermented to yield ethyl alcohol. The alcohol, which is combustible, can be mixed with gasoline to produce gasohol, an automotive fuel. There are questions however concerning the economic advantage of this energy route as well as the idea of using food to produce fuel in a world where hunger exists.

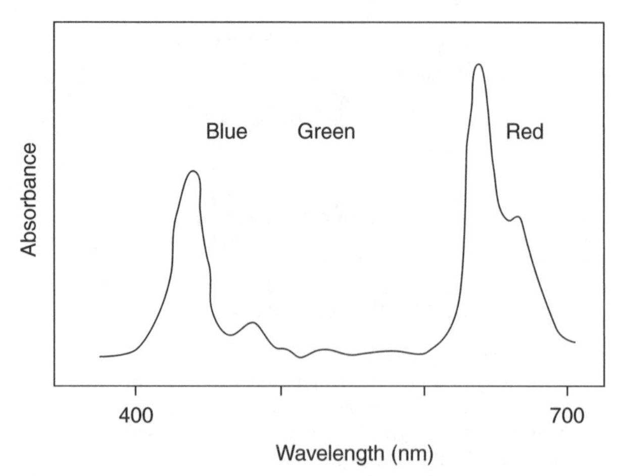


**Figure 10.11** Absorption spectra of chlorophyll.

A greater understanding of the nature of fossil fuel can be gained by examination of its origin.

## Photovoltaics

Photovoltaics, or solar cells, are devices made from the element silicon, which contain no moving parts and are able to convert direct solar energy to electrical energy at an efficiency of about 10%. Photovoltaic electrical energy is more expensive than energy generated by more conventional methods. It offers, however, convenience and freedom. Sometimes, as in geographically remote regions, solar cells represent the best option for an electrical energy source.

While photovoltaics, hydropower, solar panels, and wind energy represent "free-energy" and are nonpolluting in terms of direct emissions, there are environmental costs. These include, as in the case of hydropower, alteration of the landscape with the attendant impact on the biodiversity. In addition, solar collectors may not be aesthetically pleasing.

Two other sources of energy deserve mention; they are geothermal energy and tidal energy. Both represent sources of energy that have a lower impact on the environment than more conventional energy sources such as fossil or nuclear fuel use. (While the impact may be lower, it is not zero, however, since there will be an influence on wildlife populations.)

# Tidal Energy

Tidal energy arises from the attraction of the moon and the sun on the earth. The periodic motion of these celestial objects and the associated periodic variation of gravitational forces at different sites on earth result in the ebb and flow of tides. This tidal motion, while a source of tremendous energy, has seen only limited use. Tidal energy has been used since ancient times in terms of determining shipping times, and in the 8th century the coastal nations of France, Spain, and England harnessed tidal energy to run mills and water wheels. While direct use of tidal energy is obviously geographically limited to shorelines, the effects are more far-reaching. That is to say, tidal energy has been used to generate electricity, which is then transmitted to more interior regions. A few tidal electrical energy plants exist worldwide.

## Geothermal Energy

Geothermal energy is literally energy from the Earth. Hot springs, volcanoes, and geysers are some manifestations of this energy, which emanates from the interior of the earth. (The temperature of the Earth increases with depth. At about 2,500 miles deep the temperature is 7,200 degrees F. It is 9,000 degrees F at 4,000 miles). The Earth's interior heat escapes in regions where the tectonic plates (plates that make up the earth's crust) pass each other. The convective forces associated with the semimolten rock, or magma, drive the motion of the plates. Sometimes this magma makes its way to the surface and is ultimately seen as volcanoes.

Sometimes rainwater is able to seep into the interior of the earth for miles, through cracks and fissures, where high temperatures are encountered resulting in the formation of steam. Large regions where such steam exists represent geothermal reservoirs, which are suitable for commercial use. Direct capture of steam energy from the Earth avoids the use of any fuel and so is an attractive energy source. Geothermal energy, while geographically limited, still supplies the electrical energy needs of about 4 million people in the United States (2,800 megawatts). Worldwide geothermal power plants are responsible for the production of more than 8,200 megawatts of power in more than 20 countries. This supplies the needs of 60 million people.

Examination of the charts in Figure 10.12 reflects the various energy sources and use in the United States.

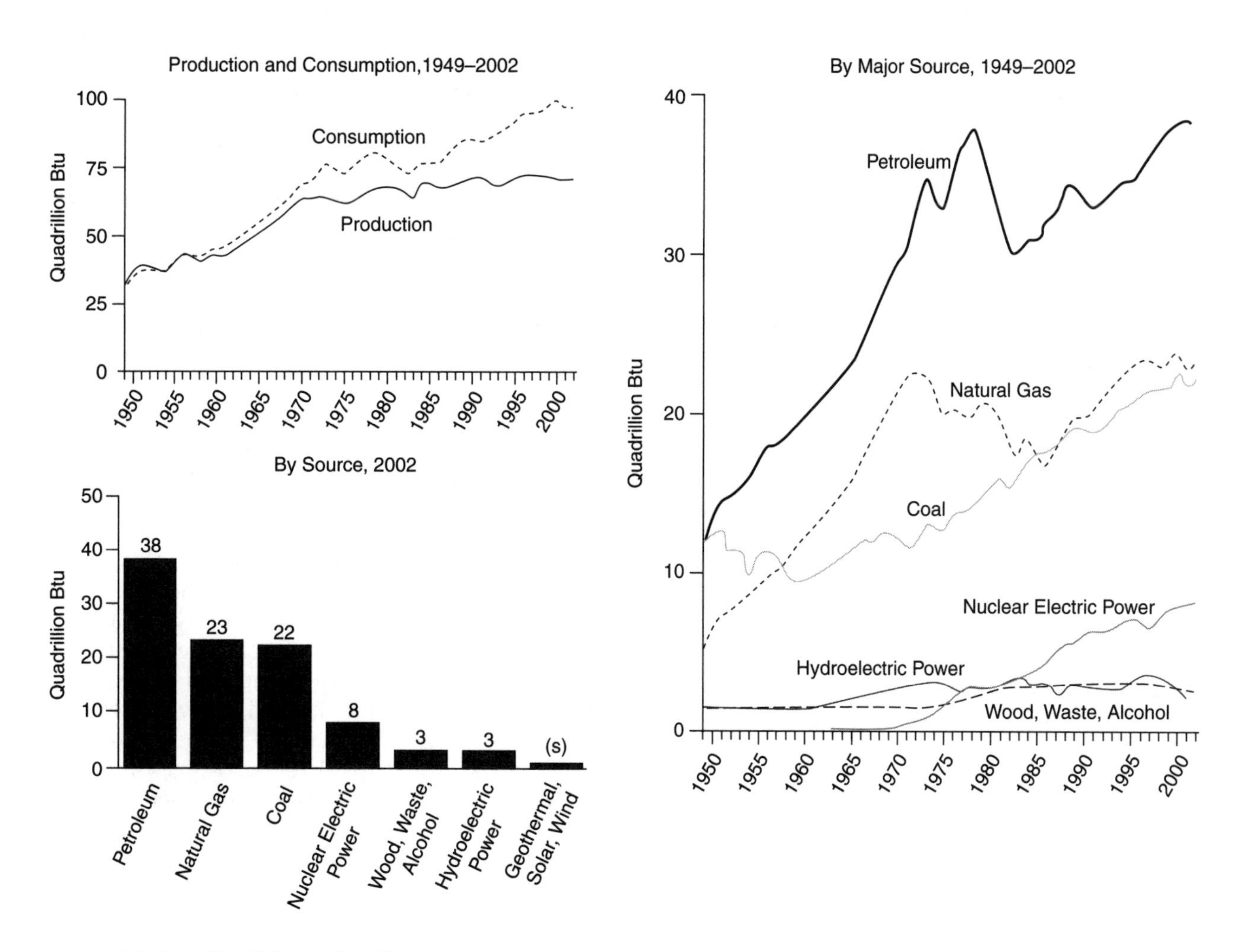


**Figure 10.12** U.S. energy use. *http://www.eia.doe.gov/emeu/aer/overview.html*

# Fossil Fuels

Another major energy resource is fossil fuel. Fossil fuel is found in nature to exist in all three states of matter. In the liquid state it is known as petroleum; examples of solid fossil fuel include various grades of coal, lignite, and peat. Natural gas represents the third form of fossil fuel. Regardless of the form in which fossil fuel exists, the fuel represents energy resources based upon the combustion process, that is, the conversion of the carbon-based compounds (the fossil fuel) to carbon dioxide. Under the best of circumstances the reactions for energy generation are:

Natural gas + Oxygen → Carbon dioxide + Water + Energy

Petroleum product + Oxygen → Carbon dioxide + Water + Energy

Coal (depending on type) + Oxygen → Carbon dioxide + Water + Energy

Examination of the first two reactions reveals that they are the reverse of photosynthesis. While photosynthesis consumes energy and carbon dioxide and water, and yields oxygen and organic matter, combustion is an oxidation; it uses oxygen and organic matter. Fossil fuels generate energy and as well as carbon dioxide. The $CO_2$, which is also formed, goes into the atmosphere. Excessive $CO_2$ becomes a pollutant.

Fossil fuels have their origin in biomass, ancient plant and animal life, which have expired, decomposed, and decayed in the absence of oxygen. The transformation of organic matter was initiated between 350 million and 50 million years ago. The decayed biotic matter remains, subjected to heat, pressure, bacterial, and fungal action, and buried by sediments decomposed in the absence of oxygen (anaerobic decomposition), ultimately yields a fossil fuel. On the other hand, if organic matter is left to rot in air, it will undergo aerobic decomposition, which is analogous to combustion.

**Aerobic decomposition**

Organic matter (contains C,H,N,S,O atoms) + $O_2$ → $CO_2$ + $H_2O$ + other products

**Anaerobic decomposition**

(pressure, heat, bacteria)
Organic matter (contains C,H,N,S,O atoms) → $CH_4$ + $H_2S$ + other products

The anaerobic equation is meant to convey the idea that decomposition in the absence of oxygen does not yield the oxygenated products ($CO_2$ and $H_2O$, end products of combustion) but rather nonoxygenated products ($CH_4$), which have fuel value.

## Coal

Depending on the nature of the organic matter, different kinds of fossil fuel were formed. Coal was formed from plant matter such as ferns, grasses, and trees. Covered by water, buried by sediments and subjected to high temperatures and pressures, chemical transformation was initiated. The transition from peat, a brown material in the earliest stage of coal formation, which may be up to 30% C, to lignite, also known as brown coal and is 40% C, to bituminous or soft coal (65% C) to anthracite coal (90% C), occurred over millions of years.

Owing to the eons of time required to produce coal (and other fossil fuels), it is understandable why these resources are called nonrenewable. Once they are used they are gone. We can't make more fossil fuel in a practical way. (You might note that coal-like material has been made in the laboratory from plant substances; however, the cost involved exceeds the fuel value of the synthetic "fossil fuel." In contrast to fossil fuels there is renewable energy, which comes from an energy source, which can be renewed, grown for example. Renewable energy is sometimes referred to as green energy.

The name fossil fuel arises from the fact that the materials from which these fuels are formed and which existed at a time when the decomposition process was initiated, are now fossils. In fact, we can see here a clear manifestation of the first law of thermodynamics: the conservation of energy. Energy from nuclear fusion reactions within the sun was converted millions of years ago to radiant energy, which streamed to earth in the form of sunlight. The sunlight was captured by early photosynthetic plants on earth and converted to chemical-bond energy trapped in the molecules that made up the plant structures. When eaten by prehistoric animals, this energy was released in the digestive and metabolic process to drive the synthesis of new animal tissue in which the energy was recaptured and, once again, stored in chemical bonds. As these prehistoric plants and animals died, were buried, and became fossilized, the energy remained in what today we know as fossil fuels. The gasoline with which we fuel our cars contains, essentially, the same energy that was generated millions of years ago in the sun's nuclear furnace.

## Natural Gas and Petroleum

The production of oil and gas also required anaerobic decomposition of plant and animal material.

**TABLE 10.1 Major Fractions of Petroleum**

| Fraction | Carbon Atoms* | Boiling Point Range (°C) | Uses |
|---|---|---|---|
| Natural Gas | $C_1$–$C_4$ | –161 to 20 | Fuel and cooking gas |
| Petroleum ether | $C_3$–$C_6$ | 30–60 | Solvent for organic compounds |
| Ligroin | $C_7$ | 20–135 | Solvent for organic compounds |
| Gasoline | $C_6$–$C_{12}$ | 30–180 | Automobile fuels |
| Kerosene | $C_{11}$–$C_{16}$ | 170–290 | Rocket and jet engine fuels, domestic heating |
| Heating fuel oil | $C_{14}$–$C_{18}$ | 260–350 | Domestic heating and fuel for electricity production |
| Lubricating oil | $C_{15}$–$C_{24}$ | 300–370 | Lubricants for automobiles and machines |

*The entries in this column indicate the numbers of carbon atoms in the compounds involved. For example, $C_1$–$C_4$ tells us that in natural gas the compounds contain 1 to 4 carbon atoms, and so on.

Raymond Chang (*Chemistry*, 7th edition, McGraw-Hill, 2002, pp. 968–969).

In addition, in order to prevent losses of gaseous fuel material to the atmosphere, or loss of the petroleum formed through the strata, certain geological rock formations were necessary to confine the evolving resource.

The nature of fossil fuel varies depending on the material that went into its formation. For example, the petroleum from oil wells varies from well to well. Some oils are dark and smelly, others are light in color and better smelling. Some oils are high in sulfur, others are not. All this relates to the initial biomass that went into its formation.

Fossil fuel, which in a sense is "buried sunshine," is of course a finite resource. The United States has large reserves of coal, a high-energy fuel. The combustion temperature of coal is high and so it is a useful fuel for a number of processes that require high temperature. On the downside, it is a solid and therefore more difficult to move around than a liquid or gas, which can be made to flow through pipes under pressure. In this sense then, natural gas and petroleum products are more convenient.

Natural gas generally refers to the lighter hydrocarbons (compounds of C and H) that exist as gases at room temperature. They include:

$CH_4$—methane,

$C_2H_6$—ethane,

$C_3H_8$—propane,

and perhaps $C_4H_{10}$—butane

As hydrocarbon molecules gain more carbon atoms, their properties change. As the number of carbon atoms increase, the substance becomes more viscous, the transition being from free-flowing liquid to oil, tar, and asphalt. Petroleum found in nature is a mixture of a number of hydrocarbon components. Petroleum is distilled (heated) so that different fractions are obtained. The process is called fractionation. The individual fractions obtained are composed of like molecules. The basic idea is that like molecules boil at about the same temperature with the boiling point increasing as the number of C atoms increases. Different fractions possess different uses aside from being an energy source. These applications are noted in Table 10.1 and Figure 10.13. How fossil fuels influence our environment will be examined in the next chapter.

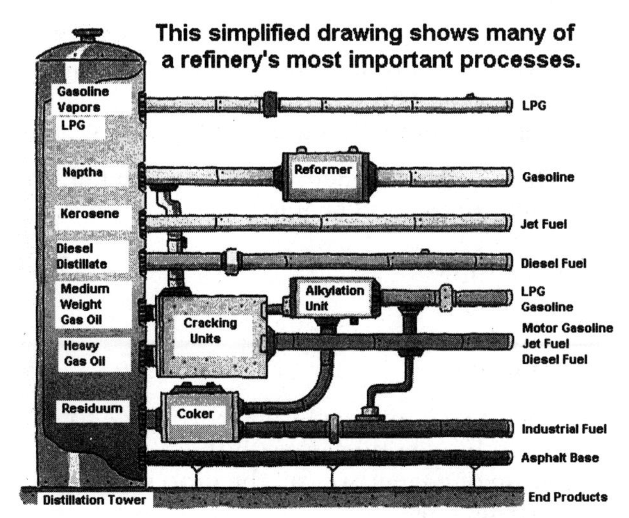


**Figure 10.13** Fractional distillation column for separating components of crude oil. *http://www.eia.doe.gov/kids/non-renewable/refinery.html*

# Review Questions

1. Discuss the origins of nuclear energy.
2. What is fission? What is fusion? What is the difference?

3. What is a chain reaction?
4. What is nuclear waste and what are the problems?
5. What is the difference between a nuclear reactor and a nuclear bomb?
6. What is "direct radiation" and how does it contribute to wind energy and hydropower?
7. What are photovoltaics?
8. What is biomass energy? How is it connected to fossil fuels?
9. What is tidal energy?
10. What is geothermal energy?

## Multiple Choice Questions

1. To paraphrase the second law of thermodynamics, _____.
   a. you can't quit
   b. you can't break even
   c. you can't win
   d. entropy always increases in a closed system
2. When a heavy nucleus splits into two lighter-weight nuclei and energy is released, we call the process _____.
   a. splitting
   b. fusion
   c. fission
   d. mitosis
3. When two light-weight nuclei coalesce to form a heavier nucleus with the consequent release of energy, we call the process _____.
   a. coalescing
   b. fusion
   c. fission
   d. mitosis
4. Most of the energy used to "run the earth" arises from _____.
   a. electricity
   b. wind
   c. the sun
   d. biomass
5. The sun primarily derives its energy from _____.
   a. coalescing
   b. fusion
   c. fission
   d. mitosis
6. In the solid state, fossil fuel can be found as _____.
   a. coal
   b. oil
   c. peat
   d. natural gas
7. An anaerobic decomposition takes place in the absence of _____.
   a. nitrogen
   b. carbon
   c. hydrogen
   d. oxygen
8. Tidal energy arises from the attraction of the moon and _____ on the earth.
   a. Mars
   b. gravity
   c. sun
   d. atmosphere
9. Photovoltaics are devices made from silicon that convert _____.
   a. analog to digital
   b. electrical energy to solar energy
   c. nuclear energy to light
   d. sunlight to electricity
10. Geothermal is energy obtained from _____.
    a. heating water
    b. sunlight
    c. the earth
    d. hydropower
11. Determine the value for x in the process:

$$^{14}_{6}\text{C} \rightarrow {}^{0}_{-1}\text{Beta particle} + {}^{x}_{7}\text{N}$$

    a. 10
    b. 12
    c. 14
    d. None of the above.
12. Nuclear fusion involves the union of _____.
    a. atoms
    b. molecules
    c. electrons
    d. nuclei
13. Nuclear fusion and nuclear fission _____.
    a. differ in that there is a net gain in weight in fusion
    b. both require high temperature to be initiated
    c. both result in a loss in a mass
    d. both have been commercially achieved
14. Fission differs from ordinary combustion in that _____.
    a. combustion involves a loss in mass
    b. fission requires fusion
    c. fission involves a loss in mass
    d. combustion involves an increase in entropy

15. An isotope of plutonium has a half-life of more than 24,000 years. If we have one gram of plutonium today, how much will remain after 92,000 years (three half-lives)?
    a. It will be all gone.
    b. 0.500 g.
    c. 0.250 g.
    d. 0.125 g.
    e. None of the above.

16. The energy associated with hydropower arises ultimately as a result of _____.
    a. the action of the moon on the tides
    b. the construction of large dams
    c. evaporation of water by the sun
    d. the production of hydroelectric plants

17. A solution of a substance is blue. This means the substance _____.
    a. absorbs light in the blue region of the spectrum
    b. transmits light in the other regions of the visible spectrum
    c. transmits light in the blue region of the spectrum
    d. None of the above.

18. Which of the following is not a fossil fuel?
    a. methane
    b. uranium
    c. kerosene
    d. coal

19. Which of the following is the cause for the seasonal variation in the Earth's temperature in the northern hemisphere?
    a. The Earth is closer in winter than summer.
    b. The Earth's motion about the sun is slower in summer.
    c. The rotation axis of the Earth is tilted so that at times the solar radiation is more direct.
    d. It is colder in winter.

20. Which of the following represents the correct sequence of boiling points of some fossil fuels (lowest to highest boiling point)?
    a. gasoline, heating oil, naphtha
    b. lubricating oil, gasoline, kerosene
    c. gasoline, kerosene, heating oil
    d. heating oil, naphtha, natural gas

21. Given light characterized by the following wavelengths,

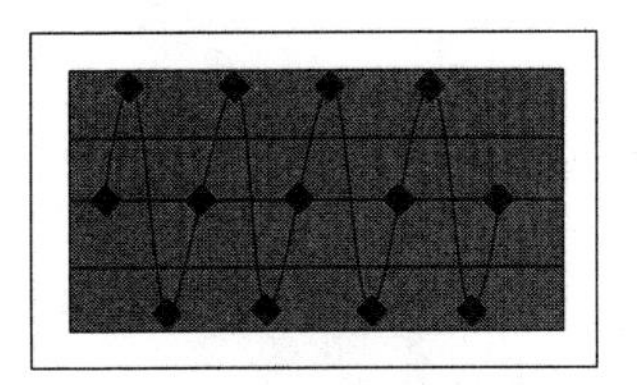

A.

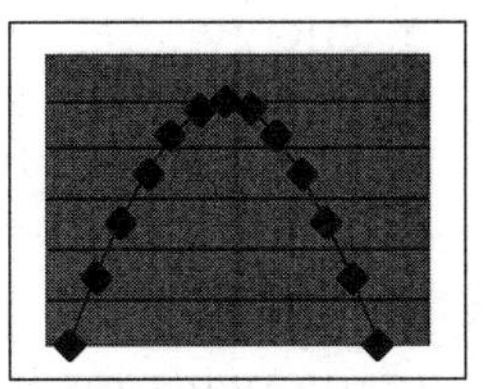

B.

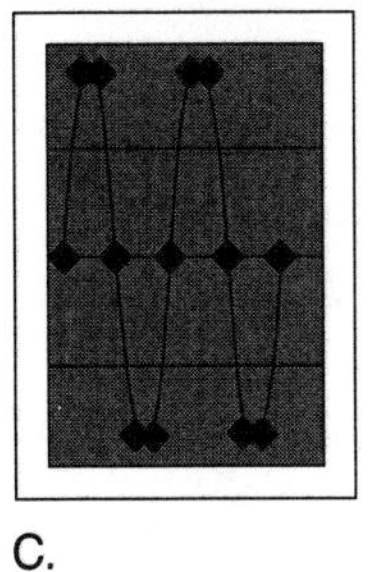

C.

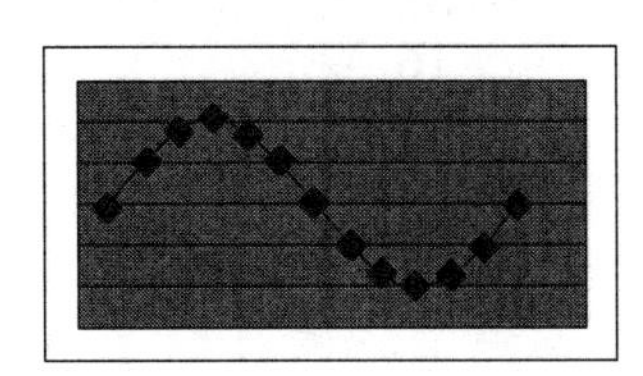

D.

Select light of the longest wavelength. _____.

22. Given this spectrum of a substance:

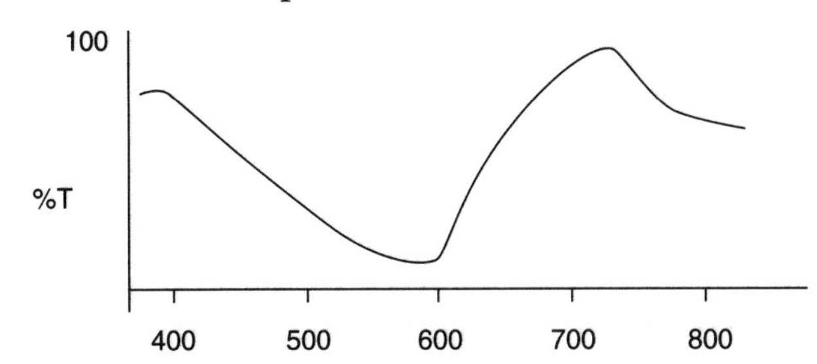


For what wavelength region does maximum absorption of light occur by the substance?
    a. 400 to 500 nm
    b. 500 to 600 nm
    c. 600 to 700 nm
    d. 700 to 800 nm

23. For radioactive materials the half-life of the material is the time it takes for one half of the radioactive nuclei present to disappear. After two half-lives, therefore, only ¼ of the nuclei remain. If 12.5% is left, then the number of half-lives it has passed through is _____.
    a. 3
    b. 4
    c. 5
    d. 6
    e. None of the above.

24. The Bohr atom, a model for a simple atomic structure, can describe _____.
    a. energy levels
    b. light absorbed by atoms
    c. light emitted from atoms
    d. All of the above.

25. Phosphorus, $^{31}_{15}P$, is an element with how many neutrons in its nucleus?
    a. 15
    b. 31
    c. 16
    d. 46

# Bibliography and Web Resources

Brief biography of co-discoverer of nuclear fission.
**http://www.sdsc.edu/ScienceWomen/meitner.html**
**http://mnmn.essortment.com/lisemeitner_rqob.htm**

Einstein Web Site.
**www.aip.org/history/einstein**

Nuclear Reactor.
**http://americanhistory.si.edu/subs/operating/propulsion**
**http://www.nrc.gov/reactors/power.html**
**http://science.howstuffworks.com/nuclear-power1.htm**
**http://www.ccnr.org/nuclear_primer.html**

Nuclear Fusion.
**http://www.jet.efda.org/pages/content/fusion1.html**

# Energy Resources and Their Environmental Effects

## The Carbon Cycle and More

*"You will die but the carbon will not; its career does not end with you. . . . It will return to the soil, and there a plant may take it up again in time, sending it once more on a cycle of plant and animal life."*

**Jacob Bronowski**

Viewed as a complete system, solar energy captured by photovoltaics is neither a source of free energy nor nonpolluting. It is only by studying a technology in terms of the systems approach that we can determine all its inputs, outputs, and processes. In this chapter we shall study various cycles such as the carbon cycle, nitrogen cycle, and oxygen cycle and their influence on our environment.

As we study the carbon cycle we shall find it is closely related to the greenhouse effect, and to other effects based on the combustion of fossil fuels such as natural gas, coal, and oil. There is interconnectedness between the use of fossil fuels and its impact on the environment and ultimately our health.

## Goals

After studying this chapter, you should be able to:

- Realize that photovoltaic sources of energy do not produce free energy.
- Be aware that a systems analysis of photovoltaic devices reveals that they also contribute to pollution.
- Understand the carbon cycle and the production of carbon dioxide.
- Understand the role of greenhouse gases in the Earth's atmosphere and the rise of the Earth's temperature.
- Understand that the combustion of fossil fuels (coal, oil, gas) is producing atmospheric carbon dioxide faster than the carbon cycle can reincorporate it into plant matter.
- Be aware of the environmental and health consequences of fossil fuel use.

## Photovoltaics

The production of energy or energy generation, generally has some environmental impact, somewhere. In isolation, a particular form of energy production may seem inexpensive and nonpolluting if we consider only dollar costs or limited environmental impact. If viewed as part of a larger system, problems may become apparent. For example, the use of photovoltaic devices seems fairly innocuous. The process seems nonpolluting. Just allow solar radiation to strike a photovoltaic plate and the solar energy will drive electrons through a wire yielding an electric current, electrical energy (see Fig. 11.1). Electrical energy is available for "free."

On closer examination from the point of view of an entire system, we find that there are environmentally negative aspects to the use of photovoltaic power (PV). There are both energy and material costs that go into the fabrication of photovoltaic devices that may appear, superficially, as transparent to us, but which have a negative impact somewhere as the following analysis suggests.

We can examine the **system** from the perspective of the energy and material inputs and outputs that are associated with the fabrication of photovoltaic devices. In addition, large photovoltaic arrays contain electrical storage facilities, batteries, which also have environmental and energy costs in terms of their fabrication and disposal. Hence a more complete analysis of PV usage requires an inventory and assessment of all inputs and outputs.

Viewed as a complete system, solar energy using photovoltaics is neither a source of free energy nor nonpolluting (see Fig. 11.2). In a similar way we can examine the environmental impact of fossil fuel use in terms of a more global perspective.

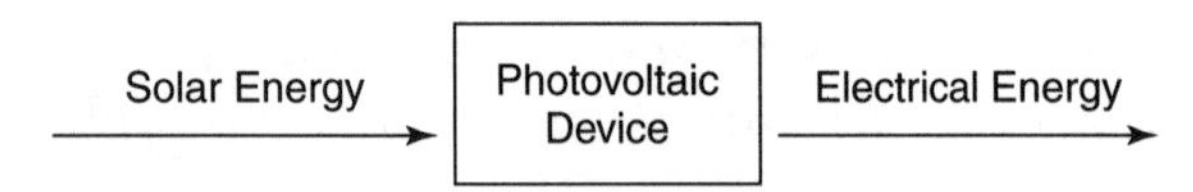


**Figure 11.1** Photovoltaic device.

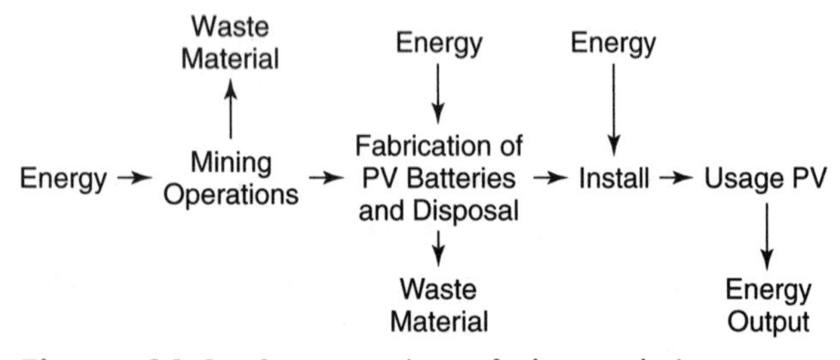


**Figure 11.2** Systems view of photovoltaic energy.

# Systems Approach

Figure 11.3 summarizes the systems connections, which we shall explore.

Briefly, it all began about 15 billion years ago with the birth of our universe through the "big bang." At that point in time, it is theorized that a massive collection of matter and energy concentrated into a relatively small space exploded outward, scattering matter and energy in all directions. Consistent with the first law of thermodynamics, there never was, nor will there ever be, a net change in the amount of matter and energy within our universe. Trapped within our sun, specifically within the atomic nuclei of the sun's matter, are vast quantities of energy. As atoms of hydrogen combine under high temperatures and pressures their nuclei fuse together, forming atoms of helium and releasing tremendous amounts of energy through this nuclear fusion process. Some of the energy released from the sun travels to earth in the form of electromagnetic radiation, which falls upon the leaves of green plants, generally referred to as photosynthetic plants. The process of photosynthesis captures this energy and stores it in the form of chemical bonds within molecules of glucose. Glucose is a molecule that contains carbon taken from the atmosphere in the form of $CO_2$ and hydrogen taken from groundwater. The plant creates glucose, a carbohydrate or carbon-containing molecule, using these materials and the sun's radiant solar energy.

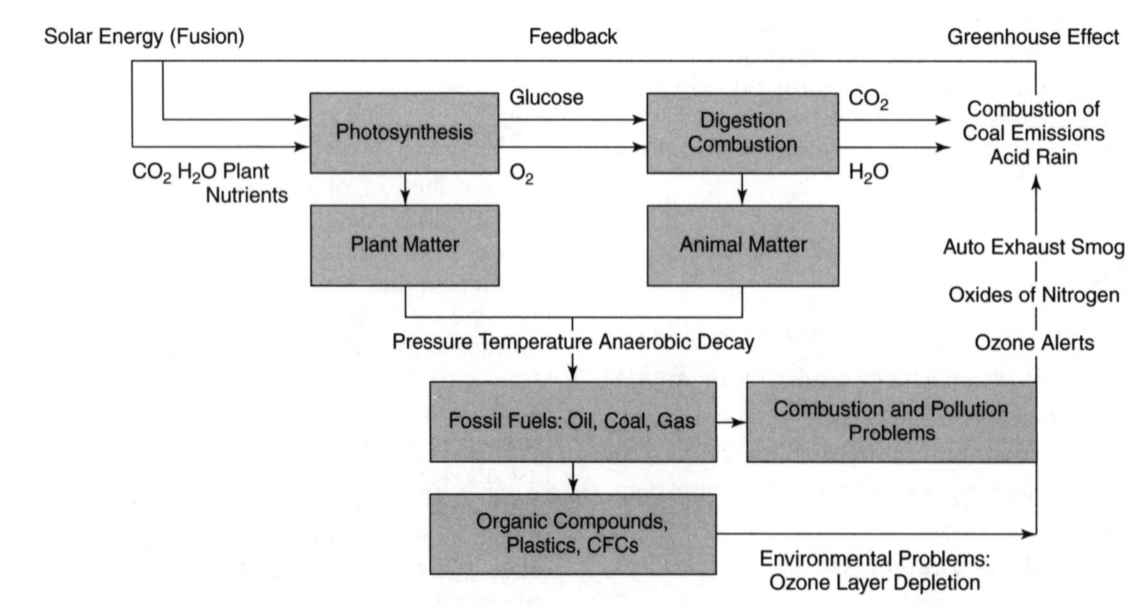


**Figure 11.3** Systems approach to energy usage.

$$6CO_2 + 6H_2O \rightarrow C_6H_{12}O_6 \text{ (glucose)} + 6O_2 \uparrow$$

When animals ingest the carbohydrates manufactured by green plants and digest, or break down, these molecules, the energy released is used to synthesize new molecules within the animal. Thus, the carbohydrates ingested by a herbivore can be broken down and the atoms rearranged to create new molecules such as muscle proteins.

In this way it is seen that energy has flowed from the sun, through the photosynthetic green plants, and into the animals. The energy flow continues as one animal is consumed by another and, when the last consumer dies, the energy trapped within its molecules will either be consumed by decomposer organisms or will remain and will ultimately become fossil fuel. Similarly, materials, namely carbon, can also be seen to flow from the atmosphere, through the photosynthetic green plants, through the animals, and back to the atmosphere when the fossil fuels are burned.

The chart can be broken down into a number of subsections, which focus on different aspects of energy and the environment.

The carbon cycle (see Fig. 11.4) summarizes the transition of carbon dioxide to plant matter to conversion to food (or fossil fuel) and carbon dioxide.

If the rate at which carbon dioxide is returned to the earth, via incorporation into plants or dissolved in water or converted into mineral carbonates or other means, matched the rate at which carbon dioxide is generated, a steady state would exist. As it develops, this is not the case. It appears that more carbon dioxide is being collected in the atmosphere than is returned to Earth. Further, the buildup of the carbon dioxide appears to increase with the growth of fossil fuel use, as noted in the graph in Figure 11.5.

This is further illustrated in the plot of carbon dioxide concentration over the last half century in Hawaii, shown in Figure 11.6. Note that there has been a continuous annual increase in carbon

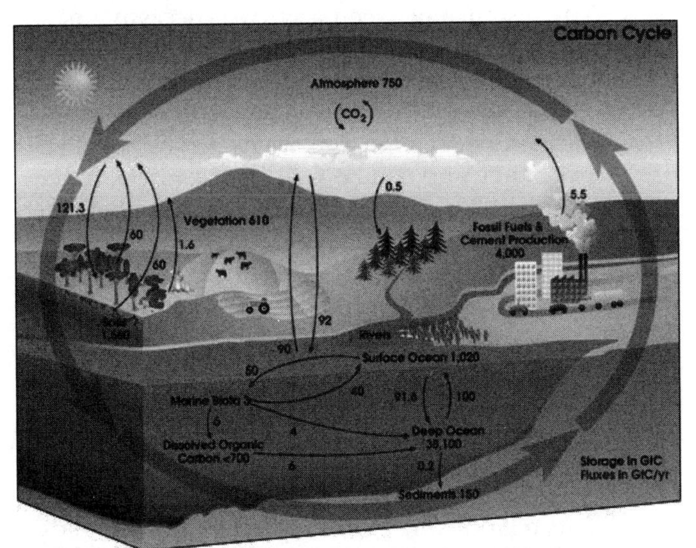


**Figure 11.4** Carbon cycle. *http://earthobservatory.nasa.gov/Library/CarbonCycle/carbon_cycle4.html*

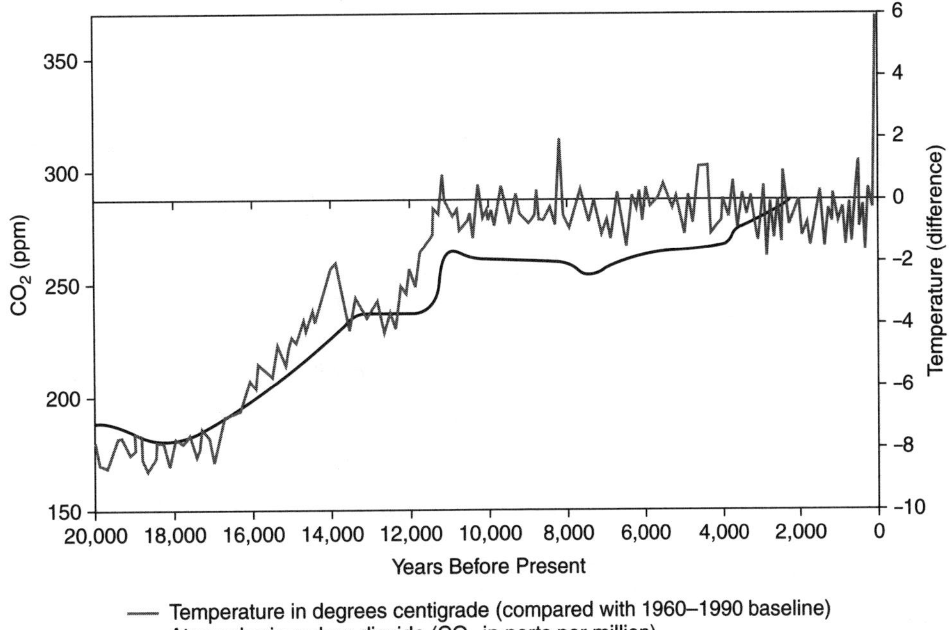


**Figure 11.5** Carbon dioxide concentration change over the centuries. *http://www.brighton73.freeserve.co.uk/gw/paleo/paleoclimate.htm*

dioxide concentration with time. (The annual cyclic variation of carbon dioxide matches variation of $CO_2$ concentration as the growing season comes and goes.)

So what's so bad about a buildup of carbon dioxide? We exhale $CO_2$ and consume the material whenever we drink carbonated soda, which contains $CO_2$ and water. It's not a poison. So what's the problem? First, let us review what constitutes the electromagnetic spectrum.

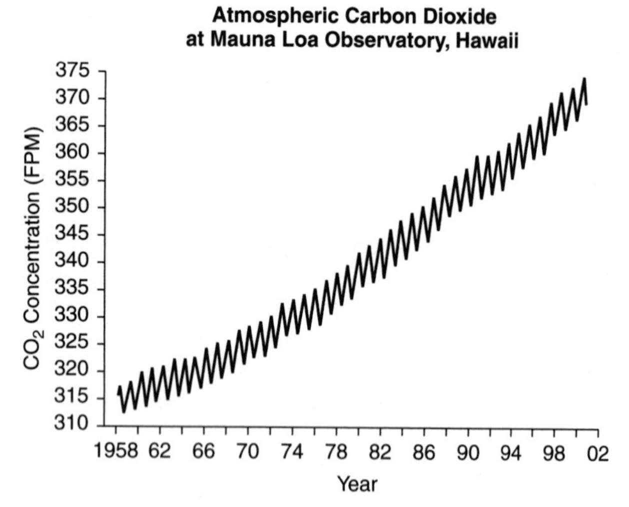


**Figure 11.6** Carbon dioxide accumulation over 45 years in Hawaii. *http://www.oar.noaa.gov/spotlite/archive/image/images/jimomedal_graph.gif*

# Electromagnetic Spectrum

Wave motion is important because waves transfer energy from one point to another without mass motion of the energy carrying particles. Recall that all waves have a wavelength and a frequency associated with them. In order to have a mechanical wave motion we need a source and a medium to carry the wave. For electromagnetic waves, which travel through a vacuum, a medium is unnecessary (see Fig. 11.7).

The distance between any two successive particles in a wave executing the identical motion we found is called the wavelength and symbolized by the Greek letter $\lambda$. The frequency or number of oscillations per second the wave makes is called the frequency and symbolized by the Greek letter $\upsilon$. The time for one oscillation (second/oscillation) is called the period T; note it is equal to $1/\upsilon$. So the velocity of a wave (rate) × time equals distance, or

$$v \times T = \lambda$$

or

$$v = \lambda/T \text{ or } v = (1/T)\,\lambda$$

or

$$v = \upsilon\lambda.$$

This relationship for the speed of a wave holds true for mechanical wave motion and for electromagnetic waves. However, electromagnetic waves travel at the speed of light, c. So for electromagnetic waves:

$$c = \upsilon\lambda$$

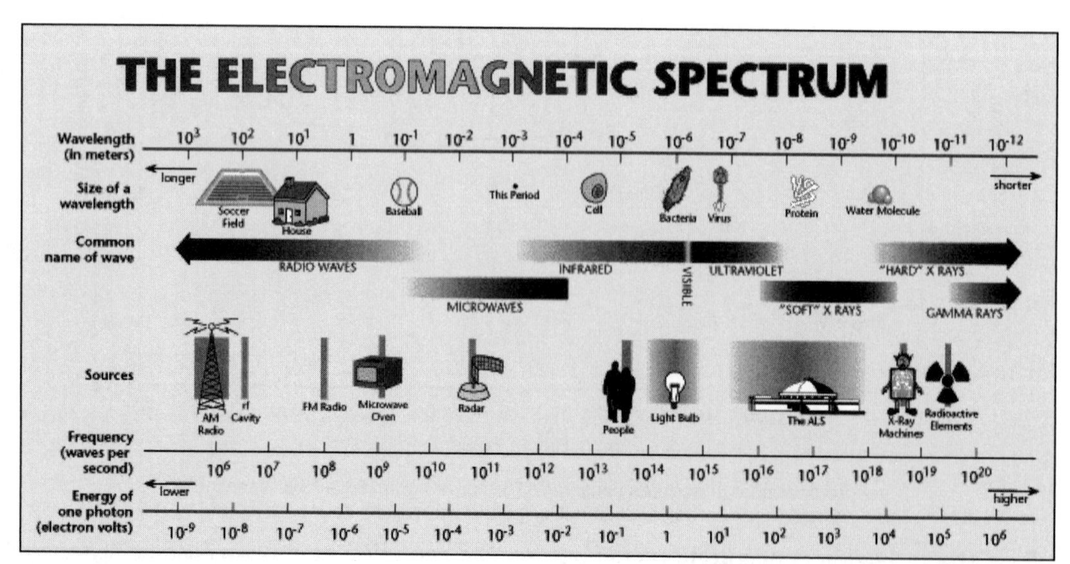


**Figure 11.7** The electromagnetic spectrum. *http://www.lbl.gov/MicroWorlds/ALSTool/EMSpec/EMSpec2.html*

Light, radio waves, TV signals, X-rays, and gamma rays are all types of electromagnetic waves and are distinguished from one another by their wavelength and frequency. For example, an AM radio wave has a wavelength of about the length of three football fields while microwaves have a wavelength of a couple of inches.

We can detect a small portion of the electromagnetic spectrum through our sense of vision. Receptors in our retina are sensitive to the portion of the electromagnetic spectrum that varies in wavelength from violet to red, or $400 \times 10^{-9}$ m to $700 \times 10^{-9}$ m. We, therefore, refer to this region as the visible portion of the spectrum. The other wavelengths in the electromagnetic spectrum are invisible.

Many cameras and TV remote controls use infrared beams. In cameras the focus of an object is adjusted according to the distance obtained from reflecting the beam off the object. Infrared beams have wavelengths greater than the wavelengths of red light in the visible spectrum.

Ultraviolet radiation has shorter wavelengths than visible light, as do X-rays and gamma rays. As a result they are highly penetrating which makes them valuable tools for diagnosis and treatment in medicine and dentistry. The shorter wavelength radiation, because they are highly penetrating, can also cause damage to human tissue and cells and therefore X-ray exposure is carefully monitored and limited.

Microwaves comprise one region of the electromagnetic spectrum and we have become increasingly reliant upon them for communications. Microwave antennas can be seen on the roofs of most tall buildings and are fixtures along our nation's highways in the form of cell phone towers (Fig. 11.8). Now, let us return to a discussion of carbon dioxide and its effects.

## Greenhouse Effect

The problem with increasing levels of atmospheric carbon dioxide is that carbon dioxide is a greenhouse gas. Greenhouse gases act in a fashion similar to the glass of a greenhouse. The greenhouse glass lets light into a greenhouse, and then keeps the heat in the greenhouse once the light strikes the ground and is radiated back as heat. Greenhouse gases act like the greenhouse glass. The greenhouse gases (carbon dioxide, methane, water, nitrous oxide, and CFCs, among others), residing above the Earth, are transparent to visible radiation and so allow sunlight to pass through and strike the Earth. Once the radiant energy strikes the Earth a portion of the light is converted to heat and is reradiated outward. While the greenhouse gases are transparent to visible radiation, they absorb the infrared radiation (heat) and so the Earth retains heat. Figure 11.9 and Figure 11.10 illustrate these points.

The graph (see Fig. 11.5) shows that the carbon dioxide concentration increase parallels an increase in the global temperature. These changes are significant. An increase in carbon dioxide concentration and global temperature can have dire consequences on the planet, which may range from melting of glaciers and polar ice caps to alteration of bio- and ecosystems, changing shorelines and rain patterns, etc.

So what does this all mean? It is pretty involved. It is reasonable to assume that greenhouse gases ought to have an effect on climate. Certainly the graph in Figure 11.5 indicates correlation. However, there have been large variations in the Earth's climate prior to the industrial era due to other causes. Changes in solar intensity, production of sunspots, and other natural variations,

**Figure 11.8** Cell phone antenna in disguise.

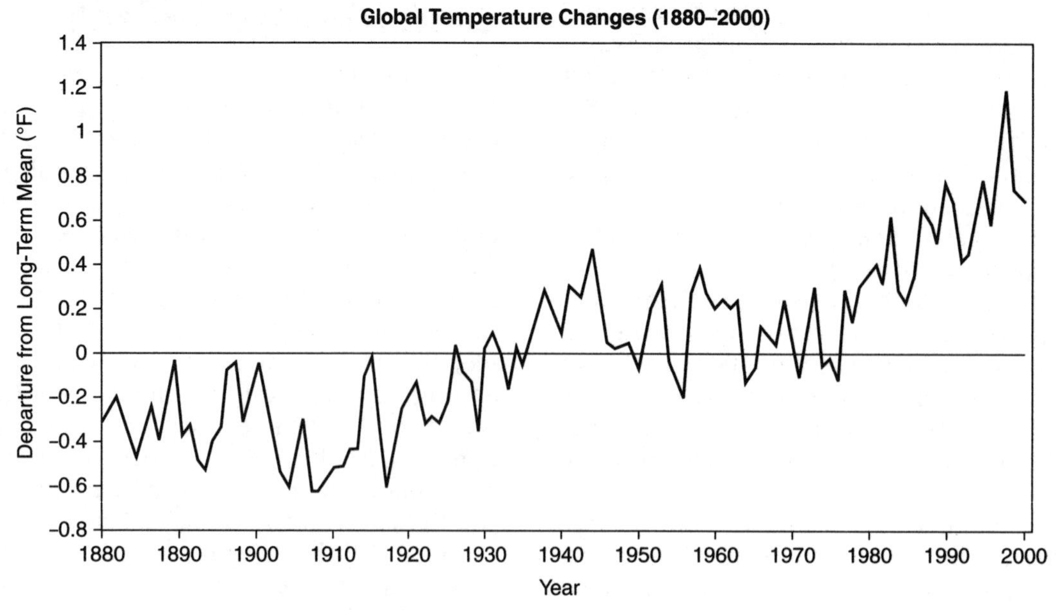


**Figure 11.9** Global temperature increases since 1860. *http://waves.marine.usf.edu/slrise_menu/greenhse.gif*

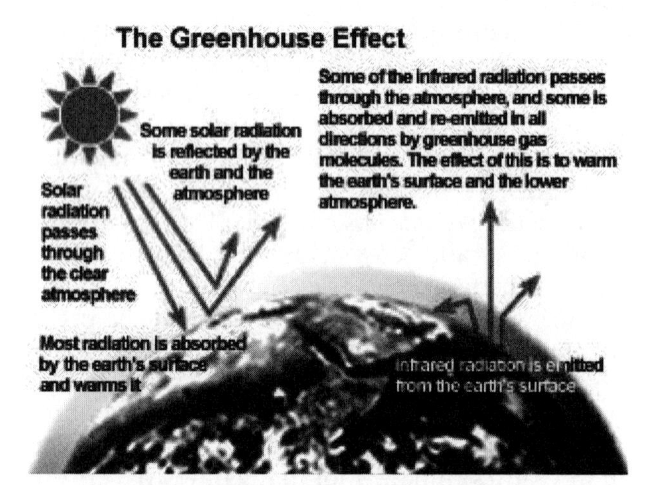


**Figure 11.10** Effects of greenhouse gases. *http://yosemite.epa.gov/oar/globalwarming.nsf/earth.jpg?OpenImageResource*

including volcanic activity and generation of aerosols and clouds, have had a profound influence on the Earth's temperature. It is logical to state that there is a natural variability associated with the Earth's temperature that depends on a number of factors.

Some of these factors are beyond our control. While there are secondary effects that can arise with greenhouse gas buildup, which mitigate its influence, greenhouse gases do appear to have an effect on the Earth's temperature and to an extent can be controlled. Since there is no model that can accurately predict the Earth's temperature based on carbon dioxide (and other greenhouse gases) buildup (because of the other uncertainties), it makes sense to monitor the influential factors. The general idea is to err on the side of caution; that is, to try to be more efficient in energy use involving fossil fuel combustion and to minimize carbon dioxide production.

The 1997 Kyoto Protocol seeks to globally mandate reduced carbon dioxide emission levels. The economic consequences to the United States for agreeing to the protocol are staggering; the United States has backed out of the agreement, which has infuriated the European Union. Now we have the confluence of energy, the environment, and politics.

Of course politics plays a significant role in access and control over energy resources, as noted by the fact that in the past decade the United States has been involved in conflicts in Kuwait and Iraq, both nations of which have considerable oil reserves.

As noted earlier, carbon dioxide, one of the products of fossil fuel combustion, is being produced faster than it is being reincorporated back into plant matter (the reduction of atmospheric carbon dioxide to a nongaseous form, carbon sequestration) owing to the burning of fossil fuel and depletion of forests and the like. Depending upon the nature of the fossil fuel (oil, coal, gas) and its use, different kinds of environmental impact are possible. To better understand this point, some appreciation of the nature of fossil fuel is necessary.

## Natural Gas

Natural gas is composed of small molecules, which have few atoms. These hydrocarbons (containing only the elements C and H) exist as gases but can

**TABLE 11.1 Natural Gas Components**

| Compound Name | Chemical Formula | Boiling Point (Celsius) |
|---|---|---|
| Methane | $CH_4$ | −162 |
| Ethane | $C_2H_6$ | −89 |
| Propane | $C_3H_8$ | −42 |
| Butane | $C_4H_{10}$ | −1 |

**TABLE 11.2 Examples of Alkanes**

| Number of Carbon Atoms | Greek Number | Formula | Name |
|---|---|---|---|
| 5 | penta | $C_5H_{12}$ | pentane |
| 6 | hexa | $C_6H_{14}$ | hexane |
| 7 | hept | $C_7H_{16}$ | heptane |

be liquefied under pressure. They are shown in Table 11.1.

While all are gases at room temperature (22 degrees C), propane and butane boil at sufficiently high temperatures so that under pressure they can be condensed into a liquid. For example, examination of a "butane" lighter allows one to see liquid butane; LP gas (or liquid propane) is propane gas that has been condensed. As you may be aware, it is used for cooking and heating purposes when a more transportable fuel is required, as in gas barbecue grills.

Natural gas delivered directly from the utility through pipes under the street to the home consists mainly of methane gas plus minor amounts of the other hydrocarbons and a small amount of a smelly substance such as tertiary butyl mercaptan added to warn of the gas's presence should there be a leak. The blue flame notes the efficient combustion of methane when it burns. A yellow flame indicates combustion, which does not efficiently convert all the methane (or other fossil fuel) to carbon dioxide; the yellow flame reflects the fact that the fuel was only partially converted to carbon dioxide, some of it becoming just the element carbon (C). The yellow glow is simply the heated carbon glowing. An example of a complete combustion is noted as follows:

Methane + Oxygen → Carbon dioxide + Water

Natural gas is an ideal fuel for home heating use since it doesn't have to be stored as coal or oil. Since it is a gas, it can be readily transported great distances via pumps and pipes.

The natural gas components just cited are also examples of alkanes, a group of hydrocarbons, which have the general formula $C_nH_{2n+2}$. (Hydrocarbons are also part of a larger group of compounds called "organic" compounds. The term organic here does not have anything to do with organic cultivation. It simply refers to the presence of carbon.)

For example, for the case where n is 3, we have $C_3H_{2(3)+2}$, or $C_3H_8$, or propane. While the names

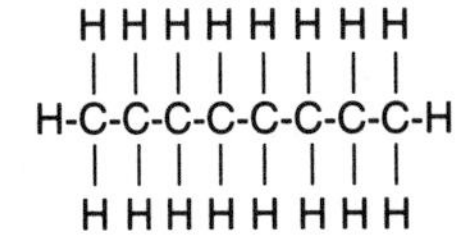


**Figure 11.11** n-octane.

```
    CH3CH3 H   H
     |  |  |   |
CH3-C - C - C —C-H
     |  |  |   |
    CH3 H  H   H
```

**Figure 11.12** Isooctane.

for the first four alkanes do not follow any particular pattern, there is a logical sequence for compounds of 5 carbon atoms or greater. For alkanes the compound name is the Greek prefix for the number of carbon atoms plus an "ane" ending. See Table 11.2. Try writing the formula for octane.

(The answer: The formula for octane is $C_8H_{18}$.)

The compounds just examined are volatile liquids; that is to say they evaporate quickly and are quite flammable. They can be used as fuels. For example, octane had wide acceptance for this purpose since volatility is just about right for ease in storage and evaporation.

Actually there are a number of compounds that have the formula $C_8H_{18}$, and all are called octanes. (Different compounds having the same formula are called isomers.) The explanation for these isomers is as follows: Because carbon atoms can form four chemical bonds, while hydrogen atoms form one chemical bond, there are a number of ways that a compound with the formula $C_8H_{18}$ can be assembled, as noted in Figure 11.11. The major component of gasoline is isooctane (Fig. 11.12).

While all of the liquid alkanes just mentioned have some potential for use as fuel for an automobile, some are better than others. Heptane, for example, is a poor fuel because it tends to ignite prematurely. After gasoline is injected into the cylinders of an automobile engine, the gasoline and air mixture are subjected to compression. If the fuel ignites too early during this period, a knocking sound will be heard and more importantly the car will perform poorly.

On the other hand, isooctane can be compressed without premature combustion. It has excellent antiknock characteristics. The octane rating of gasoline is related to the antiknock characteristic of the fuel. An octane rating of 100 is assigned to isooctane. An octane rating of 0 is assigned to heptane. So a fuel with an octane rating of 85 behaves like a mixture that is 85% isooctane and 15% heptane.

In the past, an additive (tetraethyl lead) was added to gasoline to increase the octane rating. This was a less expensive route to antiknocking enhancement, rather than altering the heptane/isooctane ratio. This is no longer the case. Law now prohibits addition of tetraethyl lead since society has recognized that lead emissions are unhealthy from a variety of perspectives. Aside from health and legal issues, combustion of gasoline with lead additives would be ruinous to the catalytic converter system and would render it inoperable.

## Fossil Fuel and Auto Emissions

So why do cars have catalytic converters? The answer, of course, is to reduce the harmful exhaust emission, but what are they?

The combustion of gasoline in air might be expected to occur as follows:

Gasoline + air → carbon dioxide + water + energy

While this equation represents an idealized situation, the combustion of gasoline occurring in an internal combustion engine of an automobile results in the formation of other species of molecules beyond those mentioned in the equation, including carbon monoxide, some hydrocarbon emissions (small fragments of unburned molecular "pieces"), and oxides of nitrogen. Part of the reason for the unburned hydrocarbon/carbon monoxide formation is that the very rapid burning of gasoline required to generate power (a large amount of energy over a short time frame) results in incomplete combustion. Carbon monoxide, an odorless gas and a deadly poison, is one of the products.

Carbon monoxide is a problem because it represents an inefficient combustion. (There is still fuel energy available) and more importantly because it reacts with hemoglobin much more strongly than oxygen reacts with hemoglobin. The net effect is that in an atmosphere of carbon monoxide, hemoglobin is not oxygenated and so cells are suffocated and die for lack of oxygen. The unburned hydrocarbons also contribute to air pollution. They are part of a pollution that is described by the term *photochemical smog* (smog mixture: smoke and fog). The automobile is the primary cause of this problem.

To gain a greater understanding of this kind of pollution consider the combustion of gasoline more completely than before.

Gasoline + Air → Carbon monoxide + Carbon dioxide + Water + Unburned hydrocarbons + Oxides of nitrogen

At first it might be a bit of a surprise to note that oxides of nitrogen are products of the combustion; however air contains both oxygen and nitrogen, and the spark plug temperature, when the gasoline is burned, is sufficient to enable oxygen and nitrogen to react to form.

$$N_2 + O_2 \rightarrow 2NO$$

Once the NO is formed it reacts in air to form nitrogen dioxide,

$$2NO + O_2 \rightarrow 2NO_2$$

Generally the level of $NO_2$ (a brownish gas) builds up as a result of the morning rush hour so by midmorning a haze may be discerned. Sunlight decomposes the $NO_2$ as noted:

$$\text{Sunlight} + NO_2 \rightarrow NO + O$$

The use of (sun) light to bring about the reaction makes the process a photochemical reaction. In regions where there is extensive automobile use and especially where it is sunny, the effects are magnified.

The atomic oxygen, O, is a free radical, a highly reactive species. (Free radicals are atoms or groups of atoms with unpaired electrons. They are very reactive. Free radicals are thought to be important factors in aging and cancer.) The reaction that ensues is

$$O + O_2 \rightarrow O_3.$$

That is, the photochemical process ultimately results in the formation of ozone, $O_3$. As we will see shortly, in the upper atmosphere ozone is a protecting molecule. However on the surface of the earth, ozone is a deadly poison and further contributes to air pollution.

Extended or repeated exposure to ozone may cause health problems, especially for the young and old and those suffering from ailments of the respiratory tract. The health problems that can arise include chest pains, bronchitis, and reduced lung capacity.

The EPA has developed an air quality index (AQI) for reporting ozone levels each day. AQI ranges from 0 to 500. Any number above 100 is deemed to be unhealthy. (In the United States the number rarely goes beyond 300.) Depending on the AQI an ozone alert may be called. During these periods those individuals who are most susceptible should remain indoors, car pooling is advised, and voluntary actions to reduce emissions are welcomed.

One of the consequences of photochemical smog is eye irritation. The ozone molecule produced goes on to react with many different substances, including car tires and hydrocarbons (like those unburned hydrocarbons emitted in exhaust gas) to produce irritating organic molecules and powerful tear-producing compounds. These lachrymators (named after the tear gland, or lachrymal gland) can cause breathing difficulties. PAN peroxyacetyl nitrate is one such example.

As noted, automobile emissions and sunlight are the cause of photochemical smog. In sunny areas with high population densities and disproportionate dependency on the automobile (like Los Angeles), the problem is significant. How can it be reduced?

Certainly one way involves the reduction of dependency on automobiles powered by fossil fuel. This could include more mass transit, hybrid cars, cars that have more efficient engines, or cars with reduced emissions. Attempts are being made to address the emissions problem. Cars are now equipped with a catalytic converter, which essentially processes the auto exhaust into emissions that are more benign.

The catalytic converter oxidizes the carbon monoxide and unburned hydrocarbons to carbon dioxide and water and reduces the oxides of nitrogen to nitrogen and oxygen, as the following illustrates.

Exhaust from engine enters the catalytic converter and is transformed:

$$CO, \text{Unburned Hydrocarbons}, NO_x \rightarrow O_2, NO_2, H_2O, CO_2.$$

The net effect of the automobile catalytic converters then is to improve the quality of air by reducing the amount of "harmful" emissions generated. A more effective way of improving air quality, however, is to drive gasoline-burning vehicles less.

Nitrogen and oxygen gas, which constitute major components of air (see Table 11.3), also enter the atmosphere through other routes, as noted in Figure 11.13.

**TABLE 11.3 Composition of Dry Air at Sea Level**

| Gas | Percentage by Volume |
|---|---|
| Nitrogen | 78.084 |
| Oxygen | 20.946 |
| Argon | 0.934 |
| Carbon Dioxide | 0.033 |
| Neon | 0.00182 |
| Helium | 0.00052 |
| Krypton, Hydrogen, Xenon, Ozone, Radon, etc. | 0.00066 |

Note: The gases listed here are the components in pollution-free dry air at ground level expressed as percent by volume.

*http://www.arm.gov/docs/education/background/compositionatmos.html*

The oxygen cycle (Fig. 11.14) illustrates some other ways that carbon dioxide may be sequestered. Note that there is ozone in the *upper atmosphere*. Its importance we will examine later.

## Coal

Another kind of fossil fuel or carbon-based air pollution also exists. This fossil fuel is coal. Coal, the solid form of fossil fuel, represents about 90% of the world's fossil fuel reserves. It tends to be a less desirable fuel for use as a source for domestic heat because delivery and automated use are less convenient, and gas and oil tend to be cleaner burning. (There are, however, processes for the gasification of coal. Such gas is called syngas, for synthetic gas.)

There are four ranks of coal, each of which represents a higher percentage of carbon present. Anthracite coal, the richest in carbon, can contain over 95% and yield the highest energy upon combustion (26 million BTU per ton).

Coal is a solid material, which as noted earlier, is composed of primarily plant material that has been subjected to anaerobic conditions involving high pressure and temperature. The resulting product contains primarily carbon and some hydrogen but also other elements as well. These other species are instrumental in the production of air pollution.

A very high-energy fuel, coal—which is plentiful—appears to be a sensible energy source to use to generate electricity especially with dwindling supplies of the natural gas and petroleum. So what's the problem?

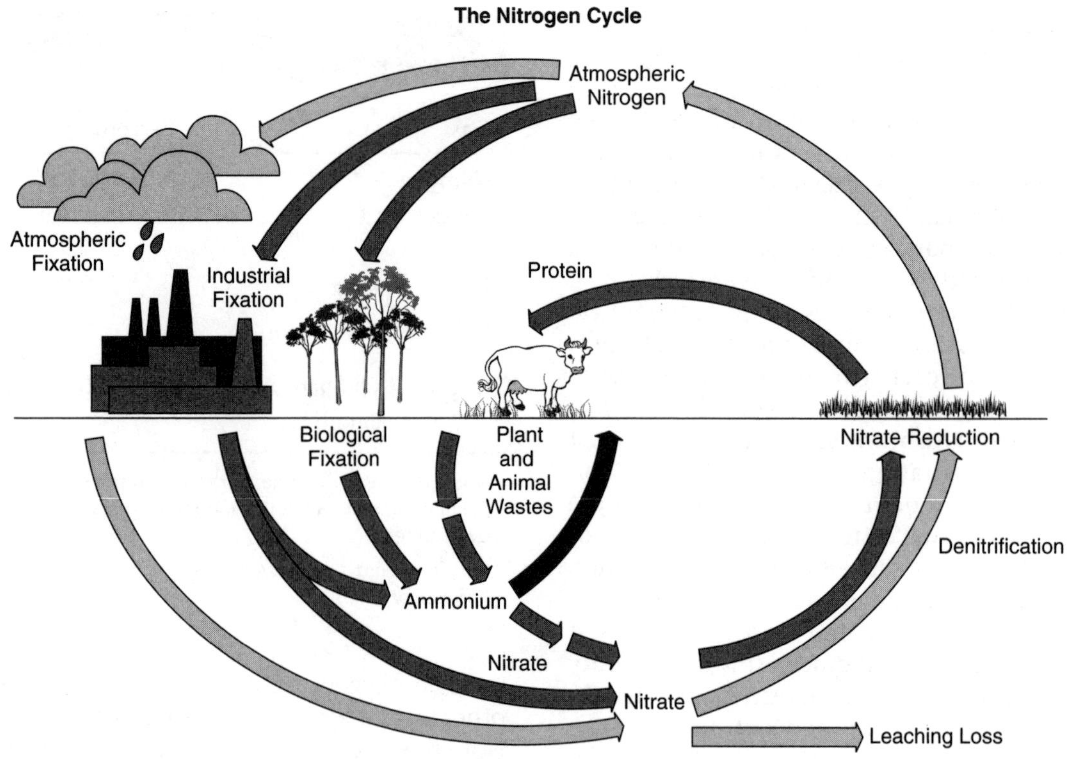

**The Nitrogen Cycle**

**Figure 11.13** Nitrogen cycle. *http://www.swrcb.ca.gov/rwqcb4/html/meetings/tmdl/calleguas%20_creek/02_0830/02_0830_Figure%204-1.pdf*

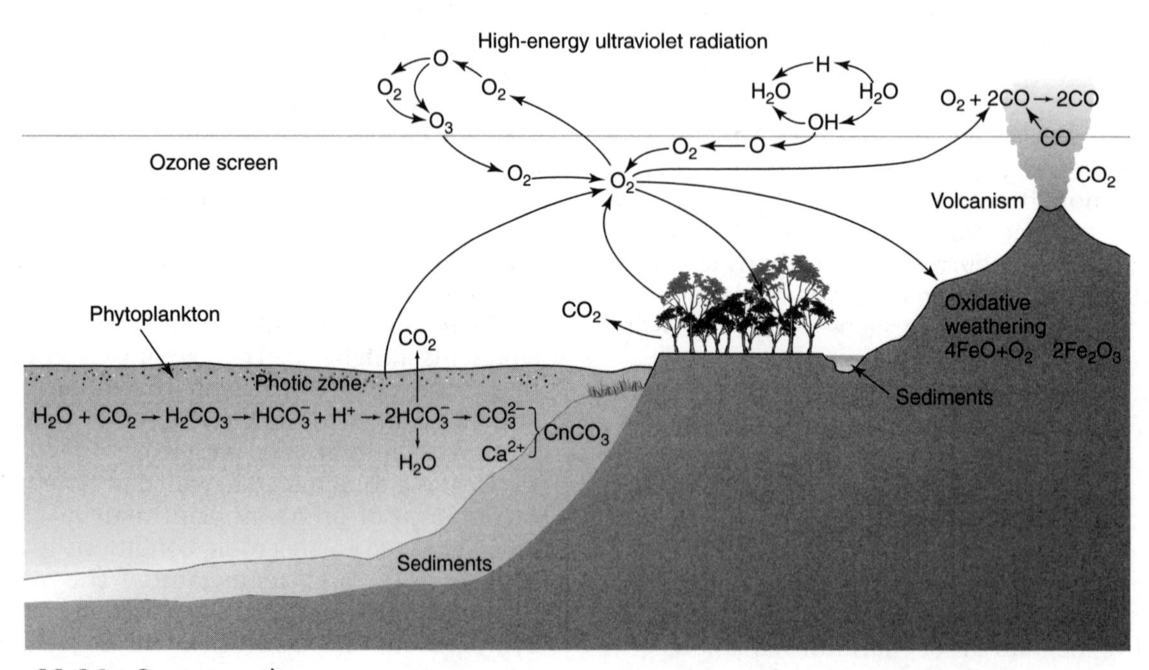


**Figure 11.14** Oxygen cycle.
(R. Chang, *Chemistry,* 7th Ed, McGraw Hill, p. 704).

Starting at the beginning, the mining of coal is a very dangerous occupation. In addition, the environmental effects of the mining operation upon the landscape can have significant deleterious impact.

Depending on the source of coal, there may be present sulfur compounds (pyrite), incombustible minerals, and mercury in some form. Combustion of coal-bearing substances is a big problem. (Coal works the same way as other energy sources. The

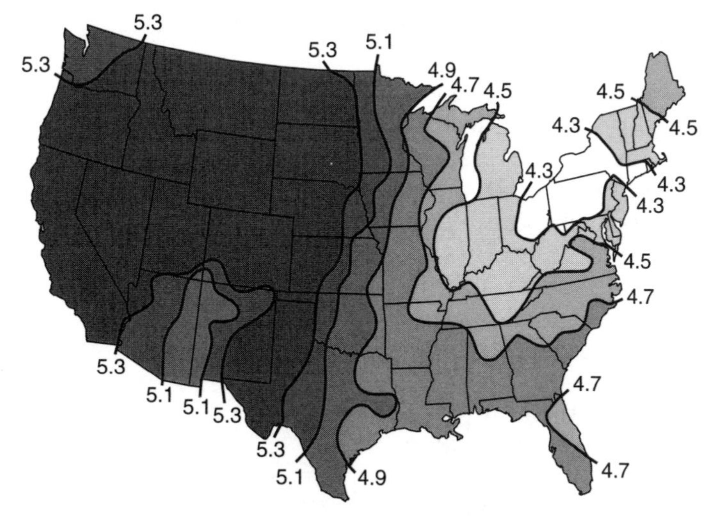


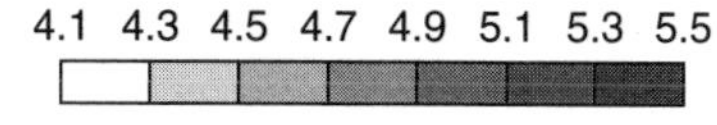


**Figure 11.15** Acid rain. *http://pubs.usgs.gov/gip/acidrain/gif/fig04.gif*

combustion generates heat, which boils water, which makes steam, which turns a turbine.)

It has been suggested that coal-fired power plants in which the coal is not "cleaned up" can produce 96% of the industries' sulfur dioxide and 99% of the industries' mercury pollution, nitrogen oxides, soot, and fly ash.

Mercury is a very poisonous substance and like lead should not be put into the atmosphere. Sulfur dioxide is an irritating and toxic substance. While it can be tolerated to a degree at very low concentrations, its effects become enhanced in the presence of fly ash (unburned ash remaining after the coal's combustion). The synergistic effect of fly ash and sulfur dioxide is especially bad on the respiratory system for the very young and very old.

Sulfur dioxide has another far-ranging effect on the environment, the production of acid rain. Acid rain is rain that is significantly richer in the hydrogen ion concentration than pure water. One measure of acidity is the pH. A pH of seven indicates a solution that is neutral, neither acidic nor basic. A pH of less than seven indicates a solution that is acidic. The lower the pH is, the more acidic the solution.

The sulfur dioxide produced via the combustion of coal-bearing sulfur compounds is lofted into the atmosphere via the industrial smokestacks, which protect the local community. The acid rain is generated via the oxidation of sulfur dioxide to sulfur trioxide in the presence of air. The sulfur dioxide in the upper atmosphere reacts with (rain) water to produce sulfuric acid ($H_2SO_4$), which has the potential to alter the pH of rivers and streams and adversely affect various ecosystems. The following equations summarize the process:

$$S + O_2 \rightarrow SO_2$$

$$SO_2 + 2O_2 \rightarrow SO_3$$

$$SO_3 + H_2O \rightarrow H_2SO_4$$

The acid rain can also react with statues and structures that are carbonate- (marble) or metal-based to leave an aesthetically unsightly coating. Various techniques may be used to make coal a more suitable and less polluting fuel, including removal of sulfur-bearing impurities and employing smokestack devices that cut down on emissions. This kind of pollution in which coal is burned and fly ash and sulfur dioxide formed is called London Smog.

In the United States acid rain is not uniformly distributed. Note that the pH is lowest in the Northeast (see Fig. 11.15). It is presumed that sulfur bearing coal burned in the more westerly regions (Illinois, Indiana, Michigan, Ohio, etc.) yield the sulfur dioxide which is blown eastward via the prevailing winds.

We see therefore that the effects of utilization of fossil fuels have a far-ranging effect on the environment. Thus far these environmental effects have been based on the combustion process. Reference to Figure 11.4 illustrates the nature of these effects in terms of the carbon cycle; however, carbon-based compounds also have an indirect environmental effect.

Petroleum, natural gas, and coal are resources which, of course, represent different forms of fossil fuel that can be used for production of energy. These resources have other uses. They can be transformed into a variety of other useful products, which range from plastics to pharmaceuticals to solvents to a host of other useful and practical materials. The element carbon, the ubiquitous element present in all fossil fuels, is responsible for these possibilities.

The element carbon has the ability to bind with itself to form stable species. In addition, it can form four chemical bonds. As a result there are huge numbers of different compounds in which carbon is present. As it develops there are more compounds of carbon (and hydrogen) than all of the other elements put together.

The field of study involving carbon and hydrogen, as noted earlier, is called organic chemistry. (The term organic chemistry was once used to mean the study of compounds from living organisms. Now we know that compounds from living organism can also be synthesized in the laboratory. More than 19 million new organic compounds have been made and hundreds of new ones are being made daily.) Fossil fuel, as well as natural products (from plants, trees, animals, etc.), serve as the resource. So how do these carbon-based compounds affect the environment? The production of plastics, which persist in the environment, and compounds, which affect the ozone layer, are two examples.

## PLASTICS

There are various kinds of plastics. They tend to persist in the environment because they are not biodegradable and as a result, the accumulation of such material is troubling since it represents visual pollution. It is instructive to examine a few cases relating to the nature of plastic since it also represents an approach to the study of ozone depletion.

One kind of plastic is formed by the reaction of ethylene (or derivatives of ethylene) to form long chained molecules.

The structure of ethylene is

```
H H
| |
C=C
| |
H H
```

Ethylene is called the monomer (mono = one, mer = part). Under suitable conditions ethylene can be made to react with itself to produce a long chained molecule or polymer (poly = many, mer = parts). Teflon is an example of a polymer where there are carbon fluorine bonds present.

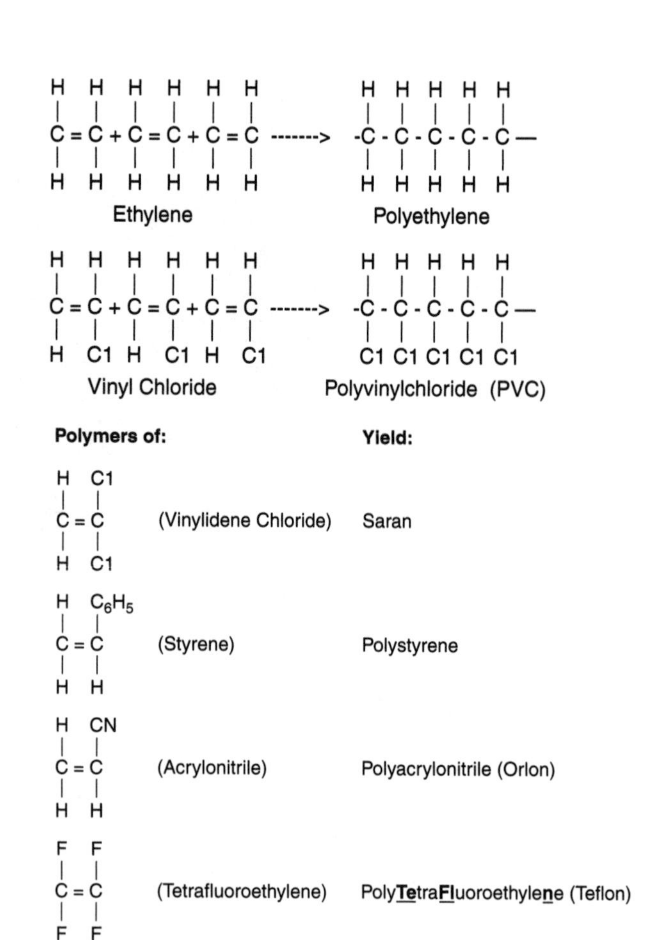


Other examples of carbon compounds where carbon fluorine bonds are present include freons. Freons, however, are much smaller molecules. Some common freons are: $CFCl_3$ (Freon 11), $CF_2Cl_2$ (Freon 12), $C_2F_3Cl_3$ (Freon113), $C_2F_4Cl_2$ (Freon 114).

These compounds are ideal refrigerants. They are readily liquefied, relatively inert, nontoxic, noncombustible, and volatile. They have been used as refrigerants, aerosol propellants, and solvents. The compounds are also known as CFCs (chlorofluorocarbons). More than one million tons of freons were produced in the late 1970s, and then the bad news came.

As noted earlier, freon is inert, so it is stable, and freon that leaks out of air conditioners or is used as a spray propellant doesn't break down—it just "hangs around." The CFCs thus produced slowly diffuse unchanged into the stratosphere, where ultraviolet radiation in the vicinity of 200 nm causes them to decompose (so the freons are not totally inert).

Freon, for example, reacts as shown by the following equations:

$$\text{CFCl}_3 \xrightarrow{(UV)} \text{CFCl}_2 + \text{Cl} \text{ (the Cl is a very reactive free radical).}$$

$$\text{Cl} + \text{O}_3 \rightarrow \text{ClO} + \text{O}_2.$$

One chlorine atom can destroy up to one hundred thousand ozone molecules before it is destroyed. (The free radical Cl atom acts like a catalyst and is regenerated in subsequent reaction steps.)

We have learned that the presence of ozone at the surface of the Earth is bad. Ozone alerts occur as a consequence of photochemical smog, etc. This is true. Ozone in the upper atmosphere, however, is a good thing. In the upper atmosphere ozone reduces the amount of ultraviolet radiation that reaches the surface of the Earth (see oxygen cycle, Fig. 11.14) by absorbing this energy. As the concentration of ozone in the upper atmosphere is diminished, more UV radiation will strike the Earth. The problem with an increase in the intensity of UV radiation is that such radiation can be cancer causing.

Ozone depletion is a serious matter. In recognition of the problem and to curtail production an international treaty, the Montreal Protocol, was signed by many of the industrialized nations. The treaty was supposed to end production of CFCs by the year 2000. Not all nations appear to be abiding by the treaty.

Examination of Figure 11.3, showing the flow of carbon energy and pollutants in the environment, now gives us greater insight in to the "interconnectedness" of fossil fuel use and its impact on the environment and ultimately on our health and the health of our planet.

In this chapter we have examined the carbon cycle and seen both the positive and negative impacts of energy use based on carbon compounds. On the positive side there is a gain in energy, creature comforts, mobility, and personal freedom, while negative aspects may include health problems associated with a variety of primary and secondary pollution effects, depletion of vital resources, and the potentially disastrous consequences of global warming. The economic and geopolitical issues are another important consideration.

One important aspect of cycles such as the carbon cycle is homeostasis, or balance between the inputs and outputs. The reason for concern regarding the greenhouse effect, as we have seen, revolves about the fact that carbon dioxide conversion into plant material lags far behind the consumption of carbon-based matter and its subsequent transformation to carbon dioxide. Hence there is a buildup of carbon dioxide and the greenhouse effect.

There are, of course, other cycles in nature that are critical to our survival and that are linked to the carbon cycle. One of these is the oxygen cycle, which has been presented earlier.

A variety of factors militate against the maintenance of homeostasis regarding the oxygen cycle. These include the reduction in oxygen production as a consequence of reduced photosynthesis arising from destruction of forests. Consumption of oxygen as more and more carbon-based fuels are burned represents another route of oxygen of reduction. As the oxygen cycle figure notes, the carbon dioxide formed can dissolve in water to form an insoluble solid in the presence of calcium. This material formed, calcium carbonate, is one route to sequestering the greenhouse gas.

While the variation of the concentration of elemental $O_2$ in the atmosphere does not appear to be a concern, the depletion of $O_3$ is worrisome. Ozone production is part of the oxygen cycle. It is produced in the upper atmosphere by the action of UV radiation on $O_2$. As noted earlier, upper atmosphere ozone is being depleted by the reaction with atomic chlorine from CFCs. This departure from homeostasis has sparked international concern. Production of ozone at the Earth's surface is related to automobile emissions, certain industrial processes, and electrical storms.

The oxygen cycle also involves other types of oxidation besides combustion of fossil fuels. These include respiration and oxidative weathering—that is, the combination of molecular oxygen with metals. Rusting, the reaction of iron and oxygen, is one such example. Its economic consequences are staggering.

Finally, oxygen is required by aerobic bacteria in order for the decay process to occur, the products of which include carbon dioxide and water. In the absence of oxygen, anaerobic bacteria are the primary decomposers and the decay process involves the production of vile-smelling $H_2S$ rather than $H_2O$. The BOD, or biological oxygen demand, is an index of the amount of oxygen required by aerobic bacteria for decomposition of matter.

The nitrogen cycle, also shown earlier, involves the circulation of nitrogen in a variety of forms through the environment. The element exists as a component of proteins, which are predominantly composed of carbon, hydrogen, oxygen, and nitrogen. The proteins or their metabolism products, plant and animal waste, is acted on by bacteria, which ultimately leads to the formation of ammonium ($NH_4^+$), nitrite ($NO_2^-$), and nitrate ($NO_3^-$) compounds. These materials may be recycled into new proteins or may undergo denitrification to yield free nitrogen. The cycle is continued as atmospheric nitrogen is converted into nitrogen compounds that can be used by plants either via industrial or biological fixation.

# Review Questions

1. Why don't photovoltaic devices produce free energy? How are they polluting?
2. Describe the carbon cycle.
3. How does carbon dioxide contribute to the greenhouse effect?
4. What is the greenhouse effect?
5. What role do the combustion of natural gas, coal, and oil play in contributing to the greenhouse effect?
6. Discuss the nitrogen cycle in detail.
7. How does the combustion of natural gas contribute to polluting the environment?
8. Why do cars have catalytic converters?
9. What is the AQI developed by the EPA?
10. What are some of the health problems caused by the combustion of fossil fuels?

# Multiple Choice Questions

1. Greenhouse gases are transparent to visible radiation but absorb _____.
   a. microwave
   b. ultraviolet
   c. infrared
   d. radio
2. The 1997 Kyoto Protocol seeks to globally mandate _____ reduced emissions.
   a. coal
   b. ozone
   c. carbon dioxide
   d. carbon monoxide
3. Natural gas contains _____.
   a. carbon and oxygen
   b. carbon and hydrogen
   c. nitrogen
   d. oxygen
4. In the presence of carbon monoxide, hemoglobin is not oxygenated and so _____.
   a. cells die
   b. carbon dioxide levels increase
   c. it turns into hydrogen
   d. it becomes gamma globulin
5. Photochemical smog results in the formation of _____.
   a. ozone
   b. nitrous oxide
   c. nitrogen dioxide
   d. carbon dioxide
6. Catalytic converters can convert _____.
   a. oxygen to carbon dioxide
   b. carbon monoxide to carbon dioxide
   c. nitrogen to nitrogen dioxide
   d. carbon dioxide to carbon monoxide
7. Coal represents _____% of the world's fossil fuel.
   a. 90
   b. 75
   c. 50
   d. 25
8. Acid rain is due to a high content of _____ in the air.
   a. carbon monoxide
   b. sulfur dioxide
   c. nitrous oxide
   d. ozone
9. UV radiation striking the earth increases as _____ decreases in the atmosphere.
   a. oxygen
   b. carbon dioxide
   c. ozone
   d. nitrogen
10. Which of the following is not a hydrocarbon fuel?
    a. natural gas
    b. ethane
    c. oil
    d. uranium
11. The greenhouse effect is caused by _____.
    a. absorption of UV radiation by greenhouse gases
    b. emission of UV radiation by greenhouse gases
    c. absorption of infrared radiation by greenhouse gases
    d. emission of infrared radiation by greenhouse gases

12. Which of the following is most responsible for the production of acid rain?
    a. the greenhouse effect
    b. nitrogen
    c. combustion of sulfur-bearing fuels and its interaction with water
    d. UV action on ozone

13. The higher the octane number, _____.
    a. the worse the antiknocking
    b. the better the knocking properties
    c. the better the antiknocking
    d. the more CO formed

14. In the system involving photosynthesis followed by combustion, which of the following best describes substance(s) involved in feedback?
    a. oxygen
    b. carbon dioxide
    c. ozone
    d. nitrogen dioxide

15. One advantage of a nuclear power plant over a conventional fossil-fuel-using plant is that _____.
    a. the nuclear plant does not contribute as much greenhouse gases (carbon dioxide)
    b. the nuclear plant generates steam
    c. the nuclear plants are inherently safe
    d. nuclear plants can be located anywhere

16. The immediate cause of depletion of the ozone layer in the upper atmosphere is _____.
    a. Cl atoms
    b. UV radiation
    c. CFCs
    d. pesticides

17. Acid rain has a pH _____.
    a. greater than 7
    b. less than 7
    c. equal to 7
    d. Can't tell, it varies.

18. Ozone is a pollutant at ground level. It is produced as a result of high temperature combustion in air of fossil fuel and _____.
    a. a catalytic converter
    b. sulfur
    c. chlorophyll
    d. sunlight

19. Which of the following is the monomer of the polymer?

    H Cl H Cl H Cl
    I—I—I—I—I—I
    –C–C–C–C–C–C–
    I—I—I—I—I—I
    H Cl H Cl H Cl

    a.
    H Cl
    I—I
    C=C
    I—I
    Cl H

    b.
    Cl H
    I—I
    C=C
    I—I
    Cl H

    c.
    Cl Cl
    I—I
    C=C
    I—I
    H H

    d.
    Cl Hl
    I—I
    C=C
    I—I
    H Cl

20. Viewing photosynthesis in terms of a system, the solar energy used in the process would be described as _____.
    a. feedback
    b. an energy product
    c. an input
    d. an output

21. If a photosynthetic process involved the input of 2000 K cal and the glucose formed yielded 700 K cal upon combustion, the efficiency of the glucose formation would be _____.
    a. 70%
    b. 35%
    c. almost 3%
    d. None of the above.

22. Given the graph shown in Figure 11.6, what is the cause of the yearly periodic oscillations?
    a. rainfall
    b. greenhouse effect
    c. air pollution
    d. rise and decline of growing activity

23. Which of the following chemical substances is most responsible for causing acid rain?
    a. coal
    b. sulfur and its oxides
    c. carbon and its oxides
    d. hydrocarbons

24. A plant cell _____.
    a. takes in carbohydrates and emits carbon dioxide
    b. uses up energy by taking in oxygen
    c. stores energy from sunlight by constructing more complex molecules
    d. stores energy from sunlight in less complex molecules

25. In the Earth's ecosystem, where does the outside energy come from?
    a. geothermal energy
    b. electrical power plants
    c. the sun
    d. decomposition of carbohydrates

26. The ecosystem in the world describes the relationship between life and the physical world. How does energy best flow through the ecosystem?
    a. sunlight to bacterial to animals to waste
    b. sunlight to animals to plants to waste
    c. plants to animals to bacteria to sunlight
    d. sunlight to plants to animals to waste

27. How can the efficiency of a solar collector be improved?
    a. Remove reflectors from outside the collector.
    b. Increase the insulation on the bottom of the collector.
    c. Remove the glass from the surface to allow more light to enter.
    d. Keep blowing air in front of the collector.

28. An experiment is done in which the effect of temperature on the rate of plant growth is investigated. The results show that there is a low temperature below which growth ceases because the plant becomes dormant. There is also a critical temperature above which there is no growth because of cell deterioration. Between these two extremes, increasing the temperature aids the process and generally increases the rate of growth. A graph of the results of this experiment might look like this: *(Please draw it.)*

29. In the conversion of its primary fuel to electricity, a cow manure plant is most similar to a _____.
    a. nuclear power plant
    b. coal-fired power plant
    c. geothermal power plant
    d. photovoltaic power plant

30. The "greenhouse effect" depends mainly upon the conversion of visible solar radiation to _____.
    a. ultraviolet and its subsequent absorption by $O_3$
    b. infrared and its subsequent absorption by $CO_2$ and $H_2O$
    c. shorter wavelength radiation
    d. sonic energy

31. In the upper atmosphere which of the following is a problem?
    a. He
    b. $N_2$
    c. Cl (atomic chlorine)
    d. $O_3$

# Bibliography and Web Resources

World pollution provides the latest world environment news from WN Network Environment.
**www.pollution.com/**

Electromagnetic Spectrum. Measuring the electromagnetic spectrum. A NASA site listing the various wavelengths of electromagnetic radiation.
**imagine.gsfc.nasa.gov/docs/science/know_l1/emspectrum.html**

The EPA Global Warming Site focuses on the science and impacts of global warming, or climate change, and on actions by governments, corporations, etc.
**www.epa.gov/globalwarming/**

CHAPTER 12

# Metabolism and Nutrition

*"Things fall apart—it's scientific."*

**Talking Heads**

Once again we shall examine the role energy plays in creating and maintaining the highly ordered structure of living systems. The second law of thermodynamics tells us that the natural tendency of systems is to degrade toward disorder. If complex systems are to be maintained, energy must continually be placed into the system; otherwise, once again, the forces of entropy will move it toward disorder. The energy for biological activities we shall find is supplied by the energy stored in the bonds of certain molecules.

The development and maintenance of a healthy, physically fit body requires a steady input of the proper and appropriately balanced nutrients. We shall find that carbohydrates are required, primarily, as an energy source in humans, fats as a secondary source, and proteins as a tertiary one. Most important is the balance among these energy sources that is required for a healthy body.

## Goals

After studying this chapter, you should be able to:

- Understand the role of energy in maintaining the body as a healthy system.
- Understand the role of chemical bonds as the source of energy for living things.
- Understand the application of the laws of thermodynamics in a cell.
- Understand how energy is extracted through metabolic breakdown.
- Define the six classes of nutrients required by the human body.
- Know the nine essential amino acids obtained outside the body.
- Understand how calories are derived from carbohydrates, fats, and proteins.
- Know the difference between HDL and LDL.
- Understand how weight control is managed.

## Entropy

In a 1980s song, *Wild, Wild Life* by the group Talking Heads, lead singer David Byrne sings the line, "Things fall apart—it's scientific." What, in essence, he is describing is the concept of entropy, a component of the laws of thermodynamics. Entropy is, simply put, the natural tendency of systems to degrade toward disorder.

In simpler terms, just as a random pile of bricks will not spontaneously organize into the

According to the laws of nature, simple arrangements (random) are naturally favored (occur naturally). Complex arrangements are not favored. Complex arrangements need to be produced using energy and need to be maintained using energy.

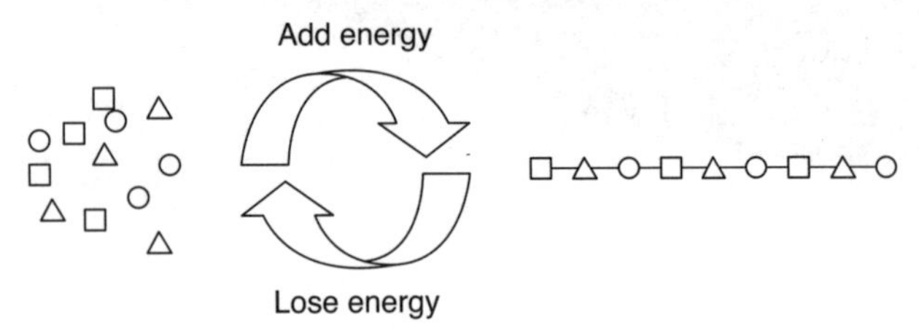


**Figure 12.1** Energy and matter.

complex system of a building, random collections of molecules will not spontaneously assemble into the complex systems of living organisms. In order for such complexity to be created from randomness, energy must be added to the system to move it in a direction against the forces of entropy. Similarly, if complex systems are to be maintained, energy must continually be placed into the system; otherwise, once again, the forces of entropy will move it toward disorder (see Fig. 12.1).

As an example, a building left abandoned and uncared for will gradually decay and fall apart in accordance with the law of entropy. In order to keep our homes and buildings intact and functional, we must constantly put effort (energy) and materials into these systems.

Similarly, living organisms require a constant supply of energy to create and maintain their highly ordered structure and systems. Photosynthetic organisms such as green plants and algae can obtain this energy directly from the sun. Organisms that do not possess this capability must obtain their energy through sources, generally through the nutrients they ingest.

Secondly, nutrients provide the physical, molecular building blocks from which organisms can synthesize new cells, tissues, and body structures required for growth and repair.

## Energy Extraction through Metabolic Breakdown

As with all systems performing work, and in keeping with the laws of thermodynamics, the energy required for the maintenance of life and for the performance of daily tasks must be supplied from some external source. Lightbulbs require energy in the form of electricity; our cars require energy stored within the bonds of gasoline molecules. The energy for biological activities is supplied by the energy stored in the bonds of the molecule adenosine triphosphate (ATP). In simple terms, ATP is the fuel that our muscles burn.

The bonds attaching the last two phosphate groups to this molecule are so-called high-energy phosphate bonds, each storing approximately 11,000 calories (11 kCal) of energy. The calorie is a unit used to measure energy, one calorie being the amount of heat energy required to raise one cubic centimeter (or 1 ml) of water one degree Celsius. When one of these phosphate bonds is split from the ATP molecule, 11,000 calories of energy (or 11 kilocalories) are available to power muscular contraction.

The removal of this phosphate group converts the ATP into ADP (adenosine diphosphate). Removal of the second phosphate group also frees 11,000 calories of energy for use in contraction and converts the ADP to AMP (adenosine monophosphate).

ATP ADP

Adenosine~P~P~P $\rightarrow$ Adenosine~P~P + P + Energy (11 kcal).

ADP AMP

Adenosine~P~P $\rightarrow \rightarrow$ Adenosine~P + P + Energy (11 kcal).

The concept of storing energy within chemical bonds is one that we should find readily understandable. For example, gasoline is a molecule in the class known as hydrocarbons. This is because gasoline consists of an arrangement of hydrogen and carbon atoms held together by chemical bonds. These atomic bonds contain energy, the same type of energy utilized by the cell. When we break apart the gasoline molecule (by burning it in the presence of oxygen so as to form new bonds and molecules), we release the energy and make it available to do work, such as move the car.

Energy is not always found in the form of chemical bonds between molecules. Energy is found in many forms such as electrical energy, mechanical energy, thermal energy, and light energy. Energy can be converted from one form to another, as we well know. Lightbulbs convert electrical energy to light energy. Your home furnace converts chemical energy (bond energy in gas or oil) to thermal (heat) energy. Your hair dryer converts electrical energy into thermal energy (heat) and mechanical energy (blower motor). The cells found in living organisms convert chemical bond energy into mechanical, thermal, and electrical energy.

The laws that govern energy use in the cell are the same laws of physics that govern energy use in any system. The first law of thermodynamics states that the total amount of energy in the universe is constant. In any physical or chemical change the total amount of energy in the universe remains constant. In other words, energy can neither be created nor destroyed; this is the law of **conservation of energy.** When cells or living organisms absorb energy from the environment to perform useful work, they must return an equivalent amount of energy to the environment, usually in the form of heat.

**Anaerobic Respiration** A relatively simple way of releasing energy is to break down ingested carbohydrates, large molecules with appreciable stores of chemical bond energy, into simpler molecules such as ethanol (a small alcohol) or lactic acid (a small acid molecule). Carbohydrates ingested in the diet are stored in the body's cells as glycogen. Through a series of biochemical steps, glycogen is broken down in a process known as glycolysis, yielding ATP and pyruvic acid. The chemical reactions of glycolysis do not require the presence of oxygen or air and are referred to as anaerobic metabolism. The ATP yield of this system is limited and is sufficient to fuel muscular activity for only an additional 30–40 seconds above the 10–15 seconds fueled by the phosphagen system.

The pyruvic acid that results from this process can itself be used to form additional ATP if oxygen is available in sufficient quantities to the muscle. In the absence of sufficient oxygen, the pyruvic acid is converted to lactic acid, which is released from the muscle into the blood. This lactic acid is believed to be partly responsible for the phenomenon of muscular fatigue. The lactic acid released into the blood can be converted by the liver into glucose, which can then reenter the pathway and supply further energy. This liver-based pathway for the recycling of pyruvic acid is known as the Cori cycle.

**The Aerobic System** The aerobic system of ATP production is one that follows the glycogen-lactic acid system. The oxidation of pyruvic acid formed during glycolysis results in a tremendous amount of ATP available for muscular activity. This system requires the presence of oxygen, or air, and is therefore called aerobic. The aerobic system is capable of supplying ATP for an indefinite time or as long as the nutrient supply lasts. Once the stored glycogen has all been used, this system can go on fueling itself with the body's fat reserves.

# Nutrition

The development and maintenance of a healthy, physically fit body requires a steady input of the proper and appropriately balanced nutrients. A nutrient is defined as any substance, obtained from food, drinks or otherwise ingested, that is used by the body for growth, synthesis, repair, or as an energy supply. There are six basic classes of nutrients required by the human body to maintain general fitness and good health as well as to provide the fuel for exercise and athletic activities (see Table 12.1).

## CARBOHYDRATES

Carbohydrates are required, primarily, as an energy source. Carbohydrates should account for between 55% and 70% of the energy supply in humans. The body's energy demands are large, even without exercise activities.

The human brain requires between 400 and 600 Calories (a measure of energy, really kilocalories) each day to drive its function. The precise amount that carbohydrates should contribute to the diet is influenced by the typical daily energy requirements (more are required for intense exercise) and dietary goals (less is required for weight loss). Each gram of carbohydrate contains 4 kcal of energy.

To a lesser extent, carbohydrates are also required as building blocks of larger molecules including the gelatin-like substance of connective tissues, cell membrane coverings, and the mucus secreted by digestive and respiratory system linings. Carbohydrates stimulate the growth of our intestinal bacteria that provide us with our main supply of vitamin K (needed for proper blood clotting), and also provide the fiber, or bulk, required in our diets.

Carbohydrates occur in two general forms. In one form, the carbohydrate occurs as small, single molecules (monomers) called **sugars,** also known as **monosaccharides,** or **simple carbohydrates.** More complex forms of carbohydrates

**TABLE 12.1 Classes of Nutrients**

| Six Classes of Nutrients Required by the Human Body |
|---|
| Carbohydrates |
| Fats |
| Proteins |
| Vitamins |
| Minerals |
| Water |

exist as long, linear, or branched chains of sugar molecules (polymers) and are alternatively known as **starches, polysaccharides,** or **complex carbohydrates** (see Fig. 12.2).

The rate at which carbohydrate-containing foods are digested and the carbohydrate monosaccharides absorbed into the bloodstream varies and is dependent upon a variety of factors. Those that are rapidly digested and absorbed and which can provide a rapid but short-term energy burst are said to possess a high **glycemic index.**

Breads, cakes, refined pastas, and bananas are examples of high glycemic index foods. Foods in which the carbohydrates enter the bloodstream more slowly and provide a longer-lasting energy supply are referred to as moderate glycemic index foods and include whole-grain breads, oatmeal, and corn.

## Fats

Fats serve primarily as a secondary energy source and, to a lesser extent, also function as building blocks for larger molecules such as those found in cell membranes and hormones. Fats provide a very concentrated source of energy, approximately two times the energy per gram as compared with carbohydrates (9 kcal/gm).

One common class of fat, lipids, also known as neutral fats, are composed of a backbone molecule of glycerol with three fatty acid molecules attached. Because of this structure, neutral lipids are often called **triglycerides.**

As seen in the illustration in Figure 12.3, the carbon atoms (C), which form the long chain backbone of the fatty acid, have four bonding sites. (The C atoms are attached to two hydrogen and two carbon atoms.) A carbon chain such as that found in a fatty acid molecule is said to be a saturated fatty acid if there are four single bonds present at each carbon. That is, if two hydrogen atoms are present at each carbon and there are no double bonds present between adjacent carbon atoms, the carbon will satisfy the requirement to form four bonds. If a double bond is present between adjacent carbon atoms, the compound is said to be unsaturated. A fatty acid with one such site would be called a monounsaturated fatty acid; one with several unsaturated sites would be a polyunsaturated fatty acid (see Fig. 12.4).

The terms saturated and unsaturated have a specific meaning when used in conjunction with chemical bonding. The carbon compound on the right is saturated. This means that each carbon atom, which is capable of forming four bonds, forms only single bonds.

The carbon compound on the left is unsaturated. Notice that there is a double bond present between the two middle carbon atoms. The degree of unsaturation is related to the number of double bonds present.

Saturated fatty acids contain no double bonds, while unsaturated fatty acids contain double bonds. Generally, saturated fatty acids tend to solidify more readily than unsaturated fatty acids.

Saturated fats, made from glycerol and saturated fatty acids and contained in foods from animal sources, tend to be very dense and solid at room temperature. Mono and polyunsaturated fats contained in foods from plants are less dense and are liquid at room temperature. The heavy, dense saturated fats are the ones that help precipitate the blockage of arteries, contribute to heart disease and stroke, and should be avoided.

Another important form of fat is based on the cholesterol molecule and is known, as a class, as **steroids.** Although most people assume that cholesterol is a "bad" fat, and it is in large amounts, it is also required for many reasons. Hormones such as estrogen, progesterone, testosterone, and aldosterone are all made from a cholesterol building

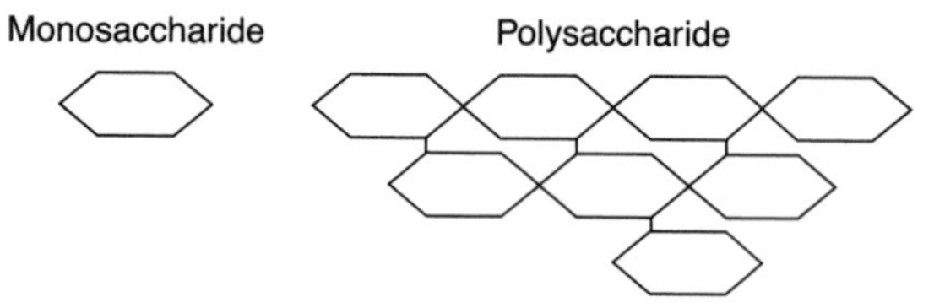


**Figure 12.2** Monosaccharide and polysaccharide.

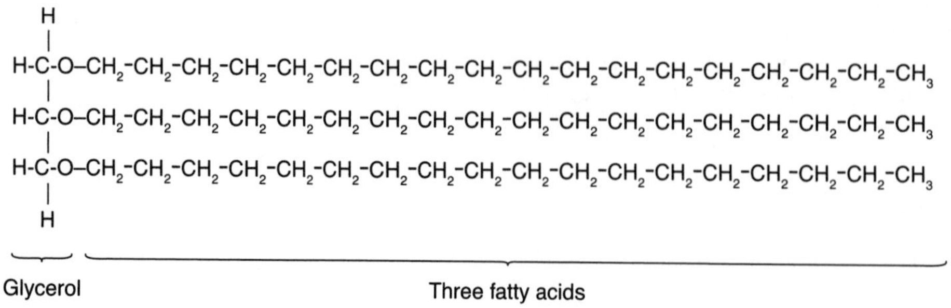

$$\begin{array}{l} \quad H \\ \quad | \\ H\text{-}C\text{-}O\text{–}CH_2\text{-}CH_2\text{-}CH_2\text{-}CH_2\text{-}CH_2\text{-}CH_2\text{-}CH_2\text{-}CH_2\text{-}CH_2\text{-}CH_2\text{-}CH_2\text{-}CH_2\text{-}CH_2\text{-}CH_2\text{-}CH_2\text{-}CH_2\text{-}CH_2\text{-}CH_3 \\ \quad | \\ H\text{-}C\text{-}O\text{–}CH_2\text{-}CH_2\text{-}CH_2\text{-}CH_2\text{-}CH_2\text{-}CH_2\text{-}CH_2\text{-}CH_2\text{-}CH_2\text{-}CH_2\text{-}CH_2\text{-}CH_2\text{-}CH_2\text{-}CH_2\text{-}CH_2\text{-}CH_2\text{-}CH_2\text{-}CH_3 \\ \quad | \\ H\text{-}C\text{-}O\text{–}CH_2\text{-}CH_2\text{-}CH_2\text{-}CH_2\text{-}CH_2\text{-}CH_2\text{-}CH_2\text{-}CH_2\text{-}CH_2\text{-}CH_2\text{-}CH_2\text{-}CH_2\text{-}CH_2\text{-}CH_2\text{-}CH_2\text{-}CH_2\text{-}CH_2\text{-}CH_3 \\ \quad | \\ \quad H \end{array}$$

Glycerol — Three fatty acids

**Figure 12.3** Triglycerides.

block. Cholesterol also plays an important role in the function of cell membranes. The body, however, can synthesize cholesterol so its importance as a nutrient is minimal.

```
  H H  H H          H H H H
  | |  | |          | | | |
H-C-C=C-C-H       H-C-C-C-C-H
  |    |            | | | |
  H    H            H H H H
 Unsaturated       Saturated
```

**Figure 12.4** Unsaturated and saturated compounds.

## PROTEIN

Protein, if necessary, can serve as a tertiary (third order) energy source but only during periods of starvation or in conditions of abnormal metabolism. For the most part, protein is required to provide building blocks, in the form of amino acids. Amino acids are the basic molecules from which proteins are built (see Fig. 12.5). The groups circled ($NH_2$, an amino group, and COOH, an acid group) are common to all amino acids. There are 20 standard amino acid building blocks that we use to build new protein. These amino acids, some obtained from the

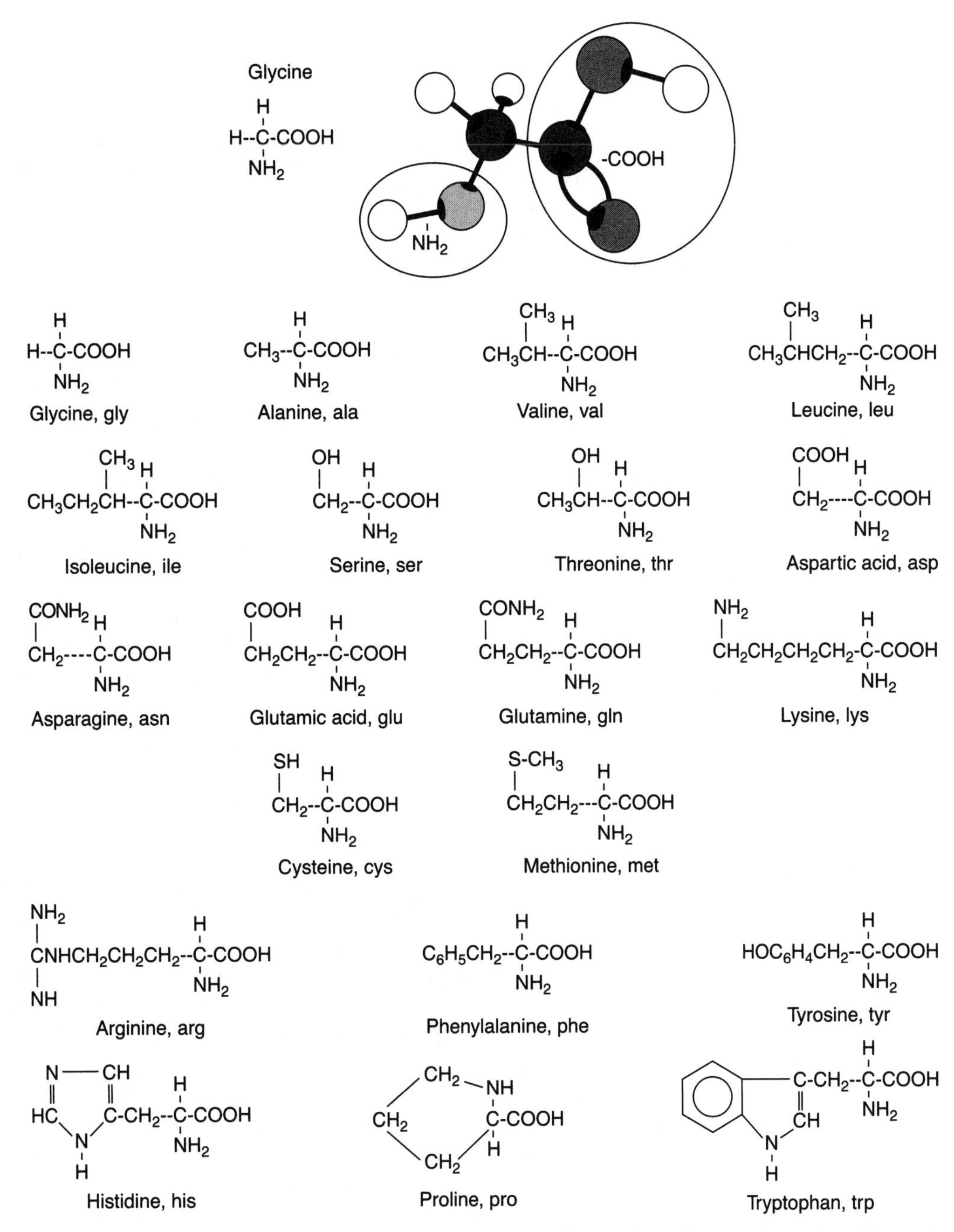


**Figure 12.5** Amino acids. *www.dep.anl.gov/S3A/S3A-amino-acids.htm* (modified to show the 20 standard amino acids)

**TABLE 12.2 Essential Amino Acids**

| Essential Amino Acids |
|---|
| Lysine |
| Histidine |
| Tryptophan |
| Phenylalanine |
| Threonine |
| Valine |
| Methionine |
| Leucine |
| Isoleucine |

protein we eat and others synthesized by the body itself, are required for the production of other proteins such as enzymes, connective tissue fibers, and the contractile elements of muscle.

Interestingly, in 1986 the nonstandard amino acid selenocysteine was found to be directly specified by the genetic code and not created by post translational modification. Now pyrrolysine has been named as the 22nd amino acid specified by the genetic code (*Science,* May 24, 2002, Vol. 296, p. 1410).

Although the body can synthesize 11 of the 20 amino acids, nine must be obtained from the diet or through dietary supplements. These are known as the essential amino acids (Table 12.2).

## Vitamins

Vitamins are organic (carbon-based) molecules, which are required by the body for use in assisting with enzyme activity or to influence certain physiological functions such as wound healing. Vitamins fall into two categories; those that dissolve in water (water soluble vitamins) and those that do not (fat soluble vitamins). (See Table 12.3.) See Table 12.4 also, which lists the functions of vitamins.

## Minerals

Minerals are inorganic elements and ions required for a host of body functions (see Table 12.5). Some minerals are required in relatively large amounts, others in only trace amounts.

## Water

Water is required to maintain the proper concentration of materials in the body, acts as a solvent, is required for body cooling, aids in joint lubrication, and serves in decomposition reactions.

**TABLE 12.3 Vitamin Classification**

| Water Soluble | Fat Soluble |
|---|---|
| B complex | A |
| $B_1$ Thiamin | D |
| $B_2$ Riboflavin | E |
| $B_3$ Niacin | K |
| $B_6$ Pyridoxine | |
| $B_{12}$ Cobalamin | |
| Biotin | |
| Folacin | |
| Pantothenic Acid | |
| C | |

**TABLE 12.4 Major Vitamin Functions**

| Vitamin | Function |
|---|---|
| $B_1$ | Energy release in mitochondria |
| $B_2$ | " " " " |
| $B_3$ (Niacin) | " " " " |
| $B_6$ | Protein and amino acid metabolism |
| $B_{12}$ | DNA & RNA metabolism |
| Biotin | ATP (energy) formation |
| Folacin | DNA & RNA metabolism |
| Pantothenic | Energy release in mitochondria |
| C | Collagen synthesis, cholesterol metabolism, synthesis of anti-inflammatory steroids |
| A | Vision |
| D | Absorption of calcium and phosphorus |
| E | Prevent $O_2$ toxicity |
| K | Blood clotting |

# Calories

The calorie (spelled with a lowercase "c") is a unit of energy measurement. Recall, one calorie denotes the amount of energy, in the form of heat, required to raise the temperature of 1 ml of water by 1°C. Admittedly, the calorie does not seem to represent all that much energy. In fact, most human activities require, and most food products contain, many thousands of calories. For this reason, when discussing human nutrition and physiology, we generally refer to calories in units of one thousand, a unit called the kilocalorie. Although the proper term for this 1,000-calorie unit, the kilocalorie, is commonly used in scientific discussions, many nutritionists, along with the general public, prefer to use the term Calorie (spelled with an uppercase "C" or capital "C"). When a slice of bread is said to contain 70 Calories, its energy con-

**TABLE 12.5 Minerals**

| Major Minerals | Main Function(s) |
|---|---|
| Na (sodium) | $H_2O$ balance, nerve function |
| K (potassium) | $H_2O$ balance, nerve function |
| P (phosphorus) | Bone and tooth structure, coenzyme |
| Cl (chloride) | acid base balance, $H_2O$ balance |
| Mg (magnesium) | Energy releasing enzymes |
| S (sulfur) | Cartilage, tendon, protein structure |
| Ca (calcium) | Bone and tooth, nerve function, clotting |
| **Trace Minerals** | **Main Function(s)** |
| Fe (iron) | Blood hemoglobin, enzyme function |
| I (iodine) | Thyroid hormone (metabolic rate) |
| Fl (fluoride) | Bone and tooth structure |
| Zn (zinc) | Enzymes |
| Se (selenium) | Enzymes |
| Cu (copper) | Enzymes |
| Mn (manganese) | Enzymes |
| Cr (chromium) | Glucose and energy metabolism |

tent is actually 70,000 calories, enough energy to raise the temperature of 1 liter of water (1000 ml) by 70° C.

An ideal diet would be one in which a person consumes only as many calories as are used in her daily activities. If the intake of calories exceeds the daily usage, the excess food will be converted to and stored as fat, as will be discussed later in this chapter. Again, in the ideal diet, approximately 48% of ingested calories should come from complex carbohydrates and naturally occurring sugars.

Processed or refined sugars should make up no more than 10% of total carbohydrates. Total carbohydrate intake should therefore account for approximately 58% of the total daily caloric intake. No more than 30% of total calories should come from fat, with less than 10% of this being saturated fats. Protein should make up the remaining 12% of total daily calorie intake.

In addition, daily cholesterol intake should be limited to 300 mg/day and sodium intake (mostly table salt) limited to 5 g/day. Cholesterol is a waxy, fat-like molecule carried in the blood by other molecules known as lipoproteins. These lipoprotein carriers come in two forms, **low-density lipoproteins (LDL)** and **high-density lipoproteins (HDL).** HDL cholesterol is referred to as "good cholesterol" because this cholesterol carrier removes cholesterol from the bloodstream. LDL cholesterol is referred to as "bad cholesterol" because too much of it can clog arteries.

Excess cholesterol circulating in the blood can contribute to the formation of fatty deposits on the walls of the arteries, narrowing their effective diameter and reducing blood flow. In the coronary arteries of the heart, this can contribute to heart attack. When cholesterol is bound to low-density lipoproteins, it tends to remain in circulation within the blood. Cholesterol bound to the high-density lipoprotein is removed from the blood and metabolized by the body's cells.

It is recommended that total blood cholesterol levels be held below 200 mg/dl (milligrams per deciliter) and an HDL over 60. Just as important as total blood cholesterol, however, is the ratio of HDL/LDL. A desirable LDL to HDL ratio would be below 3. Cholesterol is found in certain foods like dairy products, eggs, and meat. Cholesterol can be reduced by watching the diet, by eating less red meat and eggs, and by eating more fiber like oat bran cereal.

The intake of dietary fiber should be at the level of between 25 to 35 grams per day. Dietary fiber is a form of polysaccharide carbohydrate (complex carbohydrate), which is nondigestible in the human digestive system. Soluble fiber is soluble in water, rapidly absorbed by the digestive tract, and may help to reduce blood cholesterol levels. Insoluble fiber remains in the digestive tract, where it increases stool volume, reduces the amount of time that wastes remain in the large intestine, and may help to prevent colon cancer.

# Nutritional Energy Balance

The nutrients ingested, specifically the carbohydrates, fats, and proteins, carry with them material that can be converted into energy. The quantity of this energy is measured in calories and, if not used in driving daily activities, will be stored primarily as fat. The normal intake of nutrients for the average American contains approximately 2,400 kcal (kilocalories). Given the recommended ratios for the various nutrient groups, a well-balanced diet should contain 6 grams of carbohydrate for each kg of body weight (2.7 g for each lb), 1.2 g of protein per kg (.55 g/lb), and 1.2 g of fat per kg (.55 g/lb). Because of the differing energy densities (4 kcal/g for carbohydrate and protein, 9 kcal/g for fat), the diet for a normal 130-lb individual (59 kg) would be as shown in Table 12.6.

**TABLE 12.6 Caloric Intakes**

| | | |
|---|---|---|
| *Intake by Weight* | | |
| Carbohydrate | 59 kg @ 6 g/Kg = | 354 g of carbohydrate |
| Protein | 59 kg @ 1.2 g/Kg = | 71 g of protein |
| Fat | 59 kg @ 1.2 g/Kg = | 71 g of fat |
| *Intake by Calories (Kcal)* | | |
| Carbohydrate | 354 g @ 4 kcal/g = | 1,416 kcal of carbohydrate |
| Protein | 71 g @ 4 kcal/g = | 284 kcal of protein |
| Fat | 71 g @ 9 kcal/g = | 639 kcal of fat |
| Total Calories | | 2,339 kcal/day |

**TABLE 12.7 Typical American Diet**

| | |
|---|---|
| Carbohydrate: | |
| Complex | 28% |
| Simple | 20% |
| Protein | 12% |
| Fat | 40% |

**TABLE 12.8 Ideal Recommended Diet**

| | |
|---|---|
| Carbohydrate: | |
| Complex | 48% |
| Simple | 10% |
| Protein | 12% |
| Fat | 30% |

The U.S. Department of Agriculture has prepared a Recommended Dietary Allowance (RDA) for each of the above nutrients, which should be followed as nearly as is possible. See Table 12.7 and Table 12.8 for a comparison of a typical American diet to the recommended ideal diet.

## Weight Management

In order to maintain a stable weight, the amount of calories consumed per day should be roughly equivalent to the amount of calories expended per day. Excess calories consumed, be they proteins, carbohydrates, or fats, will be broken down through the metabolic process into the component parts and recombined into various forms of stored energy.

Initially, excess calories will be stored in the muscle and liver as glycogen; however, since the body's glycogen storage capacity is limited (about 1,400 kcal worth of stored glycogen), most of the excess calories will be stored as fat (about 100,000–130,000 kcal of energy are stored as fat in the normal, fit individual). Conversely, if daily calorie intake falls below the level of daily calorie expenditure, breaking down energy storage molecules, namely glycogen and fat, will make up the energy deficit.

Simply watching calories is not the best way for losing fat and reducing weight. Research has shown that the most effective method of weight reduction is to combine calorie restriction with a program of aerobic and strength training exercise. Aerobic exercise consumes a tremendous amount of energy that, in general, is first supplied by free glucose and stored glycogen and, after about 20 minutes of continuous aerobic exercise, by stored fat.

By restricting calorie intake in combination with aerobic exercise, one can compound the effect of quickly depleting glycogen reserves and catabolizing stored fat. Since, pound-for-pound, muscle is a metabolically active tissue and consumes more calories than fat, increasing the body's muscle content through strength training exercise will increase the resting daily energy requirement and help prevent excess calories from being converted to stored fat.

A recent study reported in the journal *Medicine and Science in Sports and Exercise* followed three groups of dieters, one group simply reducing calorie intake, one group engaging in aerobic exercise and calorie reduction, and the third group participating in both strength and aerobic exercise with a reduced calorie intake. After a 12-week period, all groups showed similar amounts of weight loss. The group that dieted *only,* lost both body fat and muscle mass; the diet and aerobic exercise group lost fat and increased oxygen consumption; and the diet, aerobic, and strength training group lost body fat only and gained both oxygen consumption and strength. Only this latter group is poised for long-term weight reduction without any compromise in strength.

Protein is also important in the diet for the growth and repair of muscle tissue. High protein diets, in general, are not recommended in that they have been shown to be associated with increased risks of cancer, heart disease, and osteoporosis. The average American diet contains more than the recommended 12% of calories from protein. If any dietary supplement is considered, it might best be one containing the essential amino acids.

Although often referred to as the culprit for all that is wrong with the American diet, fat does play an essential role in the average diet and an even more significant role in the diet of endurance athletes. Stored muscle glycogen, the primary fuel for intense aerobic activity, can sustain only limited periods of activity, typically 20 to 30 minutes. Longer-term aerobic activities that exceed this timeframe become increasingly reliant on muscle triglyceride reserves—in other words, stored fat.

Low-fat diets in which less than 20% of total calories come from fat can compromise this energy storage reservoir and result in premature fatigue. In addition, the absorption of the important fat-soluble vitamins, A, D, E, and K, depend on adequate levels of fat in the diet, typically a total fat intake of about 30% of total calories.

Water is, of course, always an important nutrient but becomes even more so on an increased protein or amino acid diet. During the metabolism of ingested amino acids and proteins, significant amounts of ammonia are released and must be flushed from the blood by the kidneys with a sufficient amount of water.

## Review Questions

1. What role does entropy play in living organisms?
2. Where do organisms obtain energy for biological activity?
3. What is anaerobic respiration?
4. What is the aerobic system?
5. Name the six basic nutrients required by the human body and define each.
6. What are the main roles of carbohydrates, fats, and proteins in providing nutrition for humans?
7. What role do vitamins play in providing nutrition?
8. What are the main guidelines for good weight management?
9. What is the difference between HDL and LDL?
10. What is aerobic exercise?

## Multiple Choice Questions

1. _____ is the natural tendency of systems to degrade to disorder.
   a. Energy
   b. Atrophy
   c. Entropy
   d. Chaos
2. _____ provide(s) physical, molecular building blocks that allow organisms to live.
   a. Entropy
   b. Nutrients
   c. Enthalpy
   d. Bacteria
3. Energy for biological activity is stored in _____.
   a. chemical bonds
   b. muscles
   c. enzymes
   d. lipoproteins
4. A(n) _____ is any substance obtained from food, drinks, or otherwise ingested, that is used by the body for growth, synthesis, repair, or energy.
   a. enzyme
   b. glycogen
   c. adenosine
   d. nutrient
5. _____ are required primarily as an energy source.
   a. Carbohydrates
   b. Fats
   c. Proteins
   d. Vitamins
6. A secondary source of energy and building block for larger molecules such as found in cell membranes and hormones is _____.
   a. carbohydrates
   b. fats
   c. proteins
   d. vitamins
7. _____ for the most part provide building blocks in the form of amino acids.
   a. Carbohydrates
   b. Fats
   c. Proteins
   d. Vitamins

8. Nine of 11 of the 20 amino acids that the body needs must be obtained from the diet or through dietary supplements. These are known as _____.
   a. amino acids
   b. essential amino acids
   c. enzymes
   d. calories

9. The unit of measurement used by the dietician in human nutrition is equal to ____ of the calories used by scientists.
   a. 100
   b. 1000
   c. 10,000
   d. 1,000,000

10. Cholesterol is a waxy, fat-like molecule carried in the blood by other molecules called ___.
    a. enzymes
    b. pantothenic acid
    c. lipoproteins
    d. complex carbohydrates

11. Good cholesterol is the _____.
    a. LDL
    b. HDL
    c. AMA
    d. HCL

12. Watching calories is not the best way for losing fat and reducing weight. In addition, a program of _____ must be employed.
    a. aerobic exercise
    b. metabolic control
    c. glycogen control
    d. enzyme reaction

13. Energy for biological activity is stored in _____.
    a. DNA
    b. DDT
    c. APT
    d. kilocalories

14. The heat liberated when ATP is converted to ADP is sufficient to raise the temperature of 11 kilograms of water how many degrees Celsius?
    a. 10
    b 1
    c. 5
    d. None of the above.

15. Glycolysis _____.
    a. yields ATP and oxygen
    b. requires oxygen
    c. yields ATP and pyruvic acid
    d. produces glycogen

16. Ingested carbohydrates are converted to potential energy, which is stored in the substance _____.
    a. lactic acid
    b. ethanol
    c. glycogen
    d. AMP

17. Eating 20 grams of a carbohydrate will supply _____ kilocalories (Calories) of energy.
    a. 40
    b. 20
    c. 80
    d. 180

18. Eating 20 grams of a fat will supply _____ kilocalories (Calories) of energy.
    a. 40
    b. 20
    c. 80
    d. 180

19. Proteins are polymers of _____.
    a. fats
    b. carbohydrates
    c. amino acids
    d. vitamins

20. The body requires 20 different amino acids, yet there are eight amino acids that are called essential. The reason for this is that _____.
    a. 20 are essential for life
    b. the body can synthesize all the required amino acids from a diet containing the 8 essential amino acids
    c. amino acids can be denatured
    d. some amino acids can be obtained via dietary supplements

21. If a 100 gram portion of food was 50% carbohydrate and 25% fat and protein, the total number of Calories associated with the food would be _____.
    a. 100
    b. 525
    c. 1,000
    d. None of the above.

22. The USDA has prepared a Recommended Dietary Allowance (RDA) for an ideal diet. Which of the following food types is listed in the highest percentage?
    a. complex carbohydrates
    b. simple carbohydrates
    c. protein
    d. fat

# Bibliography and Web Resources

Cholesterol and LDL level (low density lipoprotein).
**medicineassociates.com/heart.html**

American Diabetes Association. The American Diabetes Association is the nation's leading nonprofit health organization.
**www.diabetes.org/**

Defeat Diabetes: New Drug Mimics HDL, "Good Cholesterol."
**www.defeatdiabetes.org/Articles/drug031118.htm**

Food & Nutrition Information Center.
**www.nal.usda.gov/fnic/**

U.S. federal guide offering access to all government Web sites with reliable and accurate information.
**www.nutrition.gov/home/index.php3**

Weight-control Information Network. How can WIN help you? WIN produces, collects, and disseminates materials on obesity, weight control, and nutrition.
**www.niddk.nih.gov/health/nutrit/win.htm**

CHAPTER 13

# Population and Growth

*"We think the human race is up to the challenge."*

**Meadows, Meadows, Randers**

There are inherent limits within our biosphere that restrict its capacity to support life. Can the earth sustain a continually growing human population into the future? As our population increases and consequently our consumption of renewable resources, what does the future hold for us?

Understanding the concept of exponential growth is central to understanding what is happening to the human population. In 1972 the Club of Rome, a nonprofit "think tank" composed of economists, scientists, businesspersons, and former government leaders from around the world, commissioned a group at MIT to build a computer model, World3, that would allow us to understand what is happening on the earth and to make predictions about the future. The results were not encouraging and the exercise was repeated once more and a report issued in 1992, which also was discouraging. What can we do as a society and as individuals to create a sustainable environment?

## Goals

After studying this chapter, you should be able to:

- Understand exponential growth.
- Understand the concept of doubling time.
- Understand that there are limits to growth.
- Know the main conclusions contained in the first 1972 report to the Club of Rome.
- Know some of the variables considered in the computer World3 model created at MIT.
- Estimate the earth's carrying capacity in terms of population.
- Know the conclusion of the 1992 report to the Club of Rome and the recommendations for sustainability.

## Population Growth

An old economic maxim states, "grow or die." When applied to the growth of the human population, some have changed the maxim to "grow and die." The carbon, oxygen, and nitrogen cycles studies previously studied require enormous amounts of energy to function. The ultimate source of this energy comes from the sun in the form of solar radiation. The source of this radiation lies in the ability of nuclear reactions in the sun to convert mass into energy according to Einstein's famous equation $E = mc^2$.

Scientists have calculated that this transformation of mass into energy on the sun will continue into the future for billions of years. So for the future it appears that the energy available from the sun will continue at a more or less constant rate. However, there are inherent limitations within our biosphere that limits its capacity to support life.

In 1972, the World3 computer model created at MIT under commission from the **Club of Rome** predicted that under the present growth trends in world population, industrialization, pollution, food production, and resource depletion, the limits to growth would be reached sometime within the next 100 years with the result that there would be a sudden and uncontrollable decline in both population and industrial capacity (see Fig. 13.1).

Englishman Thomas Malthus (1766–1834) in "An Essay on the Principle of Population" (1798) considered the world's population to be increasing in a geometric ratio (e.g., 1, 2, 4, 8, . . . or by a multiplicative factor) while the food supply was increasing at a much slower rate defined by an arithmetic progression (1, 2, 3, 4, 5, . . . or by an additive factor). He predicted that the human population would soon outstrip the food supply and it would lead to mass starvation. He doubted the ability of science and technology to improve the food supply since any improvement would probably be offset by an increase in population.

Charles Darwin tells us that every organism "strives to the utmost to increase in number," limited by food supply, predators, environment, and disease.

The introduction of the potato in Ireland during the late 17th century provided more food for the population, which increased from about 2 million to 8 million by 1845. When the potato crop failed that year, 1 to 2 million people died from starvation and about the same number left the country. Afterwards the population stabilized, as marriage and children were relegated to later in life.

Malthus seriously underestimated the ability of demographics, social innovation, and technological innovations to control population growth. As economies shifted from an agricultural base, where cheap labor in the form of children encouraged farmers to reproduce, to an industrial base where workers lived in small apartments in cities that tended to limit reproduction, populations seemed to stabilize or even decrease.

The debate Malthus started still goes on. Are there limits to growth? Can the Earth sustain a growing human population into the future?

# Exponential Growth

All five elements—population, food production, industrialization, pollution, and consumption—of renewable resources are increasing. The mathematical model that describes the increase is called exponential growth. Many people think of growth in a linear fashion. A quantity is growing linearly (see Fig. 13.2) when it increases by a constant

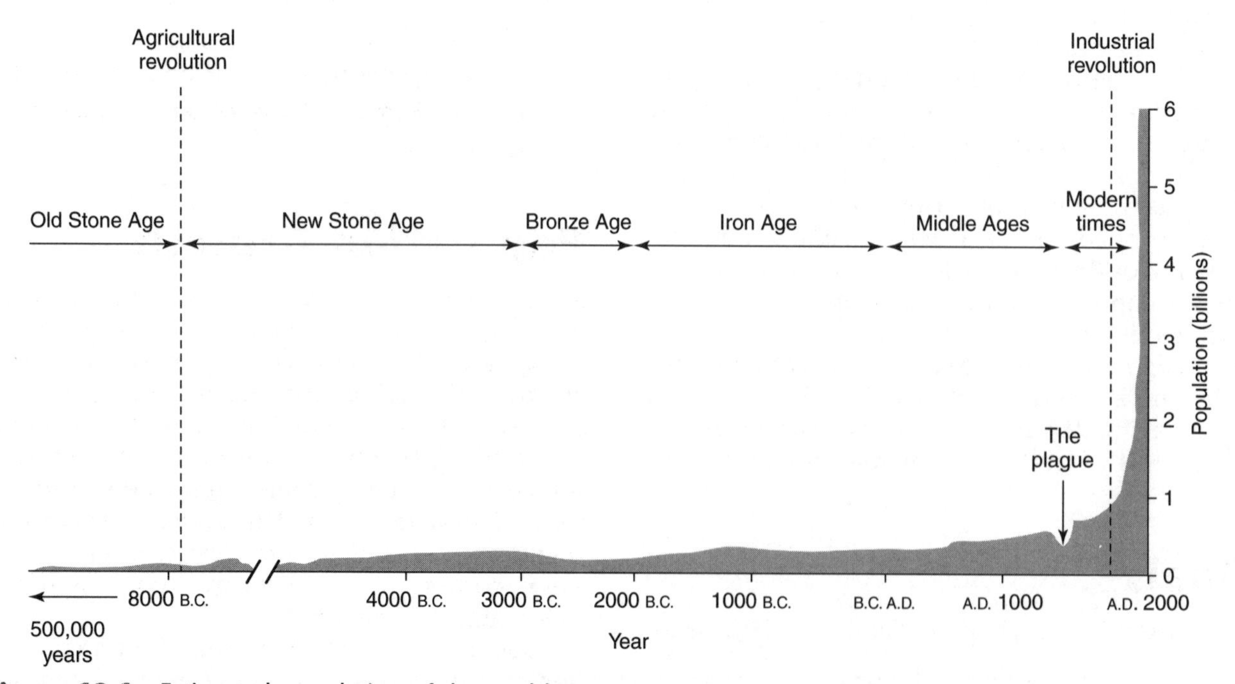


**Figure 13.1** Estimated population of the world. *http://carbon.cfr.washington.edu/esc110/2003Winter/projects/040/WritingOfHung.html*

amount in a constant time. Saving 10 dollars per week in a savings account increases the total amount in a linear way. The amount of increase each week is not affected by the amount of money already saved.

In exponential growth, Figure 13.3, the quantity increases by a constant percentage of the whole in a constant time period. It is useful to think of exponential growth in terms of doubling time or the time it takes a growing quantity to double in size.

It is known that positive feedback loops without any constraints produce exponential growth. In the world system there are two positive feedback loops in evidence producing exponential growth of population and industrial capital. The constraints that act to stop exponential growth are negative feedback loops, which become stronger as we approach the carrying capacity of the earth. These negative feedback loops involve famine, pollution, and depletion of natural resources.

In the human system there is a delay before the negative feedback loops take effect, allowing the population and capital to overshoot, which decreases the carrying capacity of the earth and intensifies decline in population and industrial capital.

Another way to control the problem of growth would be to weaken the positive feedback loops. How do we control growth deliberately? One way would be to lower the birthrate, and/or increase the death rate.

Throughout human history the human population has been growing in essentially an exponential manner. However, today there seem to be limits to growth in world population presented by industrialization, pollution, food production, and the depletion of natural resources. To understand fully how these limits come into play, one must have some understanding of the exponential function.

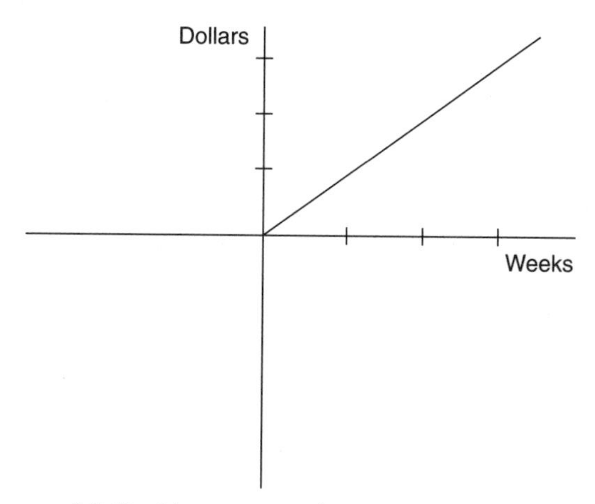


**Figure 13.2** Linear growth.

## Mathematical Review

In our basic mathematics courses we learned that a function could be a set of ordered pairs of real numbers (x, y) where there was a rule for obtaining the y values from the x values and if for each x there was one and only one y determined, then that set of ordered pairs could be called a function. Normally when denoting a function symbolically we write only the equation $y = f(x)$, where f represents the rule for obtaining the y values from the x values. For example, suppose $y = f(x)$ or $y = x^2$. This means substitute for x a real number and square it to calculate y, which will give an ordered pair of the function. If $x = 2$, then $y = 4$ and $(x, y) = (2, 4)$. Having done this for several values of x, one can plot a graph of y versus x to represent part of the function pictorially or graphically, as in Figure 13.4.

In ancient times a courtesan in a royal court in Persia supposedly invented the game of chess. The king was so pleased by the work that he asked the courtesan what would be a good reward. The

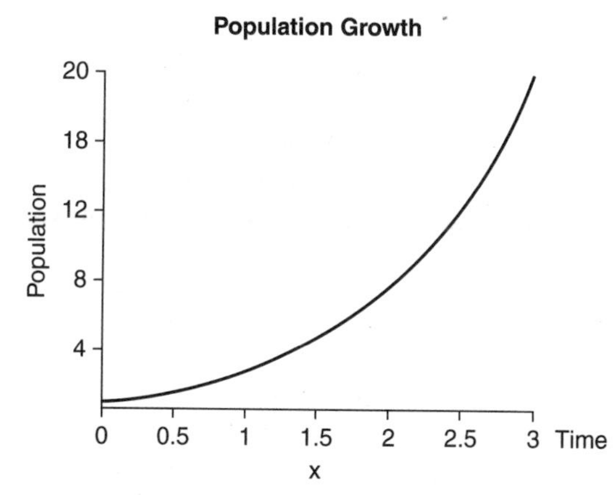


**Figure 13.3** Exponential growth in population.

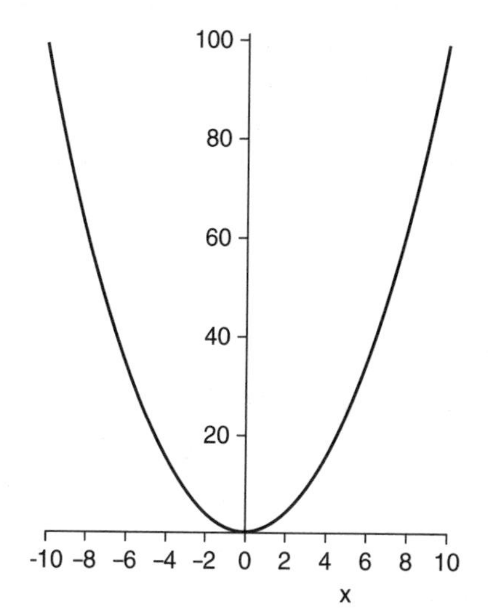


**Figure 13.4** Parabolic function.

**TABLE 13.1 Grains of Rice on Chessboard Squares**

| Square | Grains | Total |
|---|---|---|
| 1 | 1 | 1 |
| 2 | 2 | 3 |
| 3 | 4 | 7 |
| 4 | 8 | 15 |
| 5 | 16 | 31 |
| . . . | . . . | . . . |
| 64 | $2^{63}$ | $2^{64} - 1$ |

courtesan asked that one grain of rice be placed on the first square of the chessboard, two grains on the second, four on the third, and so on. So each successive square had double the amount of the previous one. The king, though somewhat embarrassed by what he thought was a minimal request for reward, agreed. If we construct a table to see how the chessboard would be filled, we discover something quite amazing (see Table 13.1). Incredibly, in the last square, the number $2^{64}$ is larger than the total amount of rice produced in single year in the entire world and may be more than the total amount ever harvested over all time.

What happened? This is an example of exponential growth. Some have characterized it as explosive growth. The secret to understanding the arithmetic is that the rate of growth (doubling for each square) applies to an ever larger *total* amount of rice, so the number of grains added with each doubling goes up, even though the rate of growth is constant.

Folding a sheet of ordinary 8½-by-11-inch sheet of paper about 1mm thick in half results in a similar phenomenon. Unfortunately, it is difficult to do more than a a few times. Starting with a piece of paper whose thickness is approximately $0.1 \times 10^{-3}$ meters, we find that after folding it seventeen times, or $2^{17} \times 0.1 \times 10^{-3}$ meters = $131{,}072 \times 0.1 \times 10^{-3}$ meters = 13.1 meters, or higher than a two-story house. After folding it 50 times, or $2^{50}$, we find it equals $1.13 \times 10^{11}$ meters, or close to the distance from the earth to the sun. This is another example of explosive or exponential growth.

## Exponential Growth and Doubling Time

Compound interest at a savings bank results in the balance on the account growing exponentially. Steady inflation causes prices to rise exponentially. Radioactive decay follows an exponential decline. The growth of populations tends to follow an exponential growth curve for a time.

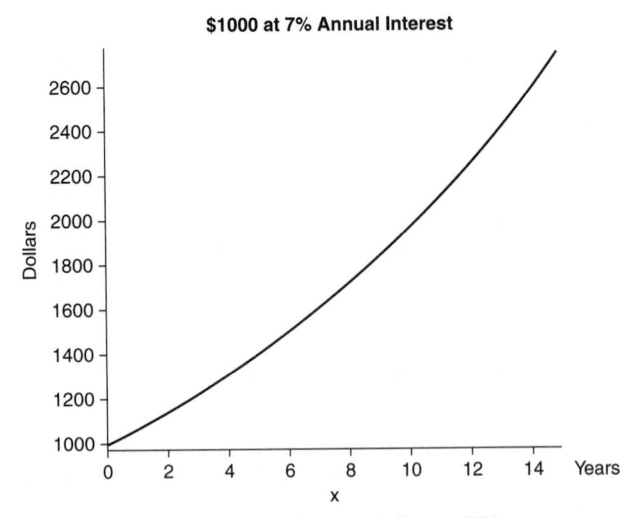


**Figure 13.5** Increase in portfolio at 7% per year.

## The Magical Formula

In exponential growth, the output value is proportional to the growth rate. In exponential growth the quantity increases by a constant percentage of the whole in a constant time period. There is a magical formula that allows us to determine the time it takes for the initial amount to double if we know the growth rate. It is T = 70/P; approximately the number of years it takes a quantity, T, to double is given by 70 divided by the growth rate percentage, P.

For example, if your stock portfolio is increasing at a rate of 10% per year, it will take seven years for it to double in value, T = 70/10. If the rate is 20% per year, you double its value in T = 70/P = 70/20 = 3.5 years. Depositing $1,000 in a bank at 7% per year annual compound interest will double to $2,000 in 10 years (see Fig. 13.5).

In the doubling process on the chess board, the number of grains placed on the next square on the chessboard was greater than the total of the previous squares. Notice square one has one grain, square two has two, and square three has four, or more than the total of squares one and two. Putting 16 grains on square five is greater than the total of 15 accumulated on the previous four squares (see Table 13.1).

The formula T = 70/P tells us that if our consumption of a natural resource doubles over a 10-year period, the rate of consumption is 7% per year. However, this means that the consumption over one decade is greater than the total of all previous consumption in the preceding years.

So, if our use of gasoline doubles over one decade, we will have used more gasoline than has been used in the entire history of the industry. If our consumption of electricity doubles over one decade, we shall have used an amount of electric-

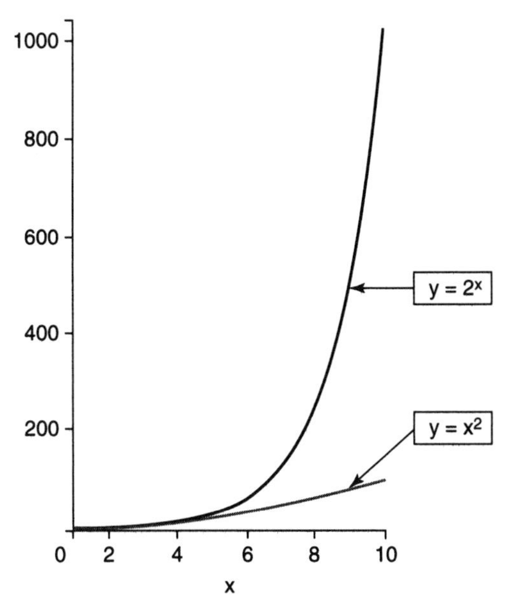


**Figure 13.6** Exponential function and power function.

ity equal to that which has been used during the entire history of the electric power industry.

In the magic formula, where doubling time T is given by 70 divided by P, the percent growth rate per unit time, $T = 70/P$, is obtained from an understanding of the mathematics underlying the exponential function (see box on p. 174). It is sometimes called the **Rule of 70.** All increasing functions are not exponential functions; for example, $y = x^a$ is a power function and behaves quite differently from $y = a^x$.

Let $a = 2$. Consider the graphs of each function. Both are increasing functions, but $2^x$ increases more rapidly, as shown in Figure 13.6.

# The Exponential Function and Population

Suppose we want to grow bacteria in a Petri dish. We start with one bacterium and find that with the appropriate amount of nutrients they divide into two about every 20 minutes. So after about 40 minutes there are only four bacteria. Twenty minutes later there are eight. The next day we come back and find $10^{24}$ bacteria, or one million billion billion. In the final 20 minutes of the 24-hour period, the same numbers of bacteria were created as were in the previous 23 hours and 40 minutes. A dramatic result has occurred.

At the end of the 24-hour period the Petri dish would contain only bacteria, including some that were dead, and all the food and nutrients would be gone. We see there is a limit to growth imposed by pollution, food, and resource depletion.

Does the human population confined to the Earth's biosphere face similar limitations?

The curve that represents the growth of bacteria in a Petri dish is termed a logistic equation, and resembles an S. That is why it is called an S-shaped curve, or a Sigmoid.

As you can see in Figure 13.8, when the population starts to grow, it does go through an exponential growth phase, but as it gets closer to the carrying capacity (approximately when the time step reaches 37), the growth slows down and it reaches a stable level.

Many natural systems exhibit such behavior. When the environment is stable, the maximum number in a population varies near the carrying capacity of the environment. However, if the environment becomes unstable, the population size can change dramatically. The use of this model is limited when dealing with real populations because of their complexity.

### Lily Pond Problem

If a lily pond doubles every day and it takes 30 days to cover the pond completely, on what day will the pond be half covered? Does the size of the pond make a difference? What happens a short time after the 30th day?

## Limits to Growth

Human population growth is similar in many respects to bacterial growth in that we are confined to our biosphere, where Earth is the environment for survival instead of a Petri dish. The rate of growth of the human population has slowed to about 1.4% per year, which when we apply the Rule of 70 gives a doubling time of about 50 years. See Figure 13.9.

The year 2000 growth rate of 1.4 percent, when applied to the world's 6.1 billion population, yields an annual increase of about 85 million people. Because of the large and increasing population size, the number of people added to the global population will remain high for several decades. This will be true even as growth rates continue to decline.

# World3 Model

In 1970, the Club of Rome, as mentioned earlier, commissioned a computer study to be done at MIT examining the questions dealing with the growth of human population and the global economy during the next century. Questions to be answered included: "What will happen if the growth in the world's population continues unchecked? (see Fig. 13.10.) What will be the environmental consequences if economic growth continues at its current

## Mathematics Underlying the Rule of 70

Consider the graphs of exponential function for three different bases, as shown in Figure 13.7.

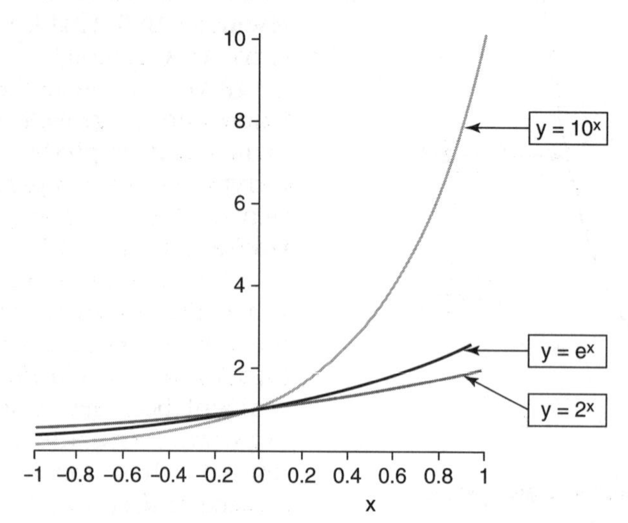


**Figure 13.7** Exponential growth.

Now the exponential function $f(t) = ba^t$, where a and b are positive real numbers and t represents time. Then if the time changes by a constant amount, T, then

$$f(t + T) = ba^{(t+T)} = b(a^t)a^T = f(t)a^T$$

where $f(t) = ba^t$. Therefore,

$$f(t + T) = f(t)a^T$$

That is, increasing t by an amount T changes the output by a constant multiple $a^T$.

When $a > 1$ (growth), choose T so that $a^T = 2$. Then for every t, $f(t + T) = 2f(t)$. The output is twice as large as before. The input that causes this is T, the time it takes to double the output. The input interval T is the doubling time of the function.

In the exponential graphs we used earlier there was one $y = e^x$. The base e is an irrational number in mathematics approximated by 2.718, the way $\pi$ is approximated by 3.14. It, e, is called the *natural exponential base.* In many applications e is the base of choice for exponential functions.

Choosing e as the base can always be done, since for any $a > 0$ we can find a k such that $e^k = a$. Then $f(x) = ba^x$ can be rewritten as:

$$f(x) = ba^x = b(e^k)^x = be^{kx}$$

where $e^k$ has replaced a. Therefore, any exponential function can be written with the natural base e if we multiply the exponent by the appropriate factor of k.

So if $a^T = 2$, then $e^{Tk} = 2$ and $f(t + T) = 2f(t)$. If we take the natural log of both sides $\ln e^{Tk} = \ln 2$, we get $kT = 0.693$, or $T = 0.693/k$. If we round off the 0.693 to 0.7, then $T = 0.7/k$.

If we multiply the top of the fraction by 100 and the bottom by 100 and get $100k = P$, the percent in the denominator changes to a whole number. We get

$$T = 70/P \text{ (Rule of 70).}$$

When the rate of using a fixed resource per year is growing at a fixed percent per year, the growth is said to be exponential. The time required for the growing quantity to increase its size by a fixed amount is constant.

For example, a rate of growth of 7% (a fixed rate) per year (a constant time interval) is exponential. The time required for the growing quantity to double in size is the doubling time T. So if the percentage of growth is 7 percent/yr, it will take 10 years for the growing quantity to double in size, $T = 70/7 = 10$.

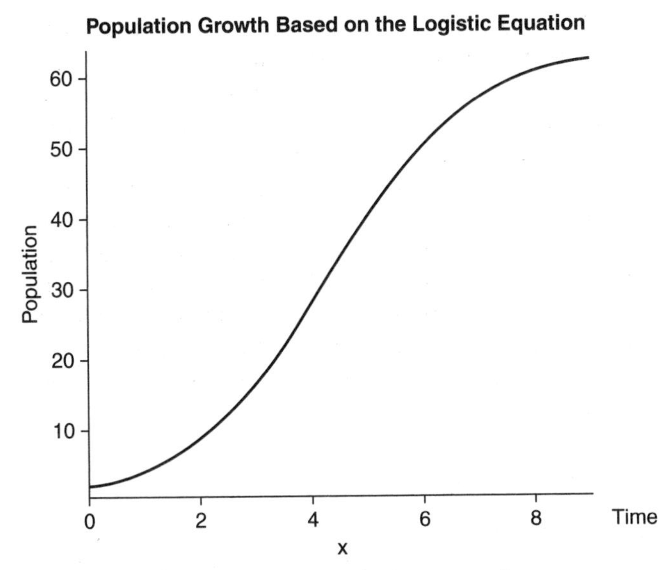


**Figure 13.8** Logistic population growth.

pace? What can be done to ensure an economy that provides sufficiently for all and that also fits the physical limits of the Earth?"

The Club of Rome is a group of scientists, economists, businessmen, international high civil servants, heads of state, and former heads of state who pool their different experiences from a wide range of backgrounds to come to a deeper understanding of world problems.

The group at MIT constructed a computerized global model, World3. Remember that the usefulness of models is that they allow a better understanding of the present as well as allowing one to make predictions about the future. The World3 model exists as equations on a computer. It was designed to understand the role of humans as our population approaches the world's carrying capacity. It provides us with insights as to what actions should be taken as we approach that limit so as to

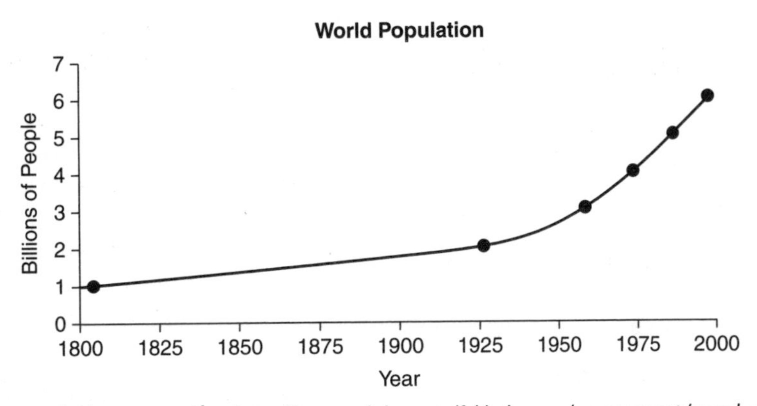


**Figure 13.9** World population growth. *http://www.globe.gov/fsl/educorn/assessment/word_pdf/ESysConnHS2student.pdf*

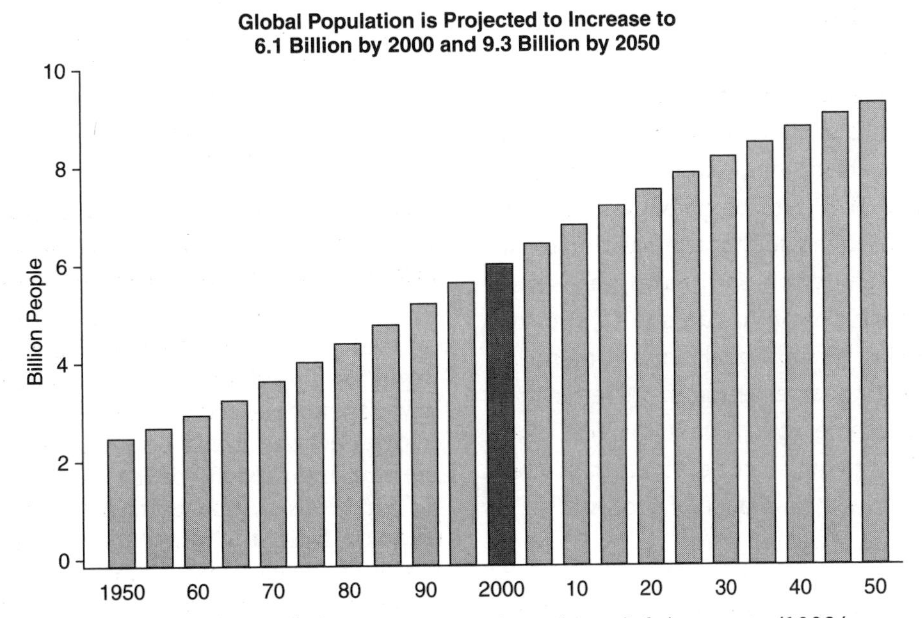


**Figure 13.10** Increase in global population. *http://www.fas.usda.gov/info/agexporter/1998/February%201998/tallying.html*

avoid a dramatic collapse as in the case of the bacteria in the Petri dish.

The World3 computer program is available at many college and university computer centers, thus allowing one to try worldwide strategic plans. Probably it would be most productive if one variable at a time was changed. The model we will examine here will be the 1992 version described in the World3 report "Beyond the Limits."

## Report to the Club of Rome (1972)

In the first study, "A Report to The Club of Rome (1972)" by Donella H. Meadows, Dennis L. Meadows, Jorgen Randers, and William W. Benders III, the model was constructed to investigate accelerating industrialization, rapid population growth, widespread malnutrition, depletion of nonrenewable resources, and a deteriorating environment. Newspaper headlines announced:

A COMPUTER LOOKS AHEAD AND SHUDDERS

STUDY SEES DISASTER BY YEAR 2100

SCIENTISTS WARN OF GLOBAL CATASTROPHE

The actual conclusions in 1972 were:

1. "If the present growth trends in world population, industrialization, pollution, food production, and resource depletion continue unchanged, the limits to growth on this planet will be reached sometime within the next one hundred years. The most probable results will be a rather sudden and uncontrollable decline in both population and industrial capacity.
2. It is possible to alter these growth trends and to establish a condition of ecological and economic stability that is sustainable far into the future. The state of global equilibrium could be designed so that the basic material needs of each person on the earth are satisfied and each person has an equal opportunity to realize his individual human potential.
3. If the world's people decide to strive for this second outcome rather than the first, the sooner they begin working to attain it, the greater will be their chance of success."[1]

[1]Meadows, Donella H., Dennis L. Meadows, and Jørgen Randers. *Beyond the Limits*. Chelsea Green Publishing, Post Mills, Vermont, 1992, pp. xiii, xv–xvi.

The study found that all five elements that were basic to the analysis (i.e., population, food production, consumption of natural resources, pollution, and industrialization) were growing and following an exponential growth pattern.

When a new report, "Beyond the Limits" was submitted to The Club of Rome in 1992 using the same World3 model and data collected over the previous 20 years, it was found that world population had slowed in growth and an estimate of 7-10 billion people was determined to be a stable population. Let us consider some variables that are dealt with in the World3 model.

## Population

How can we obtain an estimate of the carrying capacity of the Earth in terms of population? How many people can live on the Earth at one time? See Fig. 13.11. Certainly food would be a limiting factor. If we calculate the limit to the Earth's food production, we can approximate the Earth's carrying capacity.

The amount of food on earth depends directly on the sun's energy reaching the Earth. The amount of the sun's radiated power per unit area is well established and is known as the solar constant. The numerical value varies with the seasons, time of day, weather, and latitude. It is expressed in units of power/area, or watts/$m^2$.

## Carrying Capacity Calculation

The solar input, or solar constant, is the amount of solar power absorbed by a given cross-sectional area when the sun is directly overhead. The accepted value of the solar input is about 1,000 watts/$m^2$ on a clear day at sea level. Although the solar input is higher above the Earth's atmosphere since there is no reflection, most of the world's food is grown at sea level.

**Problem: How large a population can the Earth sustain based on the total amount of calories per day available as food?**

To solve this problem we compute the total calories per day available to people, or the total world food production, in terms of calories per day. If we then divide this number by the number of calories per day that a person needs, an estimate of the carrying capacity of the earth will be obtained.

First let's compute the total amount of power received from the sun each day in watts. If the radius of the earth is $r = 6.38 \times 10^6$ meters, then its cross-sectional area is $A = \pi r^2$. The cross-sectional area represents the area of the earth perpendicular to the sun's rays.

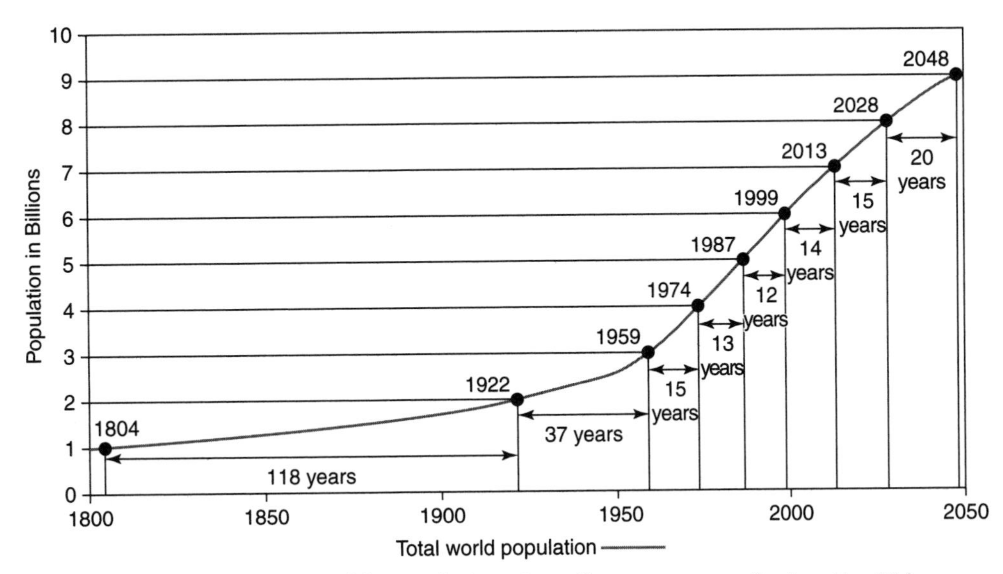


**Figure 13.11** Growth estimates of world population. *http://www.census.gov/ipc/prod/wp02/wp-02003.pdf*
Source: United Nations (1995) US, Census Bureau, International Programs Center International Data Base and unpublished tables.).

$$A = 3.14(6.38 \times 10^6)^2 \text{ m}^2 = 1.28 \times 10^{14} \text{ m}^2$$

Total Power = Solar input × cross-sectional area of Earth

$$\text{Total Power} = 1{,}000 \text{ W/m}^2 \times 1.28 \times 10^{14} \text{ m}^2 = 1.28 \times 10^{17} \text{ watts}$$

*The photosynthesis process is not 100% efficient and the chlorophyll of plants in the photosynthesis process captures approximately 0.076% of the solar energy that reaches the earth.* Therefore,

$$\text{Total Potential Photosynthesis Power} = 1.28 \times 10^{17} \text{ watts} \times 0.00076 = 9.72 \times 10^{13} \text{ watts.}$$

*Now only 10% of the Total Photosynthesis Power is ultimately converted to food for humans.*

$$\text{Solar Power Ultimately Converted to Food} = 0.10 \times 9.69 \times 10^{13} \text{ watts} = 9.72 \times 10^{12} \text{ watts}$$

To obtain the caloric equivalent of this food we use the fact that 1 watt = 0.24 cal/s to obtain the amount of solar power converted to food each day.

$$\text{Amount of Food/Day} = 9.72 \times 10^{12} \text{ watts} \times 0.24 \text{ cal/s/watt} = 2.33 \times 10^{12} \text{ cal/s}$$

Now convert this to calories per day:

$$\text{Amount of Food/Day} = (2.33 \times 10^{12} \text{ cal/s}) \times (86{,}400 \text{ s/day}) = 2.01 \times 10^{17} \text{ cal/day}$$

The average American may have a daily food intake of 2,200 Calories (plant and animal foodstuff). The number rises to approximately 11,000 primary Calories per day when the conversion of plant to animal food is considered, since five or more plant calories are required for every calorie generated of animal food. Now since the dietician's Calorie = 1,000 calories (of the scientist), *the food requirement for one well-fed person becomes 11,000,000 calories per day.*

Therefore, the carrying capacity of the earth is simply the world food production per day divided by the calorie requirement of a well fed person, or Carrying Capacity = (world food production per day)/(food requirement for a well-fed person).

*So 18 billion people, or Carrying Capacity =* $(2.01 \times 10^{17}$ *cal/day)/(11,000,000 cal/day/person)* $= 1.83 \times 10^{10}$ *persons, can theoretically be sustained on the surface of the Earth each day.*

## Population Growth and Resources Impact

Population growth drives the requirements for food and water. In reality there is a negative impact generated by population growth. Many will say that population growth is necessary for developing countries to create more producers, thus enlarging the economy, but in reality the larger population generates greater demands for food, housing, and consumer goods. See Figure 13.12.

**Food.** The World3 global model predicts that greater food production will be necessary to sustain the larger population. This means more land must be used in food production and the yield increased. Otherwise, there will be a world food shortage.

In the past 20 years cultivated land has decreased because of urbanization, erosion, salt buildup, and desertification. Interestingly, vegetarian diets require

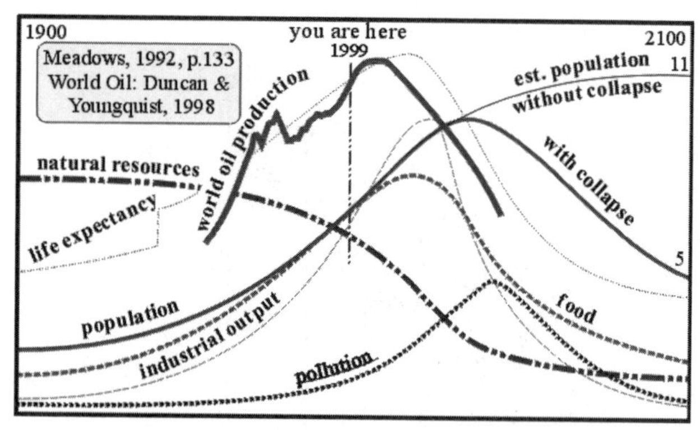


**Figure 13.12** Population, industrial output, pollution, food projections.

(Chelsea Green Publishing Company, 1992. ISBN 0-930031-62-8. Phone: 800-639-4099 or 603-448-0317; FAX: 603-448-2576.)

much less cultivated land. This is so because animal-based diets require a lot more plant matter to support the animals that feed one person.

**Water.** Although water usage is stabilizing in industrial countries, it is still found to be growing in developing countries. Within a few decades many countries will confront limits on their water supplies. Desalinization and detoxification may be a solution for some and conservation for others. The ultimate solution, of course, would be a smaller population.

**Forests.** The world's tropical forests will be used up in 30 years if tropical deforestation increases exponentially at the same rate as population growth. In many cases deforestation is followed by erosion, which prevents regrowth.

**Fossil Fuels.** There is only so much easily accessible oil, natural gas, coal, etc., available underground. Usage again is growing exponentially and the resource is limited. A solution is to convert to more efficient energy use to slow down the growth and to move to renewable energy sources such as solar and wind power.

**Materials.** Again our exponentially growing extraction of various materials from the ground such as iron, copper, aluminum, and lead results in a drop-off in quality. Recycling may help conserve this resource.

**Pollution.** Pollution is rising worldwide. It clearly has had an impact on human and animal health. Some species have disappeared as a result, and pollutants decrease the amount of clean water available. Pollution could cause the large-scale breakdown of ecosystems and it is in our best interests, therefore, to control it as much as possible.

**Industry.** Just as population is growing exponentially, so is industry. It uses energy, materials, water, and wood. It creates pollution. Either we have to slow its growth or come up with technological innovations to decrease its needs.

## Conclusions: The 1992 Report

As a result of the global data collected, the conclusions of World3 model, and the knowledge gained over the 20 years between reports, the three conclusions from 1972 would still be considered valid but strengthened as follows:

1. Human use of many essential resources and generation of many kinds of pollutants have already surpassed rates that are physically sustainable. Without significant reductions in material and energy flows, there will be in incoming decades an uncontrolled decline in per capita food output, energy use, and industrial production.
2. This decline is not inevitable. To avoid it two changes are necessary. The first is a comprehensive revision of policies and practices that perpetuate growth in material consumption and population. The second is a rapid drastic increase in the efficiency with which materials and energy are used.
3. A sustainable society is still technically and economically possible. It could be much more desirable than a society that tries to solve its problems by constant expansion. The transition to a sustainable society requires a careful balance between long-term and short-term goals and an emphasis on sufficiency, equity, and quality of life rather than on quantity of output. It requires more than productivity and more than technology; it also requires maturity, compassion, and wisdom.[2]

In the animal world an increase in population results in a natural regulation, which avoids large-scale starvation. Sea birds when faced with a decrease in plankton distribute themselves over a wider area. Guppies after reaching their maximum numbers within a given environment are found to remove excess numbers through cannibalism. In the case of mice, overpopulation results in a reduction in ovulation as density increases. In the animal kingdom it seems there is a certain limit to population density that exercises a regulatory control.

---

[2]Meadows, Donella H., Dennis L. Meadows, and Jørgen Randers. *Beyond the Limits*. Chelsea Green Publishing, Post Mills, Vermont, 1992, pp. xiii, xv–xvi.

In humans, control of population may result through war, starvation, and disease. There is a delay in feedback that may allow the population to overshoot its limit. As a society becomes more technologically advanced and wealthy, members of the society find that the only way to retain their comforts is to limit family size. It is a phenomenon we observe in highly industrialized societies like Europe, Japan, and the United States.

In less-developed countries we find the population tends to increase and the food supply diminishes to starvation levels. Clearly the solution is to provide for the poor of the world in terms of food, health, and sustainability, and to provide effective and humane controls on population increase.

# Review Questions

1. What is exponential growth? How does it differ from linear growth?
2. What role do positive and negative feedback loops play in growth?
3. What are the possible limits to growth?
4. What was the purpose of creating the World3 computer model?
5. What role does delay play in positive and negative feedback loops?
6. How can we control growth in human systems?
7. How can we work to prevent the collapse of our global systems?
8. What do you think we as a society should do?
9. What should we do as individuals?

# Multiple Choice Questions

1. The transformation of mass into energy on the sun is expected to continue for _____.
   a. thousands of years
   b. millions of years
   c. billions of years
   d. trillions of years
2. Thomas Malthus considered the world's population to be increasing _____.
   a. exponentially
   b. linearly
   c. geometrically
   d. arithmetically
3. In exponential growth the quantity increases by a constant percentage _____.
   a. regardless of the total amount
   b. of the total amount
   c. and results in a linear increase
   d. None of the above.
4. It is known that positive feedback loops without any constraints produce _____.
   a. linear growth
   b. constant growth
   c. geometric growth
   d. exponential growth
5. In exponential growth if the percentage of growth is 5%, the doubling time is _____.
   a. 10 years
   b. 14 years
   c. 15 years
   d. 20 years
6. The Rule of 70 is given by the formula _____.
   a. $T = 70P$
   b. $T = 70/P$
   c. $P = 70T$
   d. $P = T/70$
7. An approximation for the theoretical carrying capacity population of the earth is _____.
   a. 6 billion people
   b. 9 billion people
   c. 12 billion people
   d. 18 billion people
8. Technologically advanced members of society find that the way to retain their comforts is to _____.
   a. save money
   b. invest in stocks
   c. limit family size
   d. invest in real estate
9. To prevent a decline in population, the World3 1992 report recommends we _____ energy usage.
   a. ration
   b. decrease
   c. perpetuate
   d. discontinue
10. The World3 1992 report says that a transition to a sustainable society emphasize quality of life and not _____.
    a. accumulation of riches
    b. overpopulation
    c. quantity of output
    d. renewable resources
11. How many mm thick would folded paper become if a 1 mm sheet were folded over five times?
    a. 32 mm
    b. 5 mm
    c. 10 mm
    d. 5,000 mm

12. How long would it take for an individual's money in the bank account to double if the interest is 3.5% per year?
    a. 5 years
    b. 14 years
    c. 20 years
    d. None of the above.

13. If the doubling time of a bacteria's growth occurs in seven minutes, what is the percentage growth?
    a. 10% per minute.
    b. 10% per hour.
    c. 70% per minute.
    d. None of the above.

14. The carrying capacity of the Earth refers to _____.
    a. the mass of the Earth
    b. the number of individuals that can be supported on Earth
    c. the effect of gravity on people
    d. None of the above.

15. The Club of Rome is _____.
    a. a soccer organization
    b. a frat from the University of Rome
    c. an organization of professionals interested in understanding world problems
    d. a large Italian mace

16. Which is probably the most critical factor in estimating the carrying capacity of the Earth?
    a. food
    b. water
    c. electricity
    d. fossil fuel

17. The solar input is _____.
    a. 5000 watt/$m^2$
    b. 500 watt/$m^2$
    c. 1 kilowatt/$m^2$
    d. 50 watt/$m^2$

18. Which of the following represents the correct units for the solar input?
    a. cal/day
    b. watts/m
    c. cal/sec-$m^2$
    d. joules $m^2$

19. Photosynthesis _____.
    a. is a highly efficient process
    b. yields mainly food
    c. is an exothermic process
    d. is rather inefficient

20. Fossil fuel use, pollution, and population growth are all _____.
    a. connected
    b. independent of each other
    c. decreasing
    d. increasing

21. In a hypothetical population, N = 1,000, the annual rate of birth *(b)* = .15 and the annual rate of death (d) = .05. What will be the size of this population after 3 years? (Note: Use the simplified model I = rN, where r = b – d.)
    a. 1,300
    b. 1,210
    c. 1,500
    d. 1,331
    e. 1,533

22. The following graph is a model of the growth pattern in the population of gypsy moths in Westchester County. Answer the questions following the graph based on your understanding that the rate of population increase is predicted by this mathematical model.

    $$I = rN(1 - N/K)$$

    I = Rate of growth of population
    N = Population
    K = Carrying capacity
    r = Proportionality constant

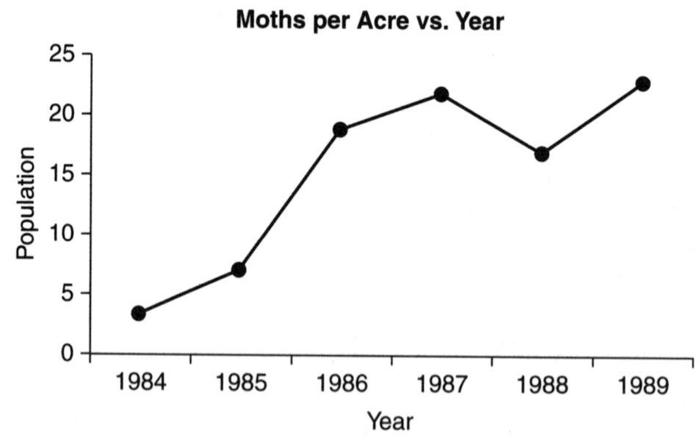


In period 1986–1987, the rate of increase in the gypsy moth population slowed because _____.
    a. K approached the value of N
    b. r approached the value of K
    c. N approached the value of r
    d. K approached the value of r
    e. N approached the value of K

## Bibliography and Web Resources

Donella H. Meadows et al., *The Limits to Growth* (New York: Universe Books, 1972).

Bartlett, Albert A. "Forgotten Fundamentals of the Energy Crisis," *American Journal of Physics,* 46(9) Sept. 1978, pp. 876–888.

Parsegian, V. Lawrence, et al. *Introduction to Natural Science Part One: The Physical Sciences,* Academic Press, New York, 1968.

Swartz, Clifford E., *Used Math.* American Association of Physics Teachers, 1993.

Limits to Growth Report.

Meadows, Meadows, and Randers, "Beyond the Limits to Growth Collapse, Envisioning a Sustainable Future," is instant must-reading. . . .

**www.context.org/ICLIB/IC32/Meadows.htm**

The "Rule of 70" for Exponential Growth. . . . The Rule of 70 is a "rule of thumb"—it is not exactly accurate. It is an approximation. . . .

**mmcconeghy.com/students/supscruleof70.html**

Exponential Growth and The Rule of 70.

**www.ecofuture.org/pop/facts/exponential70.html**

The Club of Rome home page.

**http://www.clubofrome.org/**

# Medical Screening Tests

*"More than you need to know?"*

**UK *Guardian Unlimited***

The purpose of screening tests is to determine an individual's risk factor, or likelihood, of developing a disease in the future. In contrast, diagnostic tests are generally performed on individuals who show symptoms of disease and are used to confirm the presence and specific nature of the disease so that the appropriate treatment may be started.

The human body is a complex system within which the condition of maintaining a stable, or static, condition within our body's fluids and within which our cells can live and function is known as *homeostasis.*

We shall find that medical screening tests are designed to distinguish between healthy people and those afflicted with disease by comparing the results of such tests with results obtained from known healthy populations and known diseased populations. Mathematical probabilities are used in interpreting the results.

## Goals

After studying this chapter, you should be able to:

- Distinguish between medical screening tests and diagnostic tests.
- View the body with its structure and systems as a living machine.
- Understand the role of homeostasis in a living body.
- Understand how medical screening tests use probabilities to identify the presence of disease.
- Understand how medical screening tests are used to detect disease.
- Understand what is meant by the sensitivity of a test.
- Understand the role of some common screening technologies such as amniocentesis, fetoscopy, newborn screening.
- Understand the role of risk-benefit analysis in genetic screening.

## Medical Screening Tests

We all share the experience of having visited our personal physician and undergone one, if not several, medical screening or diagnostic tests. The most likely would have involved an evaluation of body temperature, blood pressure, heart rate, vision, hearing, and blood work (cholesterol, electrolytes, white blood cell counts, etc.).

These tests generally fall into one of two categories: screening tests or diagnostic tests. Screening tests are those that are performed on healthy people who have no symptoms of disease. The

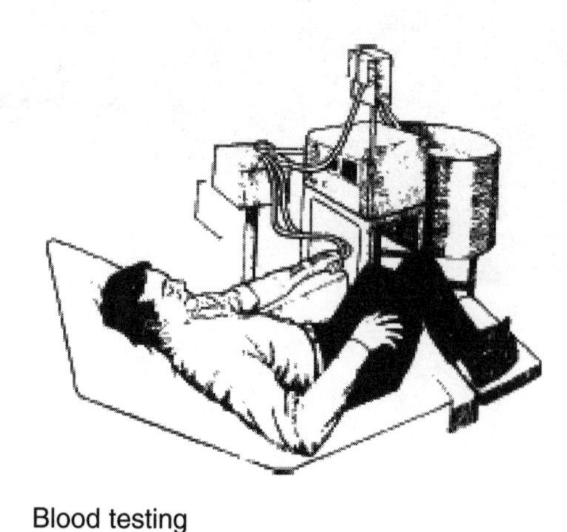

Blood testing

Blood pressure testing

**Figure 14.1** Diagnostic tests. *http://www.cdc.gov/diabetes/pubs/images/laytest.gif*

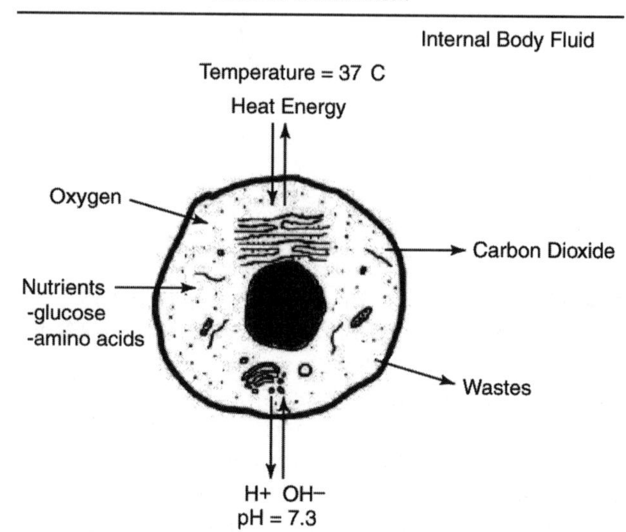


**Figure 14.2** Illustration of cells in fluid.

purpose of screening tests is to determine an individual's risk factor, or likelihood, of developing a disease in the future. They are often performed in order to establish a baseline for an individual's health status against which future screening tests can be compared.

Diagnostic tests (see Fig. 14.1) are generally performed on individuals who show symptoms of disease and are used to confirm the presence and specific nature of the disease so that the appropriate treatment may be started. Once treatment has been started, further diagnostic tests can be used to follow the progress of treatment and to determine when normal health is restored.

To understand the design and interpretation of medical screening and diagnostic tests we first must understand something about the normally functioning human body and what, exactly, goes wrong in those conditions with sickness or disease.

The human body is a complex system comprised of approximately 75 trillion individual cells. Each of these cells, like the simplest amoeba, requires a watery environment that provides it with a stable temperature (about 37° C or 99° F), a stable pH (about 7.3), a stable concentration of oxygen and carbon dioxide (partial pressures of approximately 104 mmHg and 40 mmHg, respectively), and a constant level of nutrient (approximately 90 mg/dl of glucose). See Figure 14.2.

"pH" provides a measure on a scale from 0 to 14 of the acidity or alkalinity of a solution (where 7 is neutral, greater than 7 is basic, and less than 7 is acid). Pressure is a measure of force per unit area. The standard atmospheric pressure is given at sea level by the height of mercury in a standard barometer and is equal to 760 mmHg. This is a measure of the weight of an air column that rises to the top of the atmosphere on a unit area on the surface of the earth. Perhaps you have heard it quoted as 14.7 pounds per square inch, or 14.7 psi. Instead of mmHg another equivalent set of units is used.

Given this stable and hospitable environment our cells do fine and, as a consequence, so do we as human organisms. Reduced to its simplest level, the human body, its anatomical structures and physiological systems, is nothing more than a living machine designed to maintain a constant, or static, set of conditions in which our individual cells may function.

The condition of maintaining a stable, or static, condition within our body's fluids and within which our cells can live and function is known as homeostasis. As detailed in the chapter on systems, homeostasis is maintained by a mechanism known as negative feedback. In actuality, such a system does not maintain a static, unchanging, condition so much as it maintains a physical

| Axis D MmHg | Axis C °F | Axis B mg/dL | Axis A °F |
|---|---|---|---|
| 130 | 101.5 | 130 | 74 |
| 120 | 100.5 | 120 | 72 |
| 110 | 99.5 | 110 | 70 |
| 100 | 98.5 | 100 | 68 |
| 90 | 97.5 | 90 | 66 |
| 80 | 96.5 | 80 | 64 |
| 70 | 95.5 | 70 | 62 |
| 60 | 94.5 | 60 | 60 |

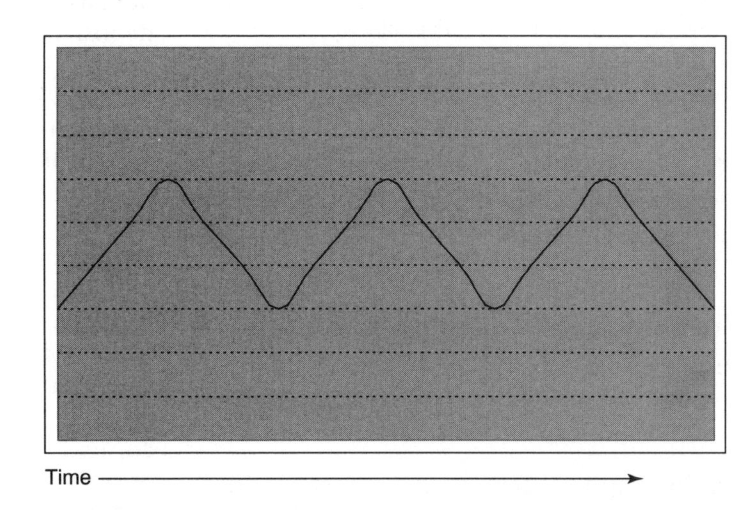


**Figure 14.3** Homeostasis.

parameter (temperature, pH, glucose level) within an acceptable range as might your home heating system, as shown in Figure 14.3. You may wish to review the sections in Chapter 6 about negative feedback and homeostasis.

If we take for a moment the example of your home heating system (y-axis "a"), we can see that the system maintains room temperature not at a constant temperature of 67° F but, rather, at an average of 67° F and within a comfortable range of, perhaps, 64°–70° F. Similarly, our body's physiological systems maintain our glucose (y-axis "b"), body temperature (y-axis "c"), and mean blood pressure (y-axis "d") within similarly comfortable ranges, not actually at an absolute static level.

Blood glucose levels, for example, will vary throughout the day, rising shortly after each meal and falling back as the body releases insulin to draw glucose into the cells for storage. At its lowest level, before a meal, blood glucose may fall to a level of 70–80 mg/dl and rise to a level of 110 mg/dl after the meal is eaten. However, it would not be normal for blood glucose levels to fall far below or to rise well above this range.

Similarly, our body temperature may vary throughout the day, falling to its lowest point during the dormancy of our sleep and rising to its highest point late in the afternoon or after heavy physical exertion. Body temperatures will generally not fall outside of the range of approximately 97–100° F in a healthy person.

Physiologically, when our body's systems are functioning properly, we refer to the measurements as being within the comfort zone, or **zone of ease,** as illustrated in Figure 14.4. When a system malfunctions, allowing some critical parameter (temperature, blood pressure, glucose level, etc.) to fall outside the zone of ease, we describe it

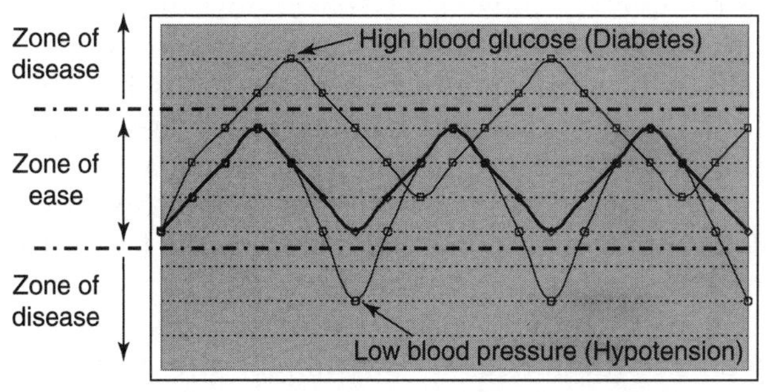


**Figure 14.4** Zones shown in medical screening.

as now being in *zone of dis-ease,* or simply, **disease.** Therefore when a person's temperature rises to 101° F, when blood pressure falls to a mean of 60mm Hg, or when blood glucose levels rise to 200 mg/dl, they are said to be sick, ill, or diseased.

Medical screening tests should measure some property associated with a person that will be shared by all people who have the condition being searched for and that is absent from all those who do not have the condition. For example, blood glucose levels higher than 150 mg/dL are common in individuals with diabetes, and body temperatures of 101° F and above are common in persons suffering from infections. Usually, measuring a particular property will give us a spread of results for a healthy population. These can be expressed as a probability distribution. Examination of those known to have the condition will also result in a probability distribution but the range will be different. An example of this is seen in Figure 14.5.

The horizontal, or x-axis, gives the value of the measured medical test. For body temperature as an example, point "A" on this axis would fall on the temperature scale at 102° F, while point "B" would indicate a temperature of about 98° F.

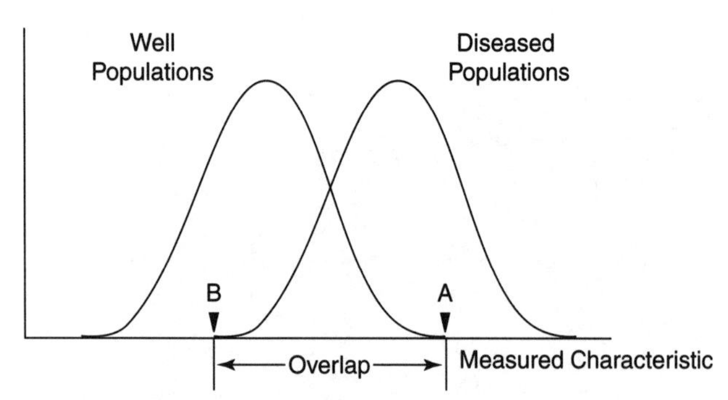


**Figure 14.5** Probability distributions for screening for disease.

From the graph we can see that a person with a temperature of greater than 102° F clearly falls within the population of diseased individuals and is completely outside of the population of well individuals. Someone with a temperature of 98° F falls within the population of well, or healthy, individuals and would be outside the area under the curve representing diseased individuals.

The people tested are either well or diseased. Two plots are drawn on the graph; one for a diseased population and one for a well population. The plots indicate the size of the diseased or well population as a function of the measurement test.

There is a range of values, however, where the well and diseased populations overlap and this may cause problems in deciding if the individual is actually diseased or not. In our example of body temperature measurements using the graph in Figure 14.5, what might be said about the health or disease state of a person with a body temperature measurement of 100° F, a point that falls midway between points "A" and "B" and within the curves of both populations?

Consider another case. Blood pressure is the force generated by blood pushing on the walls of the arteries. Two numbers record it, the systolic pressure over the diastolic pressure. The systolic pressure measures the pressure of blood in the arteries as the heart beats and the diastolic measures the pressure as the heart relaxes between beats. Normal blood pressure occurs when the systolic reading is less that 120 mmHg and the diastolic less than 80 mmHg.

Blood pressure readings for human beings vary throughout the day but when the readings are elevated above normal for a period of time, it is termed high blood pressure or, in medical terminology, hypertension. High blood pressure is dangerous because it makes the heart work too hard. The inner lining of the coronary arteries become damaged. The arteries can become narrowed as deposits build up and reduce blood flow to the organs. The risk of heart disease, congestive heart failure, kidney disease, blindness, and arteriosclerosis is increased.

Similarly, while it is easy to determine that a person with a blood pressure of 118/78 mmHg is normal with respect to this measurement and that a person with a blood pressure measurement of 220/110 suffers from serious high blood pressure, can we comfortably state that someone with a blood pressure of 140/85 suffers from hypertension? Perhaps not. This value for blood pressure measurement falls at a point that may coincide with either the higher end of the range of normal values and the lower limit of the range indicating hypertensive disease. Similarly, is someone with a body temperature of 100° F sick with infection or just at the high end of normal? The borderline individuals in both of these examples would fall in the area of overlap in Figure 14.5.

The **sensitivity** of a screening or diagnostic test refers to its ability to accurately identify the presence of disease in people who have the disease (a **true positive** result). False test results can fall into one of two categories, false positives and false negatives.

$$\text{Sensitivity} = \frac{TP}{TP + FN}$$

Sensitivity can be adjusted so that it is less likely that disease conditions will be missed. However, when we increase the sensitivity, we also increase the likelihood that the test will determine the presence of disease in a person who is actually healthy. This event, as just stated, is known as a **false positive (FP)** result. Reducing the sensitivity of the test can reduce the chance of false positives, but if the sensitivity is reduced too far, the test may fail to detect disease when it is actually present. This is known as a **false negative (FN)** result. A **true negative (TN)** result would be one in which the test correctly reported no disease in people who are actually healthy.

The specificity of a test relates to its ability to accurately indicate no disease among all healthy people and to accurately indicate the presence of disease among those who are afflicted. Specificity can be represented as:

$$\text{Specificity} = \frac{TN}{TN + FP}$$

The results from screening tests are not absolute in the sense that testing positive auto-

**TABLE 14.1 Health Status and Tests**

| CASE I | | | CASE II | | |
|---|---|---|---|---|---|
| **Health Status** | **Test +** | **Test–** | **Health Status** | **Test +** | **Test–** |
| Ill | 180 | 20 | Ill | 3,600 | 400 |
| Healthy | 990 | 18,810 | Healthy | 800 | 15,200 |

Fraction testing positive who are actually ill = 180/(180 + 990) = 180/1,170 = 0.15

Fraction testing positive who are actually ill = 3,600/(3,600 + 800) = 3,600/4,400 = 0.81

matically means that an individual is ill. The certainty of illness arising from a screening test depends upon the specificity, sensitivity, and prevalence of the illness in the test population. Consider the following two cases.

In both cases a test population of 20,000 is examined. The screening test has a sensitivity of 90% and a specificity of 95% in both cases. In Case I the prevalence of the illness is 1% and in Case II the prevalence of the illness is 20%. In each case, what are your chances of really being ill if you test positive?

To answer this question you must first determine the number of T+, T–, F+, and F– present and then determine the number of T+ found compared to all those who actually tested positive. The following table shows a useful way of organizing this information:

| Health status | Test + | Test– |
|---|---|---|
| Ill | T+ | F– |
| Healthy | F+ | T– |

***Case I*** / ***Case II***

Number ill = Population × Prevalence

| Case I | Case II |
|---|---|
| Number ill = 20,000 × 0.01 | Number ill = 20,000 × 0.20 |
| Number ill = 200 | Number ill = 4,000 |

Total Population = Number ill + Number healthy

| Case I | Case II |
|---|---|
| Number healthy = 19,800 | Number healthy = 16,000 |

**Sensitivity** = **T+/Total ill**

| Case I | Case II |
|---|---|
| .90 = T+/200 | .90 = T+/4,000 |
| T+ = 180 | T+ = 3,600 |

Number ill = (T+) + F–

| Case I | Case II |
|---|---|
| F– = 200 – 180 | F– = 4,000 – 3,600 |
| F– = 20 | F– = 400 |

**Specificity** = **T–/Total healthy**

| Case I | Case II |
|---|---|
| .95 = T–/19,800 | .95 = T–/16,000 |
| T– = 18,810 | T– = 15,200 |

Number healthy = (T–) + F+

| Case I | Case II |
|---|---|
| F+ = 19,800 – 18,810 | F+ = 16,000 – 15,200 |
| F+ = 990 | F+ = 800 |

Completing the table results in Table 14.1. Although the sensitivity and specificity of the test are the same, there is only a 15% chance of actually having an illness when testing positive in the first case but an 80% chance in Case II, or about 4.5 times as likely when the prevalence was greater. How can this be? It turns out that when the prevalence of an illness is low there will be a higher proportion of false positives than when the prevalence is high. This will reduce the fraction that test positive and are really ill, as our example has shown.

One should be aware of this effect when examining the results of screening tests with low prevalence of illness. The number of false positives can be reduced to a degree by making tests that have a greater sensitivity and specificity. However, better tests may be more costly in time and money and that tends to run counter to the idea of a quick and cheap screening test. To completely eliminate false results would require sensitivity and specificity to be 100%.

Table 14.2 indicates several medical screening/diagnostic tests that are commonly used.

# How Effective Are Medical Screening Tests?

*The Guardian,* a UK publication, printed an investigative report (Saturday, November 15, 2003) on the value of medical screening tests in early detection. They found that there was little evidence to

**TABLE 14.2 Medical Screening**

| Measured Parameter | Normal (Reference) Range | Disease Condition Indicated |
|---|---|---|
| Body Temperature | 97–99° F | Infection; hyper-, hypothermia |
| Blood Pressure | 120/80 mmHg (systolic/diastolic) | Hypertension |
| Glucose level in blood | 70–100 mg/dl | Diabetes |
| Hematocrit (red blood cell volume) | 37–48% female; 45–53% male | Anemia |
| Cholesterol level in blood | <225 mg/dl | Heart Disease / Stroke risk |
| Triglycerides (fat) level in blood | 40–200 mg/dl | Heart Disease / Stroke risk |
| Bilirubin (bile biproduct) in blood | 0–1.0 mg/dl | Liver Disease |
| Oxygen Pressure | 83–100 mmHg | Lung Disease |
| Oxygen Saturation (level in blood) | 96–100% | Lung Disease |
| Platelet Count (# in blood) | 150,000–350,000/ml | Hemophelia |

support the premise that early detection saves lives. The newspaper questioned whether the risks, financial and emotional, were worth it. Medical evidence, according to *The Guardian,* seems to indicate that medical screening tests are essentially "fishing expeditions."

# Genetic Screening Technologies

There are more than 3,000 genetic diseases and chromosomal abnormalities that are known to afflict human beings. Although most of these pose little risk of serious difficulty, those relatively few that are serious often have a significant negative impact on both the individual and society. These 3,000 genetic diseases result in the hospitalization of 12 million Americans each year at a cost to the economy of about $2.5 billion.

Among the more serious and well-known of these disorders is Tay-Sachs, Huntington's, hemophilia, Down's syndrome, sickle cell, cystic fibrosis, and beta thalassemia. Those mentioned, could all be diagnosed within the fetus by screening techniques.

Sickle cell anemia, for example, affects 1 in 400 African Americans in the United States. Thirty states presently require the screening of school children and applicants for marriage licenses for sickle cell anemia.

Although the future holds promise, there is little that can presently be done to correct most genetic diseases. Prevention of genetic disease would therefore seem to be the more effective route. There are a number of screening techniques available to diagnose the presence of genetic disease at various levels. Techniques of carrier screening are available to test for the presence of genetic defects in parents and to allow for counseling if the likelihood exists for genetic disease being transmitted to offspring.

Fetal screening techniques (also known as prenatal screening) allow for diagnoses to be made on the fetus, *in utero,* at various stages of development. Newborn screening techniques provide the diagnostic ability to test for the presence of disease in children following birth.

# Fetal Screening

## Amniocentesis

Amniocentesis involves the chromosomal analysis of cells shed from the developing fetus, and obtained from the amniotic fluid surrounding it. The procedure requires penetrating the abdominal wall and uterus with a hypodermic needle and withdrawing about 50 ml of the amniotic fluid in which the fetus develops. Fetal cells withdrawn with the fluid are then grown in laboratory culture and analyzed for chromosomal and biochemical abnormalities.

Amniocentesis is performed at 16 weeks into the pregnancy (see Fig. 14.6) and results typically require three weeks to obtain. Information about the fetus is therefore available generally after the 19th week of pregnancy.

Although the procedure is performed with the visual aid of ultrasound, the fetus does move and is sometimes stuck with the needle. On rare occasions, infection results and spreads to the fetus. With this procedure, there is a risk of about 1 in 200 (or 0.5%) that the fetus will be destroyed and spontaneously aborted.

About 50,000 amniocentesis procedures are conducted each year in the United States with costs ranging from $1,000 to $3,000 per test.

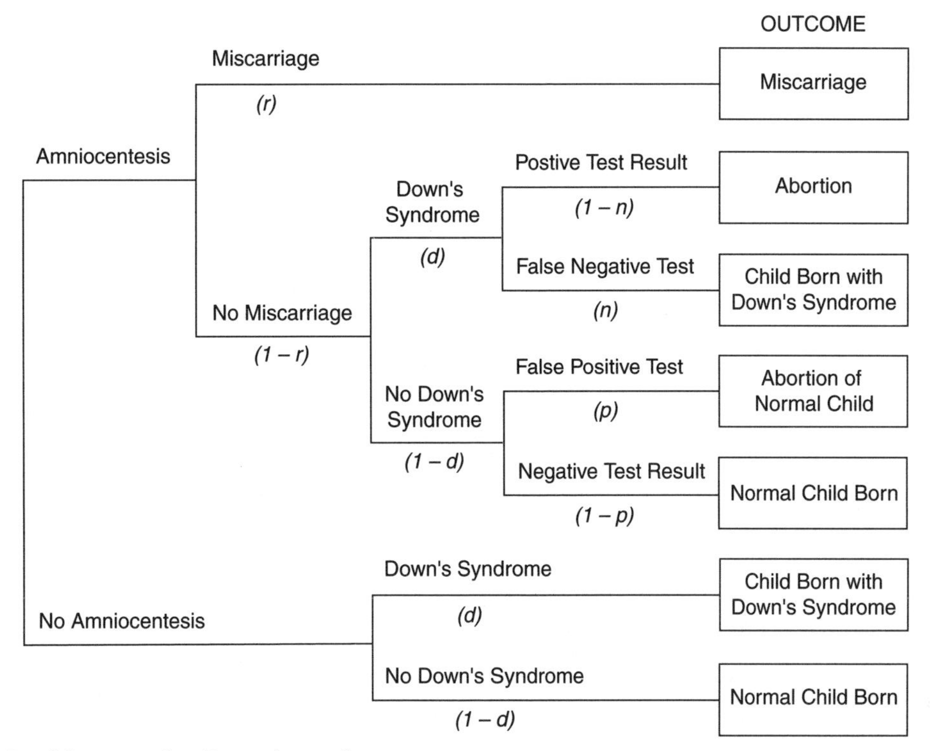


**Figure 14.6** Decision tree for Down's syndrome.

### FETOSCOPY

The procedure of fetoscopy allows for the direct visualization of the developing fetus by the physician and the sampling of fetal blood, skin, or other tissues for testing. A small incision is made through the abdominal wall, uterus, and amniotic sac. Through this opening, a thin tube is inserted, which carries fiber-optic bundles for illuminating and viewing (or photographing) the fetus. The tube also contains miniaturized instruments for withdrawing blood samples or cutting and removing small skin samples (1 $mm^3$) for laboratory analysis.

Fetoscopy may be performed no earlier than 15 weeks into the pregnancy. It is a higher risk procedure than is amniocentesis, inducing fetal death and miscarriage in 4% of the cases.

## Newborn Screening

Every state in the United States has some type of mandated newborn screening program. In New York state, mandated newborn screening was put into effect in 1965. Every child born in New York state is tested for the presence of eight genetic diseases. The cost of this screening program is not charged to the parents but is paid for from funds provided by the New York state and federal governments. The tests can all be performed on a single drop of blood, which is dried onto a piece of filter paper. A central lab located in Albany, New York, performs all tests. The specific diseases tested for are:

**Phenylketonuria (PKU).** In this disease, the body lacks the enzyme needed to break down the amino acid phenylalanine. As phenylalanine builds up in the blood, the process of normal brain development is hindered and severe mental retardation will result. Affects 1 in 15,000 newborns.

**Sickle Cell Disease (anemia).** An errant amino acid inserted into the hemoglobin molecule of red blood cells. This results in an abnormally shaped red blood cell (sickle shaped), which interferes with normal capillary blood flow. Consequences of this disease can include the increased risk of infections and the presence of anemia, or a reduced capacity of the blood to carry oxygen. The condition occurs in 1 of every 500 black newborns and to a lesser extent in Hispanics, Greeks, Italians, and Indians.

**Sickle Cell Trait.** Individuals with sickle cell trait are carriers for sickle cell anemia although they will not necessarily contract the disease.

**Branched Chain Ketonuria.** Also due to the absence of an amino acid digesting enzyme. This condition can lead to mental retardation. Affects 1 in 268,000 newborns.

**Galactosemia.** The body lacks the enzyme to break down the sugar galactose. This condition can lead to mental retardation. Affects 1 in 57,000 newborns.

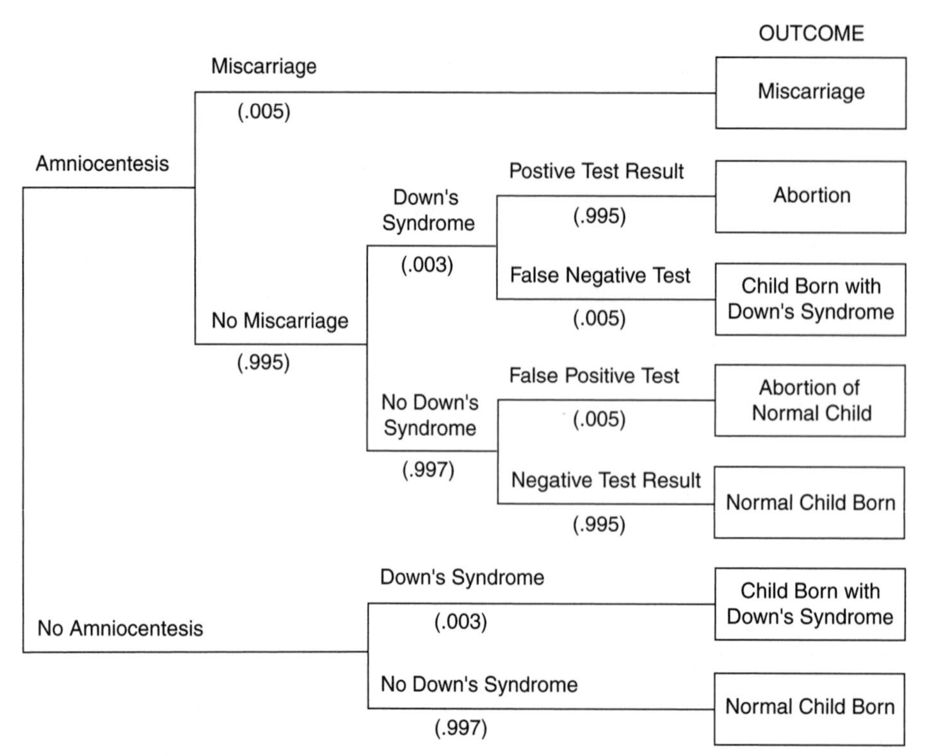


**Figure 14.7** Decision tree for amniocentesis.

**Homocystinuria.** Lack of a liver enzyme, which can result in mental retardation. Affects 1 in 225,000 newborns.

**Hypothyroidism.** Inadequate production of the hormone thyroxine by the thyroid gland can lead to stunted growth and to mental retardation. Affects 1 in 4,000 newborns.

**Biotinidase Deficiency.** The lack of the enzyme biotinidase prevents the body from adequately absorbing the vitamin biotin. This can result in skin rash, vision and hearing impairment, and brain damage. Affects 1 in every 80,000 newborns.

# Risk-Benefit Analysis in Genetic Screening

The decision tree in Figure 14.6 shows the outcomes that might be expected to occur as the result of the decision to undergo prenatal genetic screening by amniocentesis. At each branching point, or node, two different outcomes are possible. If there exists an 80% chance of one outcome (.8), then the chance of the alternative outcome must be 20% (.2). We can see, therefore, that the sum of the two individual probabilities (.8 + .2) must be equal to one. It is therefore possible to represent the probability (or risk) of one outcome as X and the other as 1 – X.

The following variables are used to represent the known probabilities (or risks) of the events associated with each branching point.

r = risk (probability) of miscarriage from the amniocentesis procedure

d = the risk (probability) of Down's syndrome occurring in the developing fetus

n = the false-negative test rate (the probability that the amniocentesis will fail to detect a case of Down's syndrome)

p = the false-positive test rate (the probability that a normal fetus will be misdiagnosed as having Down's syndrome)

Now, assume that for a 35-year-old woman the risk (probability) of conceiving a child with Down's syndrome is 1 in 328 (1/328 = .003). In this case then, d = .003. The probability of having a normal child is therefore .997 (determined as 1 – d).

The risk of miscarriage due to amniocentesis (r) is 1 in 200 (r = 1/200 = .005).

The incidence of a false-negative test result (n) is 1 in 200 (n = .005).

The incidence of a false positive test result (p) is also 1 in 200 (p = .005).

With these values inserted, the decision tree appears as shown in Figure 14.7.

From the decision tree in Figure 14.7 we can see that the risk to a 35-year-old woman of delivering a child with Down's syndrome, should she choose not to undergo amniocentesis, is .003 (or .3%). The probability of her delivering a normal child is .997 (or 99.7%). Should she choose amniocentesis, her risk of delivering a child with Down's syndrome is reduced to .000015, or .0015% (.995 × .003 × .005). This reduces the risk of a child being born with Down's syndrome 200 times. Her probability of delivering a normal child, however, is also reduced, in this case to .987 (or 98.7% [.995 × .997 × .995]), a reduction of only 1%.

# Exercises

1. If abortion is an option, what is the risk of mistakenly aborting a normal fetus due to a false-positive test?
2. Draw a decision tree; insert the proper values and determine the outcome probabilities for a woman 40 years of age (risk of Down's syndrome is 1 in 100); for a woman 28 years of age (risk of Down's syndrome is 1 in 1,000).
3. If you were a physician or public health service counselor, what recommendation would you make to each of these women and why?
4. **Fetoscopy**
   The decision tree in Figure 14.8 shows the outcomes that might be expected to occur as the result of selecting to undergo the genetic screening technique of fetoscopy. The probability values have been left out.

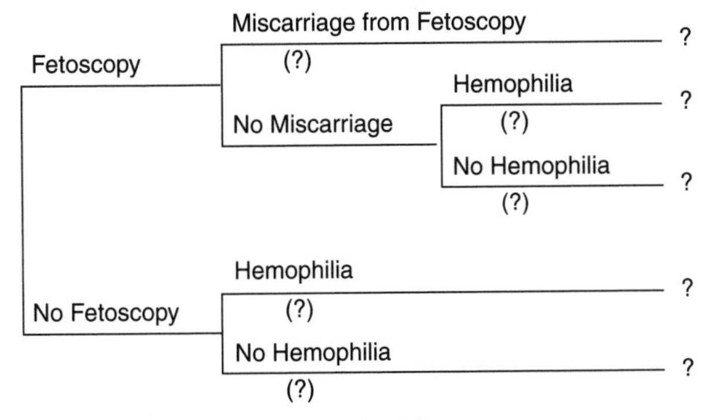


**Figure 14.8** Fetoscopy decision tree.

The technique of fetoscopy carries a 4% (.04) risk of inducing miscarriage. For women who are carriers of the disease hemophilia, the risk of male children inheriting the disease is 50%. The risk to female children is 0%. For women who are not carriers, the risk of hemophilia in offspring is 0% for both sexes.

a. What outcomes should be placed at the end of each line?
b. Determine the reduction in risk to a female carrier, of giving birth to a son with hemophilia should she choose to undergo fetoscopy.

# Review Questions

1. Distinguish between medical screening and diagnostic screening.
2. What is homeostasis?
3. What is the "zone of ease"?
4. What is meant by the sensitivity of a test? Distinguish between true positive, false positive, and false negative.
5. What is the specificity of a test?
6. Discuss fetal screening.
7. What is amniocentesis?
8. What is sickle cell anemia?
9. What is hypertension and what causes it?
10. What is the role of risk-benefit analysis in medical screening?

# Multiple Choice Questions

1. The purpose of a medical screening test is to determine a person's _____, or likelihood, of developing a disease in the future.
   a. risk factor
   b. homeostasis level
   c. disability
   d. zone of ease
2. Homeostasis has to do with maintaining _____ in a system.
   a. stability
   b. instability
   c. feedback
   d. control
3. Equilibrium is maintained in a living system using _____.
   a. control
   b. positive feedback
   c. negative feedback
   d. hypertension
4. The purpose of a screening or diagnostic test refers to its ability to detect _____.
   a. emotional disturbance
   b. small readings
   c. disease
   d. health

5. Blood pressure is the force in the arteries when the heart beats and is called _____.
   a. systolic pressure
   b. diastolic pressure
   c. neurotic pressure
   d. atrial pressure

6. When the heart is at rest, the blood pressure is called _____.
   a. systolic pressure
   b. diastolic pressure
   c. neurotic pressure
   d. atrial pressure

7. High blood pressure is defined in an adult as a blood pressure greater than or equal to 140 mm Hg systolic pressure or greater than or equal to 90 mm Hg diastolic pressure. It is called _____.
   a. high tension
   b. diastolic tension
   c. hypertension
   d. arrhythmia

8. The procedure in which the abdominal wall and uterus is penetrated with a hypodermic needle and amniotic fluid withdrawn is called _____.
   a. fetoscopy
   b. phenylketonuria
   c. amniocentesis
   d. homocystinuria

9. A medical screening for triglycerides measures the _____ level in blood.
   a. cholesterol
   b. fat
   c. bilirubin
   d. glucose

10. An abnormal level of glucose in the blood can indicate the presence of _____.
    a. hypertension
    b. anemia
    c. liver disease
    d. diabetes

11. Individuals who test positive upon administration of a screening test but are not ill are _____.
    a. true positives
    b. true negatives
    c. false positives
    d. false negatives

12. Individuals who test negative upon administration of a screening test but are ill are _____.
    a. true positives
    b. true negatives
    c. false positives.
    d. false negatives

13. If the specificity of a screening test is 90% and 20% of a test population of 5,000 are ill, the number of false positives will be _____.
    a. 4,000
    b. 400
    c. 3,600
    d. 50
    e. None of the above.

14. If the sensitivity of a screening test is 95% and 20% of a test population of 5,000 are ill, the number of false negative will be _____.
    a. 4,000
    b. 400
    c. 3,600
    d. 50
    e. None of the above.

15. Using the data from the two previous questions, suppose an individual tested positive on the screening test. What are the chances of the individual really being ill?
    a. about 70%
    b. about 100%
    c. about 20%
    d. None of the above.

16. Glucose levels in the blood are given in mg/dL; mg stands for _____.
    a. mmHg
    b. degrees Celsius
    c. mega gigs
    d. milligrams

17. Glucose levels in the blood are given in mg/dL; dl stands for _____.
    a. delayed activity
    b. direct link
    c. deciliter
    d. None of the above.

18. The presence of such substances as cholesterol, glucose, triglycerides, and bilirubin in blood is expressed as mg/dl; mg/dl represents _____.
    a. a unit of pressure
    b. a unit of concentration
    c. a unit of saturation
    d. any other kind of personal index

19. Human blood is _____.
    a. very slightly acidic
    b. very slightly basic
    c. neutral
    d. None of the above.

20. Cholesterol levels and triglyceride levels in the blood both in excess 300mg/dL represent an enhanced potential associated with _____.
    a. diabetes
    b. anemia
    c. heart disease/stroke
    d. living for tomorrow

21. Given the following diagram:

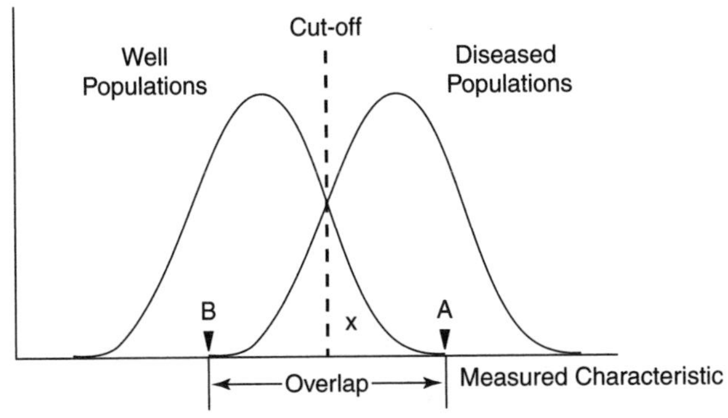


The segment of the population located at X would be described as _____.
    a. negative negative
    b. false positive
    c. false negative
    d. true negative

22. A certain screening test was found to yield 20 true positives and 40 false positives out of a test population of 10,000 people. What is the chance of having the disease if you test positive?
    a. 33%
    b. 25%
    c. 3%
    d. 6%

# Bibliography and Web Resources


Your Guide to Lowering High Blood Pressure, NHLBI. This site is for people who want to learn about preventing and controlling high blood pressure. Based on clinical guidelines and research studies.
**www.nhlbi.nih.gov/hbp/**

UMHS—Dangers of High Blood Pressure. For more information about high blood pressure, contact your doctor.
**www.med.umich.edu/1libr/heart/highbp03.htm**

High Blood Pressure. Hypertension. There is no "ideal" blood pressure reading. However, there is a range of "normal" blood pressure reading.
**sln.fi.edu/biosci/healthy/pressure.html**

Many medical screening tests may be unnecessary. The University Record, November 12, 1997.
**www.umich.edu/~urecord/9798/Nov12_97/test.htm**

Just How Effective are Medical Screening Tests? Essentially, medical screening tests are experimental fishing expeditions.
**www.researchprotection.org/infomail/03/11/17.html**

*Guardian Unlimited* special reports. Just how effective are medical screening tests?
**www.guardian.co.uk/medicine/story/0,11381,1084393,00.html**

# Technology and Risk

*"What man's mind can create, man's character can control."*

**Thomas Edison**

In this chapter we shall discuss the role of technology as a force for improving our lives and the associated risks that accompany every technological innovation. Historically, most new technologies have been embraced enthusiastically and utilized for their benefits without too much attention being paid to the associated risks.

In the early part of the last century the automobile was thought to be the solution to the pollution of our streets by horses depositing tons of manure and millions of gallons of urine each day on the streets of New York City. However, when we realize that about 45,000 people lose their lives each year and about 250,000 people are injured because of the automobile, we may think that the price we paid for this solution was too much.

In this chapter we shall examine a number of technologies and the risks that accompany them. We shall learn to evaluate and judge them in terms of the benefits derived and the risks they pose.

## Goals

After studying this chapter, you should be able to:

- Be conscious of the fact that every technology has associated risks.
- Weigh the risks in terms of the benefits.
- Identify sources of technological risk.
- Quantify risk.
- Perform elementary cost-benefit and risk-benefit analyses in making decisions.
- Control risks.

## Technology and Risk

By definition, technology is best described as the application of scientific knowledge and methods to the solution of societal problems. In short, we tend to think of technology as a positive force for improving our lives and the world in which we live and, in general, we would be correct in such thought. However, no technology, in fact no human undertaking, is without some degree of risk. Although intended to be of benefit, all technologies have risks and costs associated with them as well.

The application of any technology is best made only after a careful consideration of the relative benefits and risks resulting from its utilization. If it is believed that the benefits to be gained from the use of a technology will far outweigh any minimal risk, the decision to proceed with the technology's development and application is an obvious one. If an analysis shows that the risks far outweigh a minimal benefit, then further development of the technology would be ill advised.

Thus, when Roentgen discovered X-rays in his laboratory at the end of the 1890s, his discovery was thought to be the greatest technological innovation ever made in medical science. The human skeletal system could be viewed without cutting the skin. In the 1950s this technology was so popular it found its way into shoe and department stores, where it was used to determine the correct shoe size for a child's fast-growing foot. Mother, child, and salesperson were able to view the child's foot in the shoe using such machines. It was at this time that the 50-year old technology was found to be able to cause damage to human cells and was promptly banned except for medical use.

From 1917 to 1926, the U.S. Radium Corporation of East Orange, New Jersey, hired workers who were mostly women to paint radium-lighted watches and instruments. The corporation suffered a major setback when a number of the dial painters died from seemingly unrelated causes. It was later found out that the deaths were due to the ingestion of radium, one of the main ingredients of the luminescent paint. The radioactivity caused the hands and dial to glow in the dark but also caused radiation sickness in the workers at the plant, as they used their mouths to form a point on the radium tainted brushes.

In New York City the former Radium Dial Company is now a ⅓-acre lot in the Ninth Congressional District of Queens contaminated with Radium 226, radon, and gamma ray emissions. It sits within a three-mile radius of over 300,000 residents and only 10 feet from the Brooklyn-Queens Expressway. In 1989 the EPA determined that the site "poses a significant threat to human health and the environment" (*Federal Register*, Vol. 54, No. 157, August 16, 1989).

Again, since radioactivity was discovered at the beginning of the 20th century, it took several decades before scientists realized the harmful effects of radiation on human cells and for safety restrictions to be imposed.

In addition to the known risks inherent in any technology, there is also the perhaps greater risk that the technology will be applied in ways that were not intended or for purposes with less than honorable intent. Contrary to Edison's belief, there are many examples of technology-induced dangers and technology-related catastrophes. Below, we will examine some examples of technology-associated risk and how such risk can be measured.

# Chlorinating Water Supplies

In underdeveloped countries throughout the world one of the primary causes of disease and death is bacteria-borne illness. Frequently, these

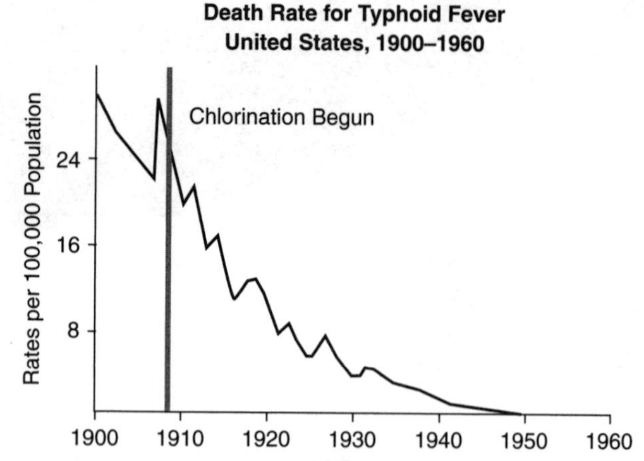


**Figure 15.1** Graph of chlorination and disease rate.

disease-causing organisms are spread from person to person through the public drinking water supply, often when it becomes cross-contaminated by sewage. According to the World Health Organization, water-borne disease is responsible for approximately 5 million deaths per year, or over 13,000 deaths every day, in developing countries.

Modern societies protect against such water-borne illnesses by upgrading both the water delivery supply and the sewage disposal system. In 1908, Jersey City, New Jersey, became the first municipality to chlorinate its public water supply system. Today, over 98% of municipal water supply systems incorporate chlorination as a means of reducing bacterial, viral, and protozoal contamination. Within public water supply systems, disinfection, the elimination of disease-causing microorganisms is most typically accomplished by initial stages of filtration and a final step of chlorination. See Figure 15.1. According to the Chlorine Chemical Council, the chlorination of public water supply and the resulting elimination of disease-causing microorganisms have played a key role in increasing the American citizens' life expectancy from 45 years in 1900 to about 76 years today.

In this process, chlorine gas is mixed with the water supply, creating chlorine concentrations that are toxic to all but the most hardy of microorganisms. The chlorine works by forming hypochlorite and hypochlorous acid when mixed with water. These molecules then react with the bacterial cell membrane lipids, chromosomal nucleic acids, and metabolic enzymes interfering with normal cell function and killing the bacterial cells by multiple means.

Although hypochlorite and hypochlorous acid in the same concentration would have a similar effect on human cells, the water we drink is typically diluted in a larger volume of gastric fluids

The supply source for United Water New Rochelle is surface water that is purchased from the New York City Department of Environmental Protection. We can pump based upon demand from five separate locations. The three sources of New York City supply that we utilize include the Croton, Catskill, and Delaware aqueducts. The Central Avenue and Little Catskill pump stations supply the day-to-day demands to the system.

| Pump Station | Source | Maximum Pumpage* |
|---|---|---|
| Central Avenue | Catskill Supply | 27 MGD |
| Croton | Croton Supply | 17 MGD |
| Little Catskill | Catskill Supply | 4 MGD |
| Troublesome Brook | Delaware Supply | 8 MGD |
| California Road | Delaware Supply | 4 MGD |

* Millions of gallons per day

## Water Quality

At United Water we take great pride in out ability to provide you with drinking water that meets or surpasses all state and federal standards. We treat the water by adding chemicals to make the water potable and pleasing to the taste. The chemicals added to your water are:

- Zinc Polyphosphate—to control corrosion within the water distribution system.
- Chlorine—for disinfection purposes and to ensure bacteriological safety.
- Fluoride—to reduce the incidence of dental cavities in children.
- Caustic Soda—to reduce acidity of the water.

The quality of your drinking water is constantly monitored—before, during and after the treatment process. We also test the treated water in each community we serve and provide the results to the Westchester County Department of Health. United Water is required to sample our supplies for over 100 water quality parameters, including inorganic and organic chemicals, coliform bacteria, chlorine concentration and turbidity. We collect approximately 2,400 bacteriological samples per year from our sources of supply and distribution system. These samples are used to monitor the effectiveness of our disinfection process. Results were well within regulatory requirements in 2000. For more detailed information see the section [below] on "Consumer Confidence Reports."

**Figure 15.2** An excerpt from the United Water of New Rochelle public information Web site.

and neutralized by food present in the stomach. Research in the early 1970s began to provide evidence that in addition to its germicidal effects, chlorine in municipal water systems also combined chemically with organic molecules in the water to form disinfection byproducts, or DBPs. Among the known DBPs are the trihalomethane compounds (bromodichloromethane, chloroform) and the haloacetic acids (monochloroacetic, dichloroacetic, and trichloroacetic acids), all suspected carcinogens or teratogens.

Municipal water chlorination (Fig. 15.2) thus provides us with a classic example of the competing risks of a) doing nothing and accepting a natural risk or, b) utilizing a technology to alleviate a problem but accepting its own inherent risks. Do water filtration and chlorination technologies work?

Absolutely. In Albany, New York, prior to the turn of the 20th century the death rate from water-borne typhoid was 110 per 100,000 of total population. When water filtration technology was added in 1900, the death rate from water-borne typhoid dropped to 20 per 100,000. After the addition of chlorination to the Albany water supply system in 1910, the typhoid death rate dropped to zero and has remained there ever since.

Although there are clearly measurable and important benefits to the use of chlorination, we must not forget that there are risks as well. In order to minimize the risk to human health caused by chlorine, its concentration in public drinking water supplies is carefully regulated. In 1996 the U.S. Congress passed the Safe Water Drinking Act (SWDA), requiring the Environmental Protection

Agency (EPA) to prepare regulations to limit human exposure to DBPs. The first of these regulations, released in 1998, was the Stage 1 Disinfectants and Disinfection Byproducts Rule, which is intended to establish regulations to "balance the risks between microbial pathogens and disinfection byproducts (DBPs)." All municipal water systems were required to comply with these regulations no later than January 2004.

According to EPA estimates, the Stage 1 Rule will result in approximately $700 million of additional costs, which, distributed among 116 million households, will cost each household approximately $6.00 per year. For this expenditure, it is estimated that total trihalomethane carcinogens will be reduced by 24%, benefiting 140 million people. Thus the EPA's cost/benefit analysis justified passage of this rule.

## Cooking Steaks

In a similar manner to the contamination spread through public drinking water supplies, the bacterial contamination of our food supply presents another route for the transmission of disease and death.

Since the advent of fire by our early human ancestors (actually, a very important technological application) people have been cooking their food largely as a means of eliminating bacterial contamination. We are well aware of the hazards of eating uncooked or even undercooked hamburger, chicken, and eggs. In each of these cases, the risk comes from the bacteria that may contaminate these food substances. Every year, as many as 30 million Americans become ill due to contaminated food, with up to 9,000 actually dying.

By cooking our food at high temperatures, particularly meat, which as an animal product is very susceptible to fecal bacteria contamination, we can kill all or the majority of any bacteria that may be present and eliminate or reduce the risk of foodborne illness (see Fig. 15.3). What is generally not known to most people is the fact that subjecting the amino acid molecules in the meats' protein to high cooking temperatures produces byproducts known as heterocyclic amines (HCAs), and these have been shown to cause a variety of cancers. We are then confronted with the Hobson's choice of either cooking our food and voluntarily subjecting ourselves to known carcinogens, or eating our food raw and subjecting ourselves to bacterial contamination. Given the almost certain risk of falling ill and perhaps dying from eating contaminated food and the minimal risk of inducing cancer through the presence of HCAs, the choice becomes clear. We accept the minimal risk to prevent the more certain risk.

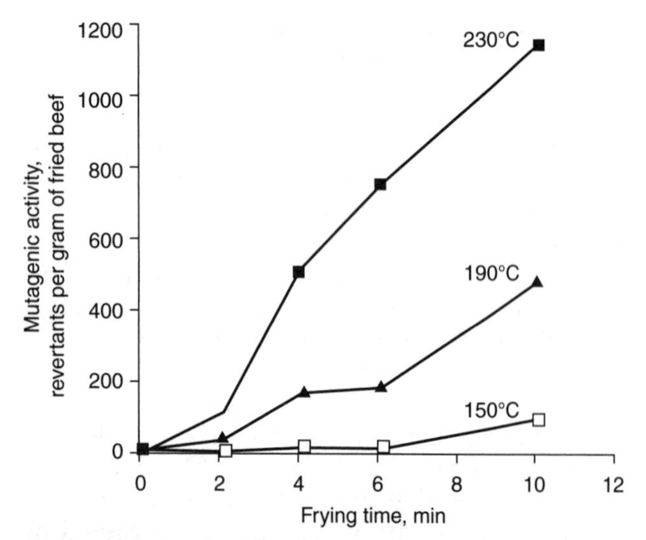


**Figure 15.3** Graph of the mutagenic activity in beef patties fried at three different temperatures. The essential point is that mutagenic activity increases with both frying temperature and time.

## Irradiation of Food

Another technology, which is now appearing in our supermarkets, involves the irradiation of food by gamma rays or electron beams within a shielded facility. When used to treat meat and poultry products it can kill most harmful bacteria. However, it is not a substitute for the proper handling of food and its manufacture but rather complements it. Irradiated but unrefrigerated meat can still spoil. The radura, Figure 15.4, is used to identify food that has been irradiated.

The Food and Drug Administration has approved irradiation of meat and poultry and allows its use for a variety of other foods, including fresh fruits and vegetables, and spices. It is found that by irradiating foods, contaminants

**Figure 15.4** Radura. *http://www.fsis.usda.gov/OA/background/irrad_final.htm*

such as bacteria, insects, and parasites may be reduced. Irradiated berries may stay fresh for three weeks versus three to five days for nonirradiated berries. The FDA has studied food irradiation for over 40 years and has found it to be safe.

However, irradiation does produce chemical changes in food and has aroused the concerns of some scientists. The substances are called radiolytic products. Cooking also changes foods, producing thermolytic products, molecules produced by the action of heat. The changes produced by irradiation are found to be minor compared to those produced by cooking. The irradiation process may cause the loss of some nutrients but no more than other processing methods. The bottom line is that food irradiation can reduce food-borne illness risk and contamination by bacteria, parasites, and insects. The choice again is should we accept minimal risk to prevent more certain risk?

# Sources of Technological Risk

In general, the risk of some unfavorable or detrimental event occurring as the result of a technological application can be attributed to one or any combination of the following sources:

Hardware factors

Human factors

Organizational factors

External Social factors

No system or component is infallible or unerring. One of the most important things to keep in mind is that engineers are trained to anticipate failure in their creations and to design against it.

## Hardware Factors

Hardware factors refer to those sources of risk that are due to the physical failure of the technological system or one of its component parts. These risks are best thought of as those that are due directly to "equipment failure." When the space shuttle Challenger was destroyed on January 28, 1986, it was found upon investigation that the physical design of the O-rings that were supposed to seal the gaps between booster rocket sections were flawed and allowed the 5,000° F exhaust gases to escape from the booster rockets and burn through the external fuel tank.

When TWA flight 800 blew up off the coast of Long Island, New York, in 1996 (see Fig. 15.5) the cause was believed to be poor design in the center fuel tank. The placement of the center fuel tank above the plane's air conditioning units subjected the fuel to high heat levels and caused the formation of high levels of fuel vapors. These vapors ignited in an explosion when poorly insulated wires leading from an in-tank fuel probe short-circuited and sparked. In both of these tragedies, a large part of the blame is placed directly on hardware components.

## Human Factors

Human factors refer to people-related rather than equipment-related sources of risk. When the Three-Mile Island nuclear power plant in Pennsylvania (see Fig. 15.6) was damaged, the root cause of the accident was "operator error." On the morning of March 28, 1979, a valve in reactor number 2 closed, turning off the cooling water supply to the reactor core. As designed, the system automatically shut down reactor 2 and alarms were sounded in the control room. In the subsequent series of steps, control room operators misinterpreted several control readings, failing to open critical valves and not realizing the extent of the problem for at least two and half hours after

**Figure 15.5** TWA flight 800 wreckage. *http://aar400.tc.faa.gov/Programs/Catastrophic/forthpage.htm*

**Figure 15.6** Three-Mile Island Power Plant. *http://www.epa.gov/history/timeline/80.htm*

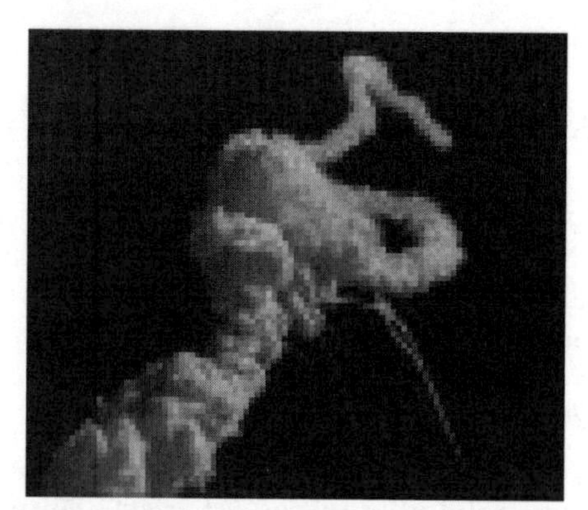

**Figure 15.7** 1986 Space shuttle Challenger disaster. *http://www.centennialofflight.gov/essay/SPACEFLIGHT/challenger/SP26.htm*

the accident began. Of course, the disaster would not have happened at all if a faulty valve had not closed. In this particular case, the accident involved a combination of both hardware and human factors.

## Organizational Factors

Organizational factors refer to sources of risk related to the operational and regulatory portions of the system. Returning to the 1986 Challenger disaster (see Fig. 15.7), NASA engineers were aware of the risk associated with such a complex technological system and had calculated in a 1983 report that the solid rocket boosters had an anticipated accident rate of 1 in every 70 launches. Knowing the risks of such a complex and dangerous system, NASA had put into place a carefully constructed launch clearance process that would delay shuttle launches, for example, when weather conditions were threatening. Knowing that the rubber O-rings were subject to losing their flexibility at cold temperatures, NASA set 40° F as the minimum temperature for the launch window. On the night of January 27, 1986, the air temperature at the Kennedy Space Center fell to 8° F. The flight was allowed to proceed, in part, because NASA managers were under pressure to prove the value and reliability of the shuttle program. The European Space Agency was becoming a competitor of NASA, and NASA had committed to an ambitious schedule of launches for 1986. This particular shuttle flight, which included the politically important flight of Christa McAuliffe, the first teacher in space, had already been canceled several times. The organizational structure of NASA, its dependence on politically influenced funding, and checks and balances in the decision-making process all represented organizational factors that made this flight particularly risky.

**Figure 15.8** World Trade Center disaster. *http://usinfo.state.gov/photogallery/index.php?album=nineeleven/wtc&image=5976663.JPG*

## External Social Factors

External social factors, completely outside of the technology itself or the human or organizational systems related to it, can also provide sources of risk. The construction of New York's World Trade Center was made possible by several technological innovations, including the novel architectural design of using external columns in the buildings' walls to help provide support and stability, the development of high-speed elevators, and the use of a motion-damping system placed in the roof to help eliminate the buildings' tendency to sway in high winds. The disaster and the unprecedented loss of life that occurred on September 11, 2001, although partially due to technological failures of the building design, escape systems, and fire control mechanisms, was precipitated however by a terrorist attack influenced by hatred and politics—external social factors (see Fig. 15.8).

# Quantifying Risk

## The Association Coefficient

Injury, financial loss, disease, and death are often thought of as consequences of risky undertakings. In order for us to make informed decisions about the relative benefits and risks, it would be helpful to have some quantitative measure of risk as well as benefit. One such quantitative measure

**TABLE 15.1 Contingency Table**

| | Outcome Present | Outcome Absent |
|---|---|---|
| Incident present | a | b |
| Incident absent | c | d |

that indicates the presence of an association between an action, such as the use of a particular technology, and an effect, is the fourfold contingency table.

The contingency table provides a model from which associations between cause and effect can be determined. The contingency table is set up as shown in Table 15.1. The values a, b, c, and d, when placed into the following equation, provide a value called the *association coefficient,* R.

$$R = \frac{(a + d) - (b + c)}{a + b + c + d}$$

The association coefficient, R, if calculated to be 1.0, demonstrates an absolute correlation between cause and effect or, for our purposes, between a technological intervention and an observed result or risk. An R value of 0 indicates no correlation whatsoever. Fractional values between 0 and 1 would indicate an increasing level of correlation.

### Relative Risk

Another way to quantify risk is to determine rate of harm that can be expected when using a technology as compared with the rate of harm expected when the technology is not employed. In other words, the relative risk is the ratio of the incidence rate of harm to those exposed to some condition, in this case a technology, relative to the incidence rate of harm for those with no exposure to the technology.

In situations where the risk associated with using the technology is no greater than the risk associated with not using the technology, the relative risk would be 1. If the technology has a greater inherent risk, that is, if using the technology increases the chance or likelihood of harm, the relative risk would be a number greater than one. For instance, a relative risk of 2 indicates that the risk of harm is doubled when using the technology. Conversely, a relative risk below 1, for example 0.5, indicates that using the technology actually reduces the risk of harm by one-half, or 50%.

## Performing Cost-Benefit and Risk-Benefit Analysis

As we have seen with municipal drinking water chlorination, the decision to utilize a technology is often made based upon an analysis of the relative benefits that we expect to gain from the technology weighed against the risks that the technology brings with it. This is typically determined through a process called **cost-benefit analysis.** Cost-benefit analysis requires that a quantitative determination be made of the cost of a technological implementation, which can be a complex but not very difficult process. It also requires a quantitative determination of benefit and this is not always so obvious. The cost-benefit ratio is calculated by dividing the total projected benefits by the total projected costs. With respect to water chlorination, the total cost of the system will include the construction, maintenance, and operating costs of the municipal water chlorination system.

In determining how to quantify the benefit, we have several options. Is the benefit a reduction in gastrointestinal illness? Is it a heightened public confidence in the water supply? Is it an overall reduction in the death rate? Actually, it is all of these things and probably several others as well.

For our purpose, the most important benefit is a reduction in the loss of life and therefore the quantitative measure of benefit is the number of lives saved. Given the known costs, and given the expected number of lives to be saved, the cost-benefit analysis can be reduced to a simple figure, the cost for each life saved.

Table 15.2 highlights some cost-benefit figures for a variety of technologies.

Alternatively, we can quantify benefit by using the **cost of illness avoided** measure. Assuming that the consumption of contaminated water is likely to result in a predictable number of illnesses per population, we can assume that the benefit of chlorination will be the elimination of such illness (such as the typhoid referenced earlier in the case of Albany, New York), and we can place a cost savings on this. For any given illness, there is an average calculable cost of medical care (doctor visits, laboratory tests, medicines, etc.) and calculable costs for lost work time. Multiplied by the expected number of reduced cases of such illness, we can readily estimate the dollars saved through "cost of illness avoided."

**TABLE 15.2 Cost-Benefit Figures**

| Technology Implementation | Cost per Life Saved |
|---|---|
| Childhood Immunizations | Less than zero |
| Adult Influenza Immunizations | $ 600 |
| Water Chlorination | $ 4,000 |
| Breast Cancer Screening | $ 17,000 |
| Kidney Dialysis | $ 46,000 |
| Heart Transplantation | $ 104,000 |
| Home Radon Control | $ 141,000 |
| Asbestos Abatement | $ 1,900,000 |
| School Bus Seat Belts | $ 2,800,000 |
| Radiation Control | $ 27,400,000 |
| Chloroform Removal in Paper Pulp Mills | $99,400,000,000 |

(*Wall Street Journal*, 7/06/1994)

# Performing Quality-Adjusted Life Year Analysis

It is possible from a purely technological view to perform a **Quality-Adjusted Life Year (QALY)** analysis on prolonging life. However, it should be noted that such an analysis does not consider the moral and religious dimensions of such a determination that must be brought into consideration if such an analysis is done. It does not define whether extraordinary steps are being applied.

Centuries of scientific research capped by several recent scientific milestones have provided us with the knowledge and techniques to treat disease and prolong life for years and even decades. Research on immune system function has provided us with immunosuppressive drugs that have made the long-term success of organ transplants, most notably the heart, a reality. Our understanding of the human circulatory system, materials science, and computer-aided design and control has further enabled us to develop fully implantable artificial hearts. While it is now possible to prolong lives through a variety of technological means, it is not always clear that we should do so. Quality of life issues may be considered.

There is disagreement as to whether we should or should not use our technological tools to prolong a human life and to what extent. Choices of intervention must be made within the context of the trauma and cost of the procedure and quality and length of extension of life. While we generally believe that extending life is desirable, it may require extraordinary means. Therefore we need to consider both the quantity and quality of life that results from the use of a medical technology and whether extraordinary means are used.

In order to make such quality of life determinations, it is helpful to calculate the QALY analysis, which is a way of taking into account the quality and length of life that will result from a technological intervention. In this quantitative model, we assign the value of 1.0 to a year of perfect health and the value of 0 to death. A year of life in which the quality of health, and therefore the quality of existence is not perfect, is assigned a fractional value of between 0 and 1.0.

As an example, the technology of hemodialysis may, on average, prolong life for 15 years, a quantity value of 15 years. Because of the need for frequent dialysis treatments (three times per week), the surgical interventions required, and the risk of frequent infections, the quality of life may be rated as 0.5. We would then calculate the QALY to be 7.5 (15 years quantity × 0.5 quality). The alternate technological option of renal transplantation would result in an average of 12 years of added life but, with the only drawback being the constant need for immunosuppressive drug therapy, the quality of life would be rated as .75, giving a QALY for transplantation of 9 (12 years × .75 quality). In calculating a value for quality of life, the issues shown in Table 15.3 are among those typically considered.

When we combine the cost of a technology's utilization with the QALY, we can derive a measure of the **cost-utility ratio** indicating a technology's cost per quality of life year added (i.e., cost per life/QALY). Table 15.4 illustrates some cost-utility ratios for various health-care procedures.

# Controlling Risk

If, indeed, risk is inescapable when using technology, what steps can and does society take to minimize these risks?

## Automobiles

Since most of us utilize this technology every day, perhaps it would be most efficient to utilize as our case study the contemporary automobile. Automobile (and truck) transportation provides for numerous social benefits related to the speed at which we can move people and materials over great distances at relatively low costs. If we make the simple comparison with transporting people and goods via horse-drawn carriage, we can

**TABLE 15.3 Quality of Life**

| | High Quality | Moderate Quality | Poor Quality |
|---|---|---|---|
| **Mobility** | Full mobility | Difficulty walking | Confined to bed |
| **Pain** | No pain | Some discomfort | Chronic extreme pain |
| **Mental State** | Happy and engaged | Anxiety and concern | Depressed |

**TABLE 15.4 Technological Intervention**

| Technological Intervention | $ Cost per QALY |
|---|---|
| Cholesterol Testing | $ 660 |
| Pacemaker Implantation | $ 3,300 |
| Hip Replacement | $ 3,540 |
| Kidney Transplantation | $14,130 |
| Heart Transplantation | $23,520 |
| Hemodialysis Treatment | $60,000 |

**Figure 15.9** A traffic control device.

immediately see the increased capability and efficiency of automobile-based transportation. While the automobile provides for significant benefits over horse-drawn transportation, it brings with it a host of additional risks and costs including accident-related injuries and death and significant environmental pollution. Rather than abandoning this technology because of its associated risks, we choose to limit the risks through regulation in design and use.

For example, today's cars contain government-mandated steel beams in the passenger compartment doors, passive restraint systems (air bags), and shatter-proof glass windows to limit the risk of injury and death in accidents. Gasoline is produced without lead additives and in different seasonal-blended formulations to minimize the production of environmentally hazardous exhaust emissions. All road and highway systems have legally-mandated speed limits and traffic flow through dangerous intersections is regulated by traffic control devices (see Fig. 15.9). What other regulatory devices or laws can you identify that have been imposed to reduce risks associated with automobile use?

## Cell Phone Regulation

Telecommunication technologies have been advancing at a rapid pace. With the development and miniaturization of cell-based mobile telephones over the past two decades, people are now able to communicate with each other, immediately and in real time, over vast distances across the planet. Mobile cell phone use has allowed parents to communicate with children, facilitated business communication, and has saved numerous lives of accident victims who, 10 or 15 years ago, might have been stranded alone and left to die.

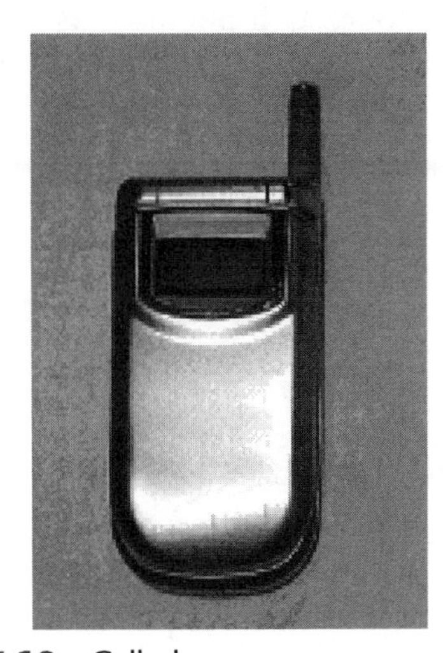

**Figure 15.10** Cell phone.

Current cellular telephones (Fig. 15.10) transmit conversations in digital form using electromagnetic radiant energy in the microwave range of the electromagnetic spectrum. Circuitry within the telephone unit converts the analog voice signal into a digital signal that is broadcast out from the telephone's antenna to a cellular base station antenna. Incoming signals are broadcast from the cellular base station antenna and picked up by the telephone handset unit. The electromagnetic energy signals travel to and from the handset's antenna at the speed of light, in all directions, and will readily pass through living tissue. Thus, a cellular telephone held to the user's ear during a conversation will transmit electromagnetic energy through the user's head.

While the energy that passes through living tissue (in this case, the user's brain) is believed to cause little harm, some of the energy is absorbed by the living cells and may cause localized changes. The quantity that is used to measure the absorbed radiation is the SAR, or Specific Absorption Rate, measured in watts per kilogram (W/kg).

The Federal Communications Commission (FCC) has established exposure guidelines for hand-held mobile cell phones that are intended to limit the user's exposure to microwave radiation. The FCC has determined that it is safe for humans to be exposed to an SAR of not more than 1.6 W/kg as averaged over 1 gram of living tissue (the European Union has set the limit at 2.0 W/kg). The FCC has required all cell phones manufactured after June 2000 to list with the FCC the SAR levels for each phone model. Cell phone users can find out about their phone's SAR by checking the FCC ID number, printed on the phone's label, at the Web site *www.fcc.gov/oet/fccid.*

Studies are now being conducted to assess the long-term risks of this relatively new technology. In the meantime some researchers recommend the use of headsets so as to keep the cell phone away from the brain.

# Review Questions

1. What are the risks and benefits associated with chlorinating the water supplies?
2. What are the risks and benefits of cooking foods?
3. Name four sources of technological risk and define them.
4. How may risk be quantified?
5. What is relative risk?
6. What is cost-benefit analysis?
7. What is quality adjusted life year (QALY) analysis?
8. Give some examples of how risk may be controlled.
9. What are the risks associated with using a cell phone?

# Multiple Choice Questions

1. _____ is best described as the application of scientific knowledge and methods to the solution of societal problems.
   a. Systems analysis
   b. Technology
   c. Research
   d. Cybernetics
2. _____ eliminates disease-causing microorganisms in the water supply.
   a. Oxygenation
   b. Fluoridation
   c. Chlorination
   d. Hibernation
3. When water is purified, disinfection byproducts, or DBPs, are formed, which are suspected to be _____.
   a. poisons
   b. carcinogens
   c. tuberculosis agents
   d. microbial pathogens
4. _____ are associated with the physical failure of the technological system.
   a. Hardware factors
   b. Human factors
   c. Organizational factors
   d. Social factors
5. The root cause of the Three Mile Island nuclear power plant disaster was due to _____.
   a. hardware factors
   b. human factors
   c. organizational factors
   d. social factors
6. When TWA flight 800 blew up off the coast of Long Island, New York, the cause was thought to be _____.
   a. hardware factors
   b. human factors
   c. organizational factors
   d. social factors
7. The Challenger space shuttle disaster can be attributed to _____.
   a. hardware factors
   b. human factors
   c. organizational factors
   d. social factors

8. The Twin Towers terrorist attack on 9/11 was thought to be precipitated by _____.
   a. hardware factors
   b. human factors
   c. organizational factors
   d. social factors

9. HCAs are _____.
   a. *E coli*
   b. fecal bacteria
   c. carcinogens caused by cooking
   d. high chlorine activation

10. A way of taking into account the quality and length of life that will result from technological intervention, or QALY, is _____.
    a. Quality Adjusted Life Yield
    b. Quality Adjusted Life Year
    c. Quantitative Applied Living Yield
    d. Quality Affirmed Living Years

11. Every technology has some associated _____.
    a. risk
    b tolerance
    c. radiation
    d. definition

12. Chlorination of water supplies _____.
    a. is 100% safe
    b. results in the formation of some suspected carcinogens
    c. has had a negative impact on water-borne typhoid
    d. is unregulated by Congress

13. Which of the following is not added to water to improve its quality?
    a. zinc polyphosphate
    b. chlorine
    c. fluorine
    d. caustic soda

14. After 1910, when chlorination of the Albany water supply was initiated, the death rate of water-borne typhoid per 100,000 dropped to _____.
    a. 110
    b. 20
    c. 10
    d. 0

15. Select the possible health risks in the absence of chlorination.
    a. ingestion of microbial pathogens
    b. ingestion of DBPs
    c. off-color water
    d. There are no health risks.

16. Select the possible health risks in the presence of adequate chlorination.
    a. ingestion of microbial pathogens
    b. ingestion of DBPs
    c. off-color water
    d. There are no health risks.

17. Select the possible health risks in the absence of cooking raw meat.
    a. ingestion of microbial pathogens
    b. ingestion of HCAs
    c. gamy food
    d. There are no health risks.

18. Gamma ray irradiation of food _____.
    a. enhances the shelf life of some fresh produce
    b. has been found to be unsafe by the FDA
    c. causes more changes to the molecular structure of the food than cooking
    d. is risk free

19. Which of the following factors generally does not influence the magnitude of risk?
    a. hardware factors
    b. human factors
    c. organizational factors
    d. social factors
    e. astrological factors

20. Operator error would be ascribed to which of the following risk factors?
    a. hardware factors
    b. human factors
    c. organizational factors
    d. social factors
    e. astrological factors

21. Engineering failures are important because _____.
    a. they show who the incompetent engineers are
    b. they show that technology advances by overcoming failures
    c. they show that engineers have large egos that must be overcome
    d. None of these.

# Essay Question

Select a technology not mentioned in this chapter.

1. Identify the problem that it was intended to solve.
2. How effectively has it addressed the problem?
3. Briefly explain the scientific principles underlying this technology.

# Bibliography and Web Resources

Food Irradiation—A Safe Measure.
**www.fda.gov/opacom/catalog/irradbro.html**

Stop Food Irradiation Project of Organic Consumers Association.
**www.organicconsumers.org/irradlink.html**

Radium Dial Corporation article.
**www.hartford-hwp.com/archives/40/046.html**

# Review of Basic Mathematics

We shall begin with a quick review of some basic concepts in mathematics, starting with definitions and operations with numbers.

**Set:** a well-defined collection of objects

**Elements of a set:** the members of the set

Example: {a, b, cow, dog, 3, 4} is a set. The *elements* of the set are the "a," "cow," "3," etc.

**Fundamental operations of algebra:** $+, -, \cdot, \div, \sqrt{}$. Let us see if we can find a set of numbers that will be closed under the fundamental operations of algebra.

**Closure:** A set is said to be closed under an operation when, operating with any two members of the set, another member of the set is obtained.

Example: a, b are members of a set. If $a \circ b = c$, where c is any member of the same set, then the set is closed under the operation "$\circ$".

**Natural numbers:** {1, 2, 3, 4, 5, . . .}
Note that adding any two natural numbers results in another natural number. Therefore, the set of natural numbers is closed under addition.

Examples: $3 + 4 = 7$

$21 + 3 = 24$

Subtracting any two natural numbers sometimes results in a number that does not belong to the set of natural numbers.

Examples: $7 - 2 = 5$

$3 - 5 = -2$, which is not a natural number.

Therefore, the set of natural numbers is not closed under subtraction.

**Consider the set of integers:** $\{. . . {}^-3, {}^-2, {}^-1, 0, {}^+1, {}^+2, {}^+3, . . .\}$. Notice that we have attached a + or – sign to each number. They are called *signed* numbers. Later we shall dispense with the signs.

**Definition of addition +:**
Let a, b, be any elements in the set of integers. Then $a + b = c$. The set is closed because when we add two integers we obtain another integer, c.

In performing the fundamental operations of arithmetic it is sometimes useful to refer to the number line. Pick any straight line. Label a point on it as zero and pick an arbitrary distance to represent one unit. A correspondence between the set of integers and points on the line may now be set up.

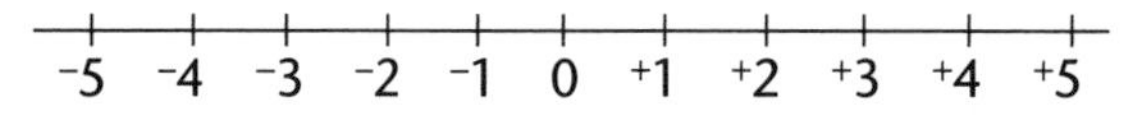


**Addition of integers:**
The operation of adding any two integers can be followed on the number line.

Examples:

a. $^{+}2 + {}^{+}3 = {}^{+}5$
(Start on the $^{+}2$ and proceed $^{+}3$ units to the right on the number line.)

b. $^{+}1 + {}^{+}3$
(Start on the $^{+}1$ and proceed $^{+}3$ units to the right to obtain $^{+}4$.)

c. $^{-}3 + {}^{+}4 = ?$
(Start on the $^{-}3$ and proceed 4 steps to the right to obtain $^{+}1$.)

d. $^{-}3 + {}^{-}2 = ?$
(Start on the $^{-}3$ and proceed 2 steps to the *left* because of the negative signed number $^{-}2$ or really two steps back to $^{-}5$.)

**Subtraction of integers:** a – b = k if and only if there exists a k such that a = b + k. In other words, what number must be added to b in order to obtain a?

Examples:

a. $^{+}5 - {}^{+}2 = ?$
or $^{+}5 = {}^{+}2 +$ ____
$^{+}5 = {}^{+}2 + {}^{+}3$
therefore, $^{+}5 - {}^{+}2 = {}^{+}3$
(Starting on the number line at $^{+}2$ how many steps to the right must be taken to obtain $^{+}5$?)

b. $^{+}10 - {}^{+}7 = ?$
or $^{+}10 = {}^{+}7 +$ ____
$^{+}10 = {}^{+}7 + {}^{+}3$
therefore, $^{+}10 - {}^{+}7 = {}^{+}3$
(Starting on the number line at $^{+}7$ how many steps to the right must be taken to reach $^{+}10$?)

c. $^{+}6 - {}^{+}8 = ?$
or $^{+}6 = {}^{+}8 +$ ____
$^{+}6 = {}^{+}8 + {}^{-}2$
therefore, $^{+}6 - {}^{+}8 = {}^{-}2$
(Starting on the number line at $^{+}8$ how many steps must be taken to the right to reach +6? Actually two steps back are needed, or $^{-}2$.)

d. $^{-}8 - {}^{-}2 = ?$
$^{-}8 = {}^{-}2 +$ ____
$^{-}8 = {}^{-}2 + {}^{-}6$
therefore, $^{-}8 - {}^{-}2 = {}^{-}6$
(Starting at $^{-}2$, how many steps to the right must be taken to reach $^{-}8$? Actually six steps back must be taken. Therefore, $^{-}6$.)

e. $^{-}3 - {}^{-}5 = ?$
$^{-}3 = {}^{-}5 +$ ____
$^{-}3 = {}^{-}5 + {}^{+}2$
therefore, $^{-}3 - {}^{-}5 = {}^{+}2$
(Starting at $^{-}5$ how many steps to the right must e taken to obtain $^{-}3$? Two steps to the right. Therefore, $^{+}2$.)

# Exercises

1. a. $^{+}16 + {}^{+}2 =$ ____
b. $^{-}20 + {}^{+}5 =$ ____
c. $^{+}7 + {}^{-}7 =$ ____
d. $0 + {}^{-}4 =$ ____
e. $^{+}12 + {}^{+}4 =$ ____
f. $^{-}9 + {}^{+}3 =$ ____
g. $^{-}6 + {}^{+}6 =$ ____
h. $0 + {}^{+}5 =$ ____
i. $^{+}4 + {}^{-}1 =$ ____
j. $^{-}5 + {}^{-}1 =$ ____
k. $^{+}9 + {}^{+}9 =$ ____
l. $^{-}22 + {}^{+}11 =$ ____
m. $^{+}6 + {}^{-}2 =$ ____
n. $^{-}16 + {}^{-}8 =$ ____
o. $^{-}3 + {}^{-}3 =$ ____
p. $^{-}6 + {}^{-}3 =$ ____

2. a. $^{+}4 - {}^{-}1 =$ ____
b. $^{-}5 - {}^{-}1 =$ ____
c. $^{+}9 - {}^{+}9 =$ ____
d. $^{-}22 - {}^{-}11 =$ ____
e. $^{+}16 - {}^{+}2 =$ ____
f. $^{-}20 - {}^{+}5 =$ ____
g. $^{+}7 - {}^{-}7 =$ ____
h. $0 - {}^{-}4 =$ ____
i. $^{+}6 - {}^{-}2 =$ ____
j. $^{-}16 - {}^{-}8 =$ ____
k. $^{-}3 - {}^{-}3 =$ ____
l. $^{-}6 - {}^{-}3 =$ ____
m. $^{+}12 - {}^{+}4 =$ ____
n. $^{-}9 - {}^{+}3 =$ ____
o. $^{-}6 - {}^{+}6 =$ ____
p. $0 - {}^{+}5 =$ ____

**Multiplication of integers:** a · b = c
Multiplication of integers can be thought of as repeated addition.

$^{+}2 \cdot {}^{+}3 = {}^{+}3 + {}^{+}3 = {}^{+}6$

$^{+}5 \cdot {}^{+}6 = {}^{+}6 + {}^{+}6 + {}^{+}6 + {}^{+}6 + {}^{+}6 = {}^{+}30$

$^{+}4 \cdot ({}^{-}2) = {}^{-}2 + {}^{-}2 + {}^{-}2 + {}^{-}2 = {}^{-}8$

$^{+}3 \cdot ({}^{-}4) = {}^{-}4 + {}^{-}4 + {}^{-}4 = {}^{-}12$

$^{+}3 \cdot ({}^{-}2) = {}^{-}2 + {}^{-}2 + {}^{-}2 = {}^{-}6$

$^{-}2 \cdot {}^{-}3 = ?$

Consider:

$$^-2 \cdot (^+3 + {}^-3) = {}^-2 \cdot (0) = 0$$

$$^-2 \cdot (^+3 + {}^-3) = {}^-2 \cdot {}^+3 + {}^-2 \cdot {}^-3 = {}^-6 + ({}^-2 \cdot {}^-3) = 0$$

implies $^-2 \cdot {}^-3 = {}^+6$

**Division of integers:** a/b = k if and only if a = bk. In other words, can a number k be found so that when it multiplies b, a is obtained? Division is the inverse of multiplication.

a. $^+12/^+4 = ?$
 $^+12 = {}^+4 \cdot$ _____
 $^+12 = {}^+4 \cdot {}^+3$
 therefore, $^+12/^+4 = {}^+3$

b. $^-6/^+2 = ?$
 $^-6 = {}^+2 \cdot$ ____
 $^-6 = {}^+2 \cdot {}^-3$
 therefore, $^-6/^+2 = {}^-3$

c. $^+8/^-4 = ?$
 $^+8 = {}^-4 \cdot$ ____
 $^+8 = {}^-4 \cdot {}^-2$
 $^+8/^-4 = {}^-2$

d. $^-10/^-5 = ?$
 $^-10 = {}^-5 \cdot$ ____
 $^-10 = {}^-5 \cdot {}^+2$
 $^-10/^-5 = {}^+2$

# Exercises

1. a. $0 \cdot {}^-4 =$ _____
 b. $0 \cdot {}^+5 =$ _____
 c. $^-22 \cdot {}^+11 =$ _____
 d. $^-6 \cdot {}^-3 =$ _____
 e. $^+7 \cdot {}^-7 =$ _____
 f. $^-6 \cdot {}^+6 =$ _____
 g. $^+9 \cdot {}^+9 =$ _____
 h. $^-3 \cdot {}^-3 =$ _____
 i. $^-20 \cdot {}^+5 =$ _____
 j. $^-9 \cdot {}^+3 =$ _____
 k. $^-5 \cdot {}^-1 =$ _____
 l. $^-16 \cdot {}^-8 =$ _____
 m. $^+16 \cdot {}^+2 =$ _____
 n. $^+12 \cdot {}^+4 =$ _____
 o. $^+4 \cdot {}^-1 =$ _____
 p. $^+6 \cdot {}^-2 =$ _____

2. a. $^-6 \div {}^-3 =$ _____
 b. $^-3 \div {}^-3 =$ _____
 c. $^-16 \div {}^-8 =$ _____
 d. $^+6 \div {}^-2 =$ _____
 e. $^-22 \div {}^+11 =$ _____
 f. $^+9 \div {}^+9 =$ _____
 g. $^-5 \div {}^-1 =$ _____
 h. $^+4 \div {}^-1 =$ _____
 i. $0 \div {}^+5 =$ _____
 j. $^-6 \div {}^+6 =$ _____
 k. $^-9 \div {}^+3 =$ _____
 l. $^+12 \div {}^+4 =$ _____
 m. $0 \div {}^-4 =$ _____
 n. $^+7 \div {}^-7 =$ _____
 o. $^-20 \div {}^+5 =$ _____
 p. $^+16 \div {}^+2 =$ _____

Now let us drop the signed number notation for more conventional notation. All numbers that are used without a sign in front are to be considered positive signed numbers.

The *absolute value* of a number is defined as
$|a| = a$ if $a > 0$
 $-a$ if $a < 0$.
In other words, $|-3| = 3$, $|5| = 5$, $|0| = 0$.

**Addition of integers:**

1. If two integers have the *same* sign, take their absolute values, add them, and put the common sign with the sum.
2. If two integers have *unlike* signs, take their absolute values, subtract them, and put the sign of the number having the larger absolute value in front.

**Subtraction of integers:** Change the sign of the number to be subtracted and follow the rules for addition.

Note that when we try to divide integers sometimes the result is not an integer. For example 8/2 = 4, an integer, but 2/8 = ¼, which is not an integer but a fraction. We say that the set of integers is not closed under the operation of division. We have seen that it is closed under the operation of +, –, and ·, which means when we perform these operations using integers we obtain integers as a result.

So, we must define a new set of numbers for which division is a closed operation. We will stop when we find a set closed under the operations of +, –, ÷, ·, and √.

**Set of rational numbers:** $\{\ldots {}^-3, {}^-2, {}^-1, 0, {}^+1, {}^+2, {}^+3, \ldots, a/b, a = \text{integer}, b = \text{integer} \neq 0.\}$

The set of rational numbers consists of all the integers and numbers that can be expressed as the quotient of two integers, (i.e., a/b, b ≠ 0).

Rational numbers include common fractions such as ⅝, $^-7/8$; and mixed numbers such as 4⅓, since it be written as $13/3$, and terminating or repeating decimals like 3.<u>01</u> since it can be written as $301/100$.

**Definition of equality:** Two rational numbers a/b and c/d are equal if and only if ad = bc.

Examples:

a. 6/9 = 10/15 since 90 = 90

b. 5/13 ≠ 2/5 since 25 ≠ 26

Note that a/b = ka/kb for any integer k, k ≠ 0, since akb = aka.

$10/25 = 2 \cdot 5/5 \cdot 5 = 2/5$
$9/15 = 3 \cdot 3/5 \cdot 3 = 3/5$

Note also from the definition of equality:

$2/-5 = -2/5$ since $2 \cdot 5 = -5 \cdot -2$ or $a/-b = -a/b$

**Definition of addition of rational numbers:** $a/b + c/d = (ad + bc)/bd$

Examples:

a. $2/5 + 5/8 = (2 \cdot 8 + 5 \cdot 5)/5 \cdot 8$
$= (16 + 25)/40 = 41/40$

b. $1/3 + -5/8 = (8 \cdot 1 + 3 \cdot -5)/3 \cdot 8$
$= (8 + -15)/24 = -7/24$

c. $-2/9 + -5/6 = (6 \cdot -2 + 9 \cdot -5)/6 \cdot 9$
$= (-12 + -45)/54 = -57/54$

**Definition of subtraction:** $a/b - c/d = k$ or $a/b = c/d + k$, k is a rational number. In other words, what fraction k must be added to c/d to give a/b? Rule: Change the sign of the fraction to be subtracted and added.

Examples:

a. $2/5 - 5/9 = 2/5 + -5/9 = (9 \cdot 2 + -5 \cdot 5)/9 \cdot 5 = (18 + -25)/45 = -7/45$

b. $-3/8 - 5/6 = -3/8 + -5/6 = (6 \cdot -3 + 8 \cdot -5)/6 \cdot 8 = (-18 + -40)/48 = -58/48$

**Multiplication of fractions**: $a/b \cdot c/d = ac/bd$

Examples:

a. $3/5 \cdot 2/7 = 6/35$

b. $-2/9 \cdot 4/7 = -8/63$

c. $-1/7 \cdot -3/8 = 3/56$

**Division of fractions:** $a/b \div c/d = k$ if and only if there exists a k such that $a/b = k \cdot c/d$, or k is $a/b \cdot d/c$

Examples:

a. $2/9 \div 4/5 = 2/9 \cdot 5/4 = 10/36$

b. $-3/8 \div 2/5 = -3/8 \cdot 5/2 = -15/16$

If we now test the rational numbers to see if the set is closed under our five algebraic operations of +, −, ·, ÷, and √, we find that we can compute $\sqrt{16}$ and $\sqrt{4}$ but we cannot express the $\sqrt{2}$ as the ratio of two integers or as a rational number. Therefore we say the $\sqrt{2}$ is an irrational number. Therefore we must extend our set of numbers and create a new number system that includes the irrational numbers.

**Irrational numbers:** a number that cannot be expressed as the quotient of two integers.

Example: $\sqrt{2}$, $\sqrt{3}$, $\pi$

**Set of real numbers:** $\{\ldots {}^-3, {}^-2, {}^-1, 0, 1, 2, 3, \ldots a/b, b \neq 0 \ldots \sqrt{{}^+a}\}$

The rules for +, −, ·, ÷, √ are analogous.

For the real numbers the following structural properties exist for + and ·. Let a, b, c be any real numbers.

1. $a + b = c$ (closure)
2. $a + (b + c) = (a + b) + c$ (associative)
3. $a + b = b + a$ (commutative)
4. $a \cdot b = c$ (closure)
5. $a(bc) = (ab)c$ (associative)
6. $ab = ba$ (commutative)
7. $a(b + c) = ab + ac$ (distributive)
8. If $a = b$, $a + c = b + c$
9. If $a = b$, $ac = bc$

# Exercises

1. Determine whether the following pairs of rational numbers are equal or unequal.
   a. $-3/7, 3/-7$
   b. $-5/12, -65/156$
   c. $3/8, 51/137$
   d. $-8/13, -11/17$
2. Find the sums:
   a. $2/3 + 3/4 =$ _____
   b. $2/5 + -1/2 =$ _____
   c. $7/-6 + 4/38 =$ _____
   d. $-3/10 + -4/15 =$ _____
   e. $2/-3 + 3/-8 =$ _____
   f. $7/6 + 5/10 =$ _____
   g. $8/-3 + 7/-6 =$ _____
   h. $-5/-7 + 3/24 =$ _____
   i. $-10/-3 + -4/-5 =$ _____
3. Find the differences:
   a. $3/4 - 2/3 =$ _____
   b. $7/8 - -3/5 =$ _____
   c. $-9/2 - 7/8 =$ _____
   d. $-4/9 - -3/8 =$ _____
   e. $3/8 - -7/3 =$ _____
   f. $13/4 - 7/3 =$ _____
4. Multiply:
   a. $2/3 \cdot -4/7 =$ _____
   b. $-3/5 \cdot -7/9 =$ _____
   c. $3/7 \cdot -4/9 \cdot 2/5 =$ _____
   d. $7/2 \cdot -1/3 \cdot -5/3 =$ _____
   e. $13/7 \cdot 7/13 =$ _____
   f. $13/7 \cdot -7/13 =$ _____
5. Evaluate:
   a. $2/3 \cdot 3/4 + 5/7 =$ _____
   b. $2/3 \cdot (3/4 + 5/7) =$ _____
   c. $3/4 \cdot (2/3 - 1/4) =$ _____
   d. $(3/4 \cdot 2/3) - 1/4 =$ _____
   e. $(2/3 + 4/7 + 3/5) \cdot 1/4 =$ _____
   f. $2/3 \cdot (4/7 + 3/5) \cdot 1/4 =$ _____

6. Find the following quotients:
   a. $2/5 \div 4/3 =$ _____
   b. $9/2 \div -3/5 =$ _____
   c. $1/4 \div 1/4 =$ _____
   d. $0/1 \div 2/7 =$ _____
   e. $-2/5 \div -5/7 =$ _____
   f. $2/3 \div 5 =$ _____

7. a. $2/3 \div 4/5 =$ _____
   b. $(3/4 + 2/7) \div -5/2 =$ _____
   c. $(2 + 3/4) \div 2/3 =$ _____
   d. $(3/2 - 1/2) \div 2/5 =$ _____
   e. $(6 - 5) \div 2/3 =$ _____
   f. $5/8 \div (2 + 4) =$ _____

**Algebraic operations:** the fundamental operations together with the process of taking roots

**Exponents:** $b^m = b \cdot b \cdot b \cdots b$ (m factors), where b is real and m is a positive integer.

Examples:

a. $3^2 = 3 \cdot 3$
   $2^5 = 2 \cdot 2 \cdot 2 \cdot 2 \cdot 2$
   $x^4 = x \cdot x \cdot x \cdot x$

b. $x^3 \cdot x^2 = x \cdot x \cdot x \cdot x \cdot x = x^5$

**Algebraic expression:** a mathematical expression resulting from the application of a finite number of algebraic operations applied to a collection of variables and real numbers

Examples:

a. $\frac{2x^2 + 6}{\phantom{x}}$

b. $x + \sqrt{2x} - 6$

**Term:** each partial expression with its sign

**Similar terms:** term with the same *literal* parts

**Monomial:** an algebraic expression written as the product of a real number and nonnegative powers of the variables

Examples:

a. $3x^3$

b. $8x^2y$

**Binomial:** sum of two monomials

Examples:

a. $x + 2y$

b. $x^2 + 6$

**Trinomial:** sum of three monomials

Examples:

a. $x^2 + 2x + 6$

b. $2x + 3y + z$

**Polynomial:** any sum of monomials

# Exercises

1. Evaluate
   a. $x \cdot x \cdot x$
   b. $-2 \cdot 4 \cdot x \cdot x \cdot x$
   c. $(-3) \cdot (-7) \cdot xxxxy$
   d. $-xxxyy$

2. Evaluate for $x = -2$
   a. $4(x - 5)$
   b. $5x(3x - 2)$
   c. $x + 3 - (2x + 8)$
   d. $3(2x + 4)$

3. Evaluate for $x = 2$, $y = -3$
   a. $3[x - (4y + 5)]$
   b. $x - [2x - (x^2 - y)]$
   c. $(x^2 - 3)(x + 1)$
   d. $(x^2 + 2x + 1)(2x - 3)$
   e. $(x - 2y)^3$

**Addition and subtraction of monomials:** Two monomials may be combined by addition or subtraction if they are similar terms.

Examples:

a. $6x + 9x = (6 + 9)x = 15x$

b. $4x - 7x = (4 - 7)x = -3x$

c. $7y^2z + y^2z + 4y^2z = (7 + 1 + 4)\, y^2z = 12y^2z$

**Multiplication and division:**

Example: $(-4xy^2)(8xy) = -4 \cdot 8xx\, y^2y = -32x^2y^3$

*Rule:* To multiply or divide, determine the numerical coefficients first and then the exponents of the variable.

Examples:

a. $(-3ab^2) \cdot (5ac) = -15a^2b^2c$

b. $(-4m^2) \cdot (-n) = 4m^2n$

c. $(6ax^2) \cdot (-3) = -18ax^2$

d. $(7x^2yz^3) \cdot (3xy^3) = 21x^3y^4z^3$

e. $\frac{14\mu^2v^3}{-2uv^3} = \frac{14u \cdot u\ v \cdot v \cdot v}{-2u\ v \cdot v \cdot v} = -7u$

f. $\frac{18ab^2}{-3ab^2} = -6$

g. $\frac{-24u^2v^2w}{-8uw} = 3uv^2$

h. $\frac{7ax^2}{a} = 7x^2$

i. $\frac{9x^2y}{-9x^2y} = -1$

# Exercises

1. Add:
   a. $(-7a) + (-4a) =$ _____
   b. $-9y + (5y) =$ _____
   c. $8k - 13k =$ _____
   d. $7ab^2c + 6ab^2c =$ _____
   e. $-4m^2n - 4m^2n =$ _____
   f. $2p^9 - 6p^9 =$ _____
   g. $9ab - 2ab =$ _____
2. Multiply:
   a. $3x^2 \cdot (-2x) =$ _____
   b. $(-6p^2)(5p) =$ _____
   c. $(8abx)(-7a^2xy) =$ _____
   d. $(-6a)(-8u^3) =$ _____
   e. $(-3b^3)^2 =$ _____
3. Divide:
   a. $(-x^7) \div (-x) =$ _____
   b. $-27k^2 \div 9k^2 =$ _____
   c. $8a^2by^2 \div (-4aby^2) =$ _____
   d. $12p^2 \div 6p =$ _____
   e. $3x^2y \div (-x) =$ _____

**Addition and subtraction of polynomials:** Group like terms and add or subtract.

Examples:

a. $(2x + 3y) + (5x - 9y) = 2x + 5x + 3y - 9y$
$= 7x - 6y$

b. $(2a - 5b - c) + (8a + 4b - 3c) = (2a + 8a)$
$+ (-5b + 4b) + (-c - 3c) = 10a - b - 4c$

c. $(7y^2 - 6) - (4y^2 - 5y + 2)$
$= 7y^2 - 6 - 4y^2 + 5y - 2$
$= (7y^2 - 4y^2) + 5y - 2 - 6$
$= 3y^2 + 5y - 8$

**Multiplication:** a direct application of the distributive property: $a(b+c) = ab + ac$

Examples:

a. $3a(a^2 - 4ab + 8b^2) = 3a^3 - 12a^2b + 24ab^2$

b. $-2uv^2(u^3 - 2u^2 + u - 6) = -2u^4v^2 + 4u^3v^2$
$- 2u^2v^2 + 12\ uv^2$

c. $(4x^2 - 7x - 5)(2x + 3) = (4x^2 - 7x - 5)2x$
$+ (4x^2 - 7x - 5)3 = 8x^3 - 14x^2 - 10x + 12x^2$
$- 21x - 15 = 8x^3 - 2x^2 - 31x - 15$

or multiplying $4x^2 - 7x - 5$
by $2x + 3$

$$\begin{array}{r} 8x^3 - 14x^2 - 10x \quad\quad\quad \\ 12x^2 - 21x - 15 \\ \hline \end{array}$$

or $8x^3 - 2x^2 - 31x - 15$

**Division of polynomials:**

Examples:

a. $\dfrac{8b^3 - 14b^2 + 12b}{2b} = 4b^2 - 7b + 6$

b. $\dfrac{x^3 - 3x^2 + x}{-x} = -x^2 + 3x - 1$

c. $(37 + 8x^3 - 4x) \div (2x + 3)$

$$\begin{array}{r|l} & 4x^2 - 6x + 7 \\ 2x+3 & 8x^3 + 0x^2 - 4x + 37 \\ & \underline{8x^3 + 12x^2} \\ & -12x^2 - 4x \\ & \underline{-12x^2 - 18x} \\ & 14x + 37 \\ & \underline{14x + 21} \\ & 16 \text{ remainder} \end{array}$$

so $(37 + 8x^3 - 4x) \div (2x + 3) = 4x^2 - 6x + 7$ with a remainder of 16.

# Exercises

1. Add:
   a. $(3x - 2) + (4x + 2) =$ ___
   b. $(2x^2 - 6xy + y^2) + (-x^2 + xy + y^2) =$ ___
   c. $(4x^2 - 2y^2) + (x^2 - y^2) + (3x^2 + 5y^2) =$ ___
   d. $7b^3 + b^2 + 6 - b - 2b^2 + 4b + 7 - 5b^2 - 4b + 11b^3 =$ ___
2. Subtract:
   a. $(2x^2 + 5x - 1) - (x^2 - 2x + 4) =$ ___
   b. $(3a^2 - 2ab + b^2) - (9a^2 - 2ab - b^2) =$ ___
   c. $(2y^2 + 5y + 9) - (y^2 - 1) =$ ___
   d. $(mn - m^3 + 3m^2) - (m^3 + 5mn + 6m^2) =$ ___
3. Multiply:
   a. $(3a - b)(2a + c) =$ ___
   b. $(y + 7)(2y - 3) =$ ___
   c. $(3u - 8v)(3u + 8v) =$ ___
   d. $(2a - 3b + c)(2a - c) =$ ___
4. Divide:
   a. $(k^3 - 3k^2) \div k$
   b. $(m^2n^2 - 7m^4m^3) \div mn^2$
   c. $(a^3 - u^2 - 14a + 24) \div (a - 2)$
   d. $(x^3 + 2x^2 - x + 2) \div (x - 1)$

**Linear equation:** any equation of the form $ax + b = 0$ where a and b are real numbers and $a \neq 0$

Examples:

a. Solve $3x - 5 = 4$ for x.
$3x - 5 + 5 = 4 + 5$ (adding 5 to both sides)
$3x = 9$
$\frac{1}{3} \cdot 3x = \frac{1}{3} \cdot 9$ (multiplying both sides by $\frac{1}{3}$)
$x = 3$

b. Solve $2(y - 4) + 12 = 3(y + 3)$ for y.

$2y - 8 + 12 = 3y + 9$

$2y + 4 = 3y + 9$

$2y + 4 - 4 = 3y + 9 - 4$

$2y = 3y + 5$

$2y - 3y = 3y - 3y + 5$

$-y = 5$

$(-1)(-y) = -1 \cdot 5$

$y = -5$

c. Solve $\frac{1}{3}(x - 4) + 2 = \frac{1}{2}(x + 3)$ for x.

$$\frac{1x}{3} - \frac{4}{3} + \frac{2}{1} = \frac{1x}{2} + \frac{3}{2}$$

$$\frac{1x}{3} + \frac{-4 \cdot 1 + 2 \cdot 3}{3 \cdot 1} = \frac{1x}{2} + \frac{3}{2}$$

$$\frac{1x}{3} + \frac{2}{3} = \frac{1x}{2} + \frac{3}{2}$$

$$\frac{1x}{3} + \frac{2}{3} + \frac{-2}{3} = \frac{1x}{2} + \frac{3}{2} + \frac{-2}{3}$$

$$\frac{1x}{3} = \frac{1x}{2} + \frac{3 \cdot 3 + 2 \cdot -2}{2 \cdot 3}$$

$$\frac{1x}{3} = \frac{1x}{2} + \frac{5}{6}$$

$$\frac{1x}{3} - \frac{1x}{2} = \frac{1x}{2} - \frac{1x}{2} + \frac{5}{6}$$

$$\frac{2x - 3x}{2 \cdot 3} = \frac{5}{6}$$

$$\frac{-x}{6} = \frac{5}{6}$$

$(-6/1)(-x/6) = (-6/1)(5/6)$

$x = -5$

# Exercises

Solve for the unknown variable.

1. $y + 5 = 4$
2. $y - 3 = 2$
3. $4y = 12$
4. $\frac{5y}{4} = 20$
5. $4x - 6 = 14$
6. $2x + 5 = 11$
7. $3(x + 5) = 21$
8. $5(x - 7) = 15$
9. $7(x - 3) = x + 9$
10. $3(x + 8) = 12 + x$
11. $\frac{3(x + 3)}{2} = \frac{2(x - 2)}{3}$
12. $\frac{1(x - 3)}{2} + \frac{1(x + 2)}{4} = 8$
13. $F = ma$, solve for m.
14. $p = a + b + c$, solve for a.
15. $A = lw$, solve for w.
16. $V = (1/3)Bh$, solve for B.
17. $V = \pi r^2 h$, solve for h.
18. $A = p + prt$, solve for r.
19. $S = 2\pi rh + 2\pi r^2$, solve for h.
20. $F = (9/5)C + 32$, solve for C.

# Scientific Notation and Significant Figures

Scientists use scientific notation as a convenient way of expressing very large numbers such as the distance from the earth to the sun, 93,000,000 miles ($9.3 \times 10^7$ miles), or very small numbers such as the wavelength of green light 0.000000560 m ($560 \times 10^{-9}$ m).

## Exponents

Recall:

Def: $b^m = b \cdot b \cdot b \cdots b$ (m factors), where b is real and m is a *positive* integer.

Examples: $3^2 = 3 \cdot 3 = 9$

$2^5 = 2 \cdot 2 \cdot 2 \cdot 2 \cdot 2 = 32$

$x^4 = x \cdot x \cdot x \cdot x$

$x^3 \cdot x^2 = x \cdot x \cdot x \cdot x \cdot x = x^5$

Also, $b^{-m} = 1/b^m$

Examples: $3^{-2} = 1/3^2 = 1/(3 \cdot 3) = 1/9$

$2^{-5} = 1/2^5 = 1/(2 \cdot 2 \cdot 2 \cdot 2 \cdot 2) = 1/32$

$x^{-4} = 1/x^4 = 1/(x \cdot x \cdot x \cdot x)$

$x^{-3} \cdot x^{-2} = 1/x^3 \cdot 1/x^2 = 1/x^5$

It is customary to write a number in scientific notation in the following form:

$$n \cdot x \cdot 10^c$$

where n is a number one or larger and less than 10, while c is a positive or negative integer (whole number).

To write a number in scientific notation, one moves the decimal point until a number between 1 and 10 is obtained. If the decimal is moved to the left, the number of times it is moved is counted and used as the power "c" for ten. It is positive when the decimal has to be moved to the left and negative if moved to the right.

Examples:

a. $287 = 2.87 \times 10^2$

b. $0.00382 = 3.82 \times 10^{-3}$

Express the radius of the earth in scientific notation: r = 3960 mi

Answer: $3.96 \times 10^3$ mi

Express the circumference of the earth in scientific notation: C = 24,860 mi

Answer: $2.486 \times 10^4$ mi

Express the distance from the earth to the moon in scientific notation: d = 240,000 mi

Answer: $2.4 \times 10^5$ mi

Express the speed of light in scientific notation: v = 299,800,000 m/s

Answer: $2.998 \times 10^8$ m/s

Express the charge on electron in scientific notation: 0.00000000000000000016 coulombs

Answer: $1.6 \times 10^{-19}$ coulombs

# Significant Figures

Experimental science is based on measurement. Any time we make a measurement there is always a possibility of an error or uncertainty. The error is the maximum difference between the measured value and the true value. When reporting a measurement the uncertainty or error is indicated by a + and – value. For example, the length of a block of wood may be reported as being 12.36 cm + or –.03 cm. This means its true value is less than 12.39 cm and greater than 12.33 cm.

Many times when measurements are made the results are expressed in terms of significant figures. For example, if the length of the block was reported as 12.36, then the error would be associated with the last digit, which resides in the hundredths place. The number has four significant figures with an error of .01. If the length of the block was given as 12 cm, then the number has two significant figures and since the last digit is a two in the units position the error would be 1.

A convenient way of determining the number of significant figures is writing the number using scientific notation. Thus the distance from the earth to the sun reported as 93,000,000 miles or $9.3 \times 10^7$ miles would have two significant figures.

If the radius of the earth was given as 4,000 mi or, in scientific notation, $4 \times 10^3$ miles, we see the number of significant figures would be one when written in this form. Note that if we wrote $4.00 \times 10^3$ then the number of significant figures is three.

# Calculations

Great care must be taken with significant figures when doing calculations using numbers obtained from measurements. Suppose you measure the sides of a rectangle and are asked to compute its area.

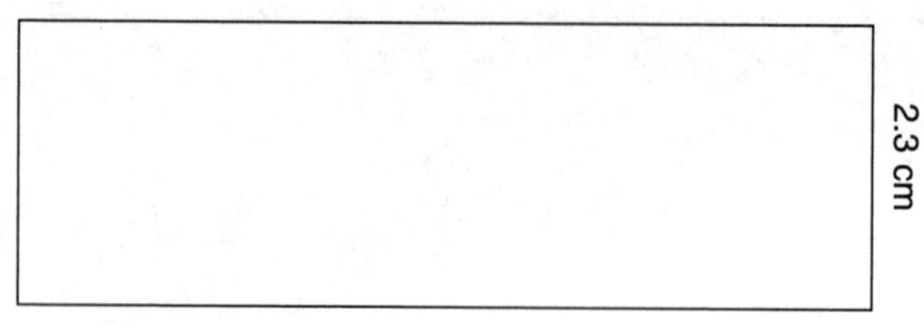


$$\text{Area} = \text{length} \times \text{width}$$

$$A = 4.28 \text{ cm} \times 2.3 \text{ cm}$$

Now you multiply these numbers on your calculator to obtain:

$$A = 9.844 \text{ cm}^2$$

This answer is *incorrect.* It implies the error is less than is possible based on the measurements that were made. The answer has four significant figures. The correct answer cannot have more significant figures then what appears in the least accurate measurement. In this case the width is 2.3 cm, or two significant figures. Therefore, the answer given by the calculator must be rounded off to

$$A = 9.8 \text{ cm}^2.$$

When numbers are added and subtracted, the location of the decimal point is what matters. The perimeter of the rectangle considered previously would be the sum of its sides: 4.28 cm + 2.3 cm + 4.28 cm + 2.3 cm, or

$$P = 13.16 \text{ cm}.$$

The possible error in the length of 4.28 cm is 0.01, and in the width of 2.3 cm it is 0.1. So the sum has an error of about 0.1 and should be written as 13.2 cm.

Thus we see great care must be taken when using the calculator so as not to make errors.

APPENDIX C

# Functions

In our basic mathematics courses we learned that a function could be a set of ordered pairs of real numbers (x,y) where there was a rule for obtaining the y values from the x values. If for each x there was one and only one y determined, then that set of ordered pairs could be called a **function.**

Normally when denoting a function symbolically we write only the rule $y = f(x)$. **f** represents the rule for obtaining the y's from the x's. A graph of $y = x$ follows here. When $x = 0$, $y = 0$. When $x = 1$, $y = 1$. This is the graph of a linear function.

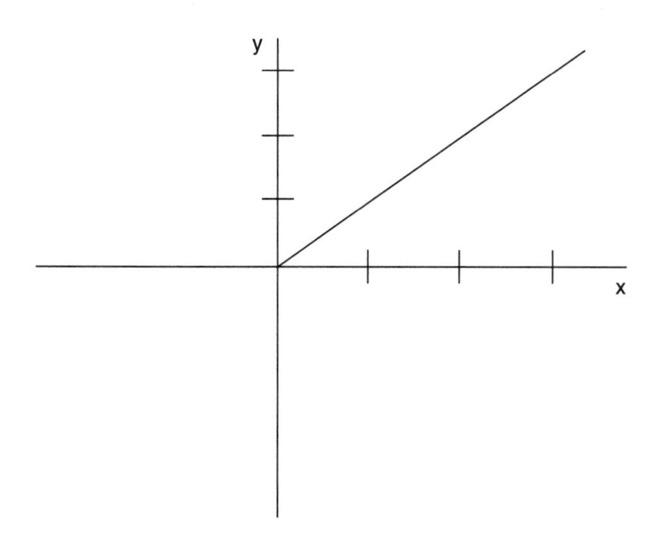


Many people think of growth in a linear fashion. A quantity is growing **linearly** when it increases by a constant amount in a constant time. Saving 10 dollars per week in a savings account increases the total amount in a linear way. The amount of increase each week is not affected by the amount of money already saved.

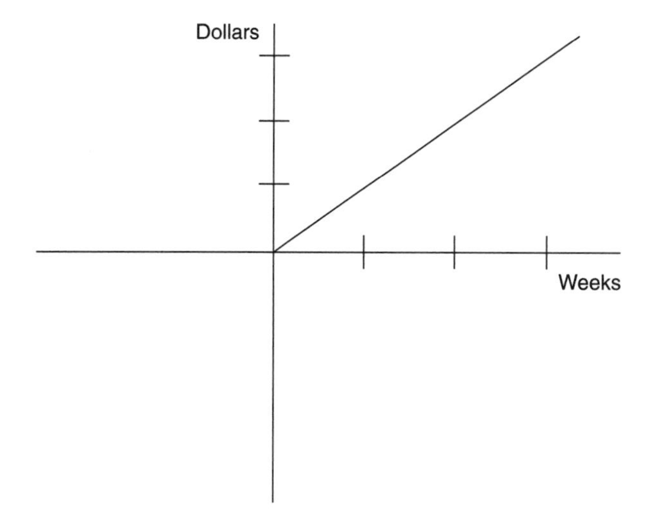


## Linear Functions

A function defined by $y = f(x) = mx + b$, where m and b are real numbers, is called a linear function. The graph of a linear function is a line.

Suppose $y = 2x + 4$. Suppose we let $x = 0$, then $y = 4$, and then let $x = 1$ and then $y = 6$. Suppose $2x + 4y = 8$. Draw the graph of this linear function.

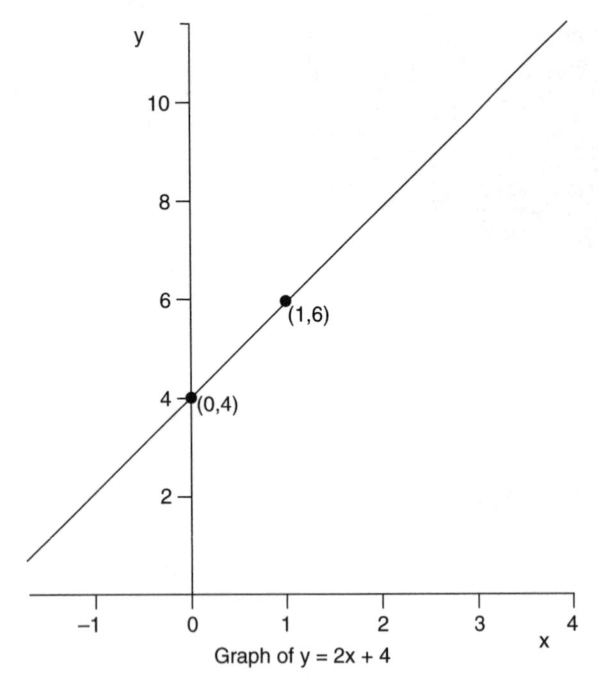

Graph of $y = 2x + 4$

Solving for y:

$$4y = -2x + 8$$
$$y = -(1/2)x + 2$$

Substituting $x = 0$ gives $y = 2$ and substituting $x = 2$ gives $y = 1$, or (0,2) and (2,1).

Therefore, the graph is:

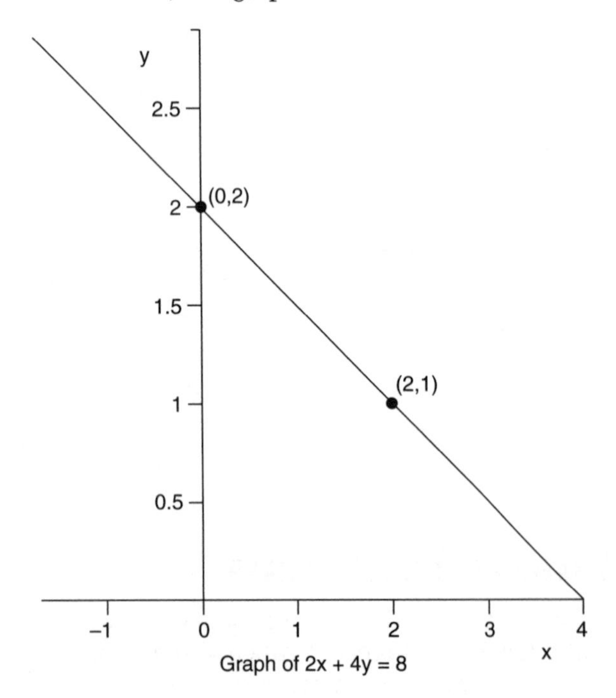

Graph of $2x + 4y = 8$

# Pythagorean Theorem

Pythagoras of Samos lived around 500 B.C. and founded a philosophical and religious school in Croton (in southern Italy) that had many followers. He is known for his famous geometrical theorem even though some believe the Babylonians knew of it 1,000 years earlier. He is given credit for being the first to prove it.

Today we know this theorem as the Pythagorean theorem for right triangles, or $a^2 + b^2 = c^2$.

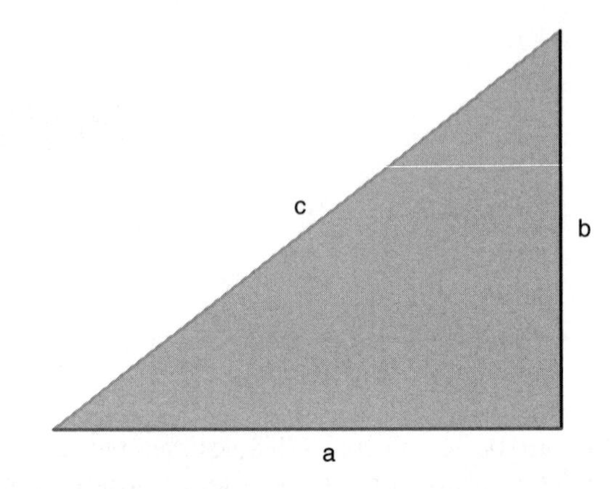


**a** is called the base of the triangle; **b** is the height; **c** (the side opposite the right angle) is called the hypotenuse. It is interesting to note that if the triangle is a right triangle (i.e., one angle is 90°) then Pythagoras' theorem is true and, conversely, if Pythagoras' theorem is true for the three sides of the triangle it has to have one right angle.

Pythagoras' theorem provides us with the means of determining the length of a line in a plane. If we take two ordered pairs on the line and apply Pythagoras' theorem, we obtain the distance between the two points.

For example two points on the line $y = (4/3)x$ are (3, 4) and (6, 8). If we plot these points on an x-y coordinate system we can construct a right triangle and apply Pythagoras' theorem to obtain the distance between them.

$$a^2 + b^2 = c^2$$
$$3^2 + 4^2 = c^2, \text{ or}$$
$$c^2 = 25$$
$$c = 5$$

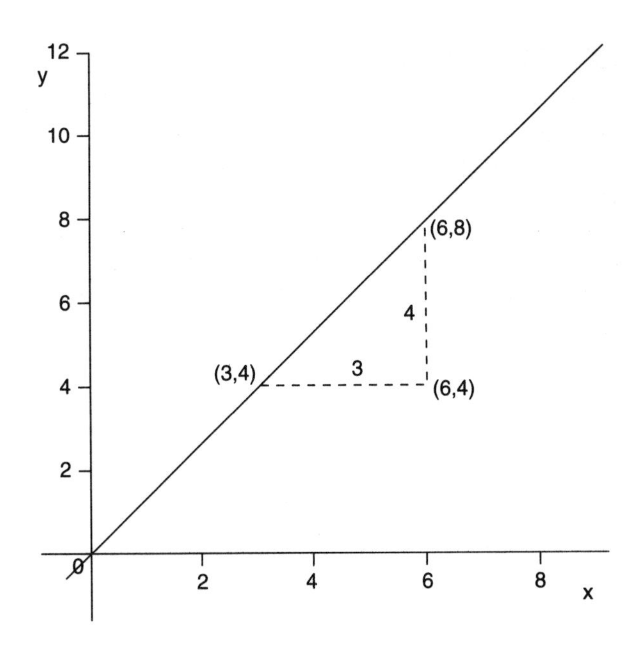


More generally, for any two points in the x-y plane we obtain $d^2 = (x_2 - x_1)^2 + (y_2 - y_1)^2$, or

$$d = \sqrt{(x_2 - x_1)^2 + (y_2 - y_1)^2}$$

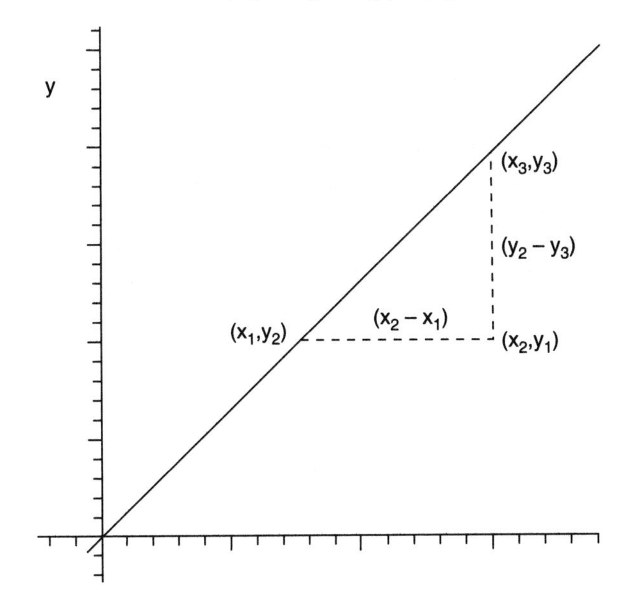


# Trigonometric Formulas

It is observed that in a right triangle, one may form the ratio of two sides at a time.

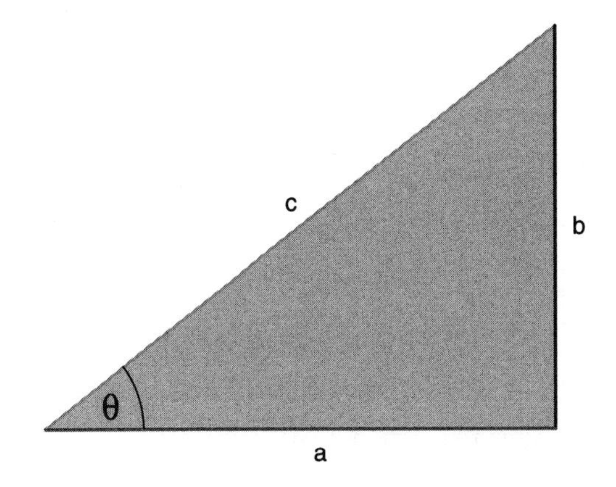


For example, b/c, a/c, b/a, c/b, c/a, a/b covers all possibilities. Each of these ratios is given a name.

Let the angle that side **c** makes with **a** be called **θ.** Then the names are:

$$\sin\theta = b/c \qquad \cos\theta = a/c \qquad \tan\theta = b/a$$

where sin is short for sine, cos for cosine, tan for tangent.

Sometimes these three are remembered as:

Sin θ = Opposite side over Hypotenuse
Cos θ = Adjacent side over Hypotenuse
Tan θ = Opposite side over Adjacent side

Some students remember the formulas by thinking of a possible foreign sounding word:

SOHCAHTOA.

These three trigonometric formulas are valuable in doing many scientific and engineering calculations.

## Slope of a Line

We are all familiar with ski slopes. In describing your skiing prowess to others you probably would give them some idea of the steepness of the hill you can ski down on. It might be a 30° slope or a 45° slope. You usually give them the angle the hill makes with the horizontal.

In mathematics instead of using the angle directly to specify the steepness of a line or slope, **m,** of a line, the tangent of the angle the line makes with the horizontal is used. So again referring to two arbitrary points on a line $(x_1, y_1)$, $(x_2, y_2)$ the slope of the line may be written in general as:

$$m = \tan\theta = (y_2 - y_1)/(x_2 - x_1).$$

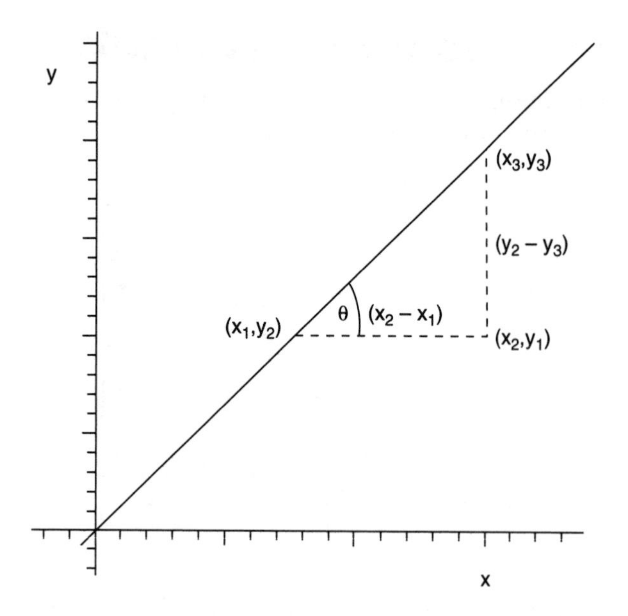


Find the slope of the line between (2,4) and (6,6).

$$m = \tan \theta = (y_2 - y_1)/(x_2 - x_1)$$
$$\text{or } m = (6 - 4)/(6 - 2) = 2/4 = 1/2$$
$$m = 1/2$$

If an arbitrary ordered pair on the line such as (x,y) is used a known point $(x_1, y_1)$ the formula for slope would be:

$$m = \tan \theta = (y - y_1)/(x - x_1)$$
$$m = (y - y_1)/(x - x_1).$$
$$\text{or } m(x - x_1) = (y - y_1).$$

Solving for y: $mx - mx_1 = y - y_1$

$$\text{or } y = mx + (-mx_1 + y_1)$$
$$\text{or } y = mx + b$$

where we let $b = (-mx_1 + y_1)$. Therefore $y = mx + b$, and the equation of the straight line or linear function is obtained. Note if we let $x = 0$ and substitute in the equation, we find $y = b$. When $x = 0$, the line hits the y-axis at b and b is known as the y-intercept.

## OTHER FUNCTIONS

Suppose $y = f(x)$ or $y = x^2$. This means: substitute for x a real number and square it to calculate y, which will give one ordered pair of the function set. If $x = 2$, then $y = 4$ and $(x, y) = (2, 4)$. Having done this for several values of x, one can plot a graph of y versus x to represent part of the function pictorially or graphically. We obtain a parabola:

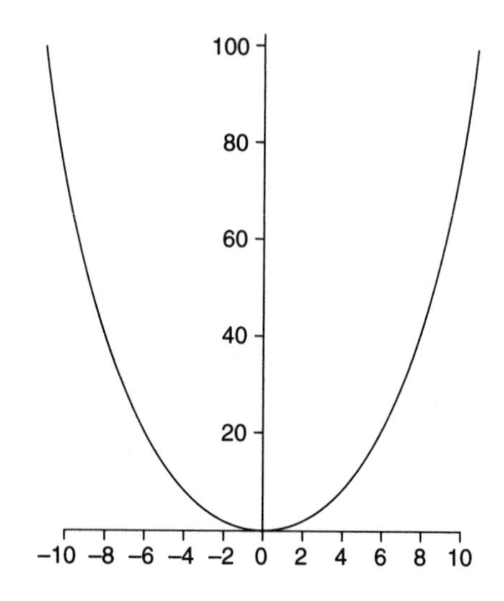


In **exponential** growth, the quantity increases by a constant percentage of the whole in a constant time period. It is useful to think of exponential growth in terms of doubling time, or the time it takes a growing quantity to double in size.

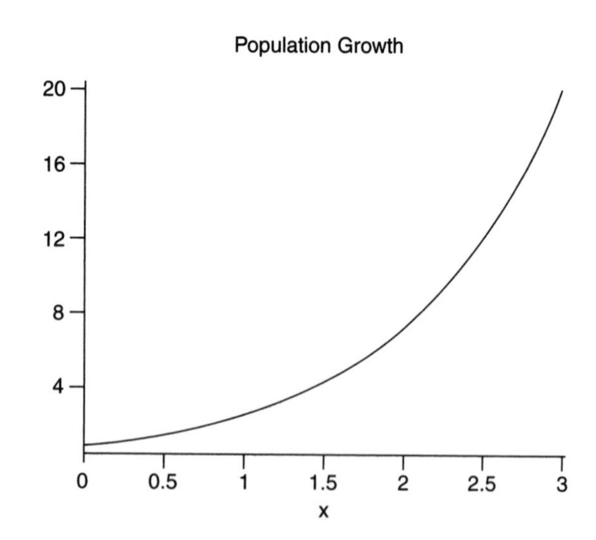